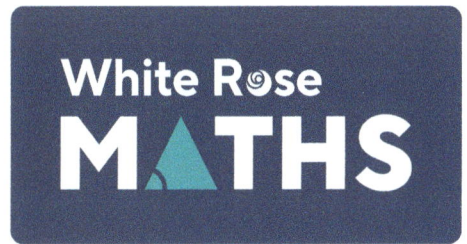

Year 1C
A Guide to Teaching for Mastery

Series Editor: Tony Staneff
Lead author: Josh Lury

Contents

Introduction to the author team	4
What is *Power Maths*?	5
What's different in the new edition?	6
Your *Power Maths* resources	7
The *Power Maths* teaching model	10
The *Power Maths* lesson sequence	12
Using the *Power Maths* Teacher Guide	15
Power Maths Year 1, yearly overview	16
Mindset: an introduction	20
The *Power Maths* characters	21
Mathematical language	22
The role of talk and discussion	23
Assessment strategies	24
Keeping the class together	26
Same-day intervention	27
The role of practice	28
Structures and representations	29
Variation helps visualisation	30
Practical aspects of *Power Maths*	31
Using *Power Maths* flexibly in Key Stage 1	33
Working with children below age-related expectation	35
Providing extra depth and challenge with *Power Maths*	37
Using *Power Maths* with mixed age classes	39
List of practical resources	40
Getting started with *Power Maths*	43

Unit 11 – Multiplication and division — 44
Count in 2s	46
Count in 10s	50
Count in 5s	54
Equal groups	58
Add equal groups	62
Make arrays	66
Make doubles	70
Grouping	74
Sharing	78
End of unit check	82

Unit 12 – Fractions — 84
Recognise and find a half of a shape	86
Recognise and find a half of a quantity	90
Recognise and find a quarter of a shape	94
Recognise and find a quarter of a quantity	98
End of unit check	102

Unit 13 – Position and direction — 104
Describe turns — 106
Describe position – left and right — 110
Describe position – forwards and backwards — 114
Describe position – above and below — 118
Ordinal numbers — 122
End of unit check — 126

Unit 14 – Numbers to 100 — 128
Count from 50 to 100 — 130
10s to 100 — 134
Partition into 10s and 1s — 138
Number line to 100 — 142
One more and one less — 146
Compare numbers — 150
End of unit check — 154

Unit 15 – Money — 156
Recognise coins — 158
Recognise notes — 162
Count in coins — 166
End of unit check — 170

Unit 16 – Time — 172
Before and after — 174
Days of the week — 178
Months of the year — 182
Tell the time to the hour — 186
Tell the time to the half hour — 190
End of unit check — 194

Introduction to the author team

Power Maths arises from the work of maths mastery experts who are committed to proving that, given the right mastery mindset and approach, **everyone can do maths**. Based on robust research and best practice from around the world, *Power Maths* was developed in partnership with a group of UK teachers to make sure that it not only meets our children's wide-ranging needs but also aligns with the National Curriculum in England.

Power Maths – White Rose Maths edition

This edition of *Power Maths* has been developed and updated by:

Tony Staneff, Series Editor and Author

Vice Principal at Trinity Academy, Halifax, Tony also leads a team of mastery experts who help schools across the UK to develop teaching for mastery via nationally recognised CPD courses, problem-solving and reasoning resources, schemes of work, assessment materials and other tools.

Josh Lury, Lead Author

Josh is a specialist maths teacher, author and maths consultant with a passion for innovative and effective maths education.

The first edition of *Power Maths* was developed by a team of experienced authors, including:

- **Tony Staneff and Josh Lury**
- **Trinity Academy Halifax** (Michael Gosling CEO, Emily Fox, Kate Henshall, Rebecca Holland, Stephanie Kirk, Stephen Monaghan and Rachel Webster)
- **David Board, Belle Cottingham, Jonathan East, Tim Handley, Derek Huby, Neil Jarrett, Stephen Monaghan, Beth Smith, Tim Weal, Paul Wrangles** – skilled maths teachers and mastery experts
- **Cherri Moseley** – a maths author, former teacher and professional development provider
- **Professors Liu Jian and Zhang Dan**, Series Consultants and authors, and their team of mastery expert authors:
  Wei Huinv, Huang Lihua, Zhu Dejiang, Zhu Yuhong, Hou Huiying, Yin Lili, Zhang Jing, Zhou Da and Liu Qimeng

  Used by over 20 million children, Professor Liu Jian's textbook programme is one of the most popular in China. He and his author team are highly experienced in intelligent practice and in embedding key maths concepts using a C-P-A approach.

- **A group of 15 teachers and maths co-ordinators**

  We consulted our teacher group throughout the development of *Power Maths* to ensure we are meeting their real needs in the classroom.

What is *Power Maths*?

Created especially for UK primary schools, and aligned with the new National Curriculum, *Power Maths* is a whole-class, textbook-based mastery resource that empowers every child to understand and succeed. *Power Maths* rejects the notion that some people simply 'can't do' maths. Instead, it develops growth mindsets and encourages hard work, practice and a willingness to see mistakes as learning tools.

Best practice consistently shows that mastery of small, cumulative steps builds a solid foundation of deep mathematical understanding. *Power Maths* combines interactive teaching tools, high-quality textbooks and continuing professional development (CPD) to help you equip children with a deep and long-lasting understanding. Based on extensive evidence, and developed in partnership with practising teachers, *Power Maths* ensures that it meets the needs of children in the UK.

Power Maths and Mastery

Power Maths makes mastery practical and achievable by providing the structures, pathways, content, tools and support you need to make it happen in your classroom.

To develop mastery in maths, children must be enabled to acquire a deep understanding of maths concepts, structures and procedures, step by step. Complex mathematical concepts are built on simpler conceptual components and when children understand every step in the learning sequence, maths becomes transparent and makes logical sense. Interactive lessons establish deep understanding in small steps, as well as effortless fluency in key facts such as tables and number bonds. The whole class works on the same content and no child is left behind.

Power Maths

- Builds every concept in small, progressive steps
- Is built with interactive, whole-class teaching in mind
- Provides the tools you need to develop growth mindsets
- Helps you check understanding and ensure that every child is keeping up
- Establishes core elements such as intelligent practice and reflection

The *Power Maths* approach

Everyone can!
Founded on the conviction that every child can achieve, *Power Maths* enables children to build number fluency, confidence and understanding, step by step.

Child-centred learning
Children master concepts one step at a time in lessons that embrace a concrete-pictorial-abstract (C-P-A) approach, avoid overload, build on prior learning and help them see patterns and connections. Same-day intervention ensures sustained progress.

Continuing professional development
Embedded teacher support and development offer every teacher the opportunity to continually improve their subject knowledge and manage whole-class teaching for mastery.

Whole-class teaching
An interactive, whole-class teaching model encourages thinking and precise mathematical language and allows children to deepen their understanding as far as they can.

What's different in the new edition?

If you have previously used the first editions of *Power Maths*, you might be interested to know how this edition is different. All of the improvements described below are based on feedback from *Power Maths* customers.

Changes to units and the progression

- The order of units has been slightly adjusted, creating closer alignment between adjacent year groups, which will be useful for mixed age teaching.
- The flow of lessons has been improved within units to optimise the pace of the progression and build in more recap where needed. For key topics, the sequence of lessons gives more opportunities to build up a solid base of understanding. Other units have fewer lessons than before, where appropriate, making it possible to fit in all the content.
- Overall, the lessons put more focus on the most essential content for that year, with less time given to non-statutory content.
- The progression of lessons matches the steps in the new White Rose Maths schemes of learning.

Lesson resources

- There is a Quick recap for each lesson in the Teacher Guide, which offers an alternative lesson starter to the Power Up for cases where you feel it would be more beneficial to surface prerequisite learning than general number fluency.
- In the **Discover** and **Share** sections there is now more of a progression from 1 a) to 1 b). Whereas before, 1 b) was mainly designed as a separate question, now 1 a) leads directly into 1 b). This means that there is an improved whole-class flow, and also an opportunity to focus on the logic and skills in more detail. As a teacher, you will be using 1 a) to lead the class into the thinking, then 1 b) to mould that thinking into the core new learning of the lesson.
- In the **Share** section, for KS1 in particular, the number of different models and representations has been reduced, to support the clarity of thinking prompted by the flow from 1 a) into 1 b).
- More fluency questions have been built into the guided and independent practice.
- Pupil pages are as easy as possible for children to access independently. The pages are less full where this supports greater focus on key ideas and instructions. Also, more freedom is offered around answer format, with fewer boxes scaffolding children's responses; squared paper backgrounds are used in the Practice Books where appropriate. Artwork has also been revisited to ensure the highest standards of accessibility.

New components

480 Individual Practice Games are available in *ActiveLearn* for practising key facts and skills in Years 1 to 6. These are designed in an arcade style, to feel like fun games that children would choose to play outside school. They can be accessed via the Pupil World for homework or additional practice in school – and children can earn rewards. There are Support, Core and Extend levels to allocate, with Activity Reporting available for the teacher. There is a Quick Guide on *ActiveLearn* and you can use the Help area for support in setting up child accounts.

There is also a new set of lesson video resources on the Professional Development tile, designed for in-school training in 10- to 20-minute bursts. For each part of the *Power Maths* lesson sequence, there is a slide deck with embedded video, which will facilitate discussions about how you can take your *Power Maths* teaching to the next level.

Your *Power Maths* resources

Pupil Textbooks

Discover, **Share** and **Think together** sections promote discussion and introduce mathematical ideas logically, so that children understand more easily.

Using a Concrete-Pictorial-Abstract approach, clear mathematical models help children to make connections and grasp concepts.

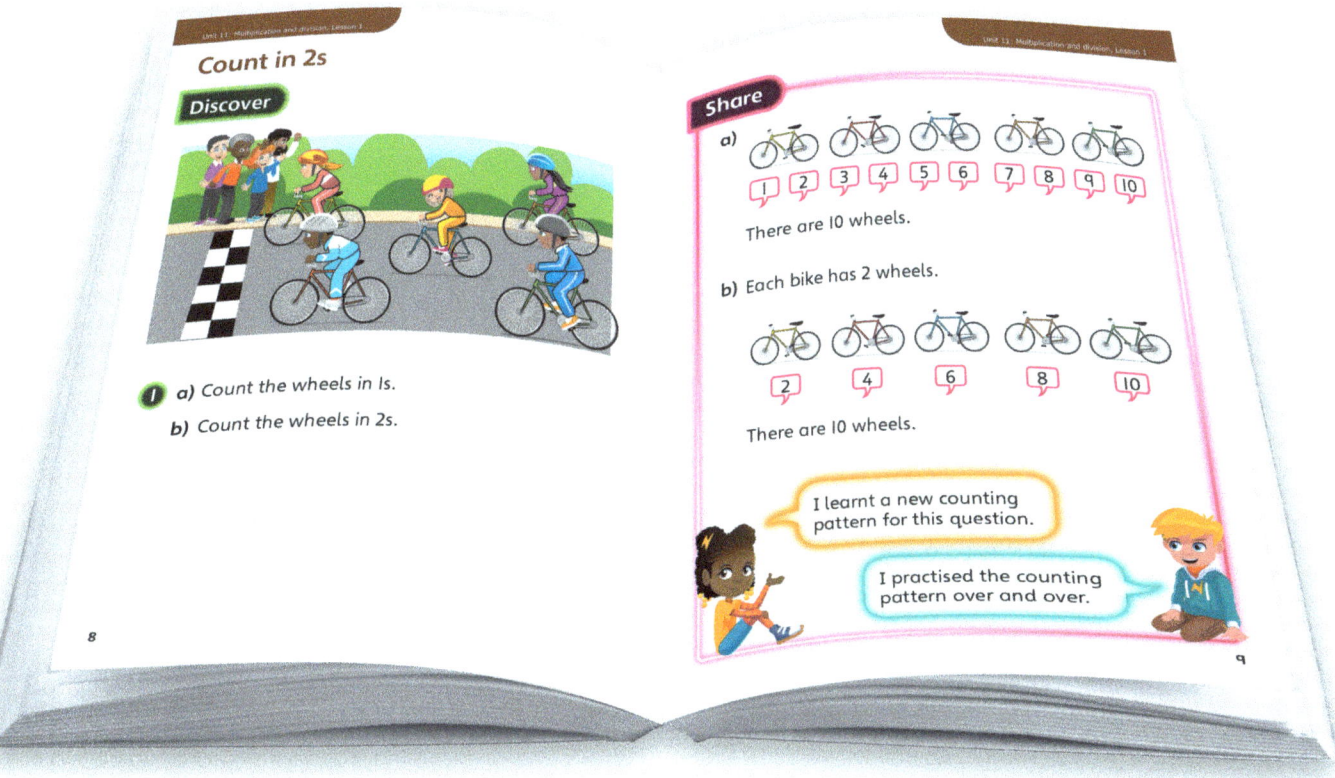

Appealing scenarios stimulate curiosity, helping children to identify the maths problem and discover patterns and relationships for themselves.

Friendly, supportive characters help children develop a growth mindset by prompting them to think, reason and reflect.

To help you teach for mastery, *Power Maths* comprises a variety of high-quality resources.

The coherent *Power Maths* lesson structure carries through into the vibrant, high-quality textbooks. Setting out the core learning objectives for each class, the lesson structure follows a carefully mapped journey through the curriculum and supports children on their journey to deeper understanding.

Pupil Practice Books

The Practice Books offer just the right amount of intelligent practice for children to complete independently in the final section of each lesson.

Practice questions are finely tuned to move children forward in their thinking and to reveal misconceptions.

The practice questions are for everyone – each question varies one small element to move children on in their thinking.

Calculations are connected so that children think about the underlying concept.

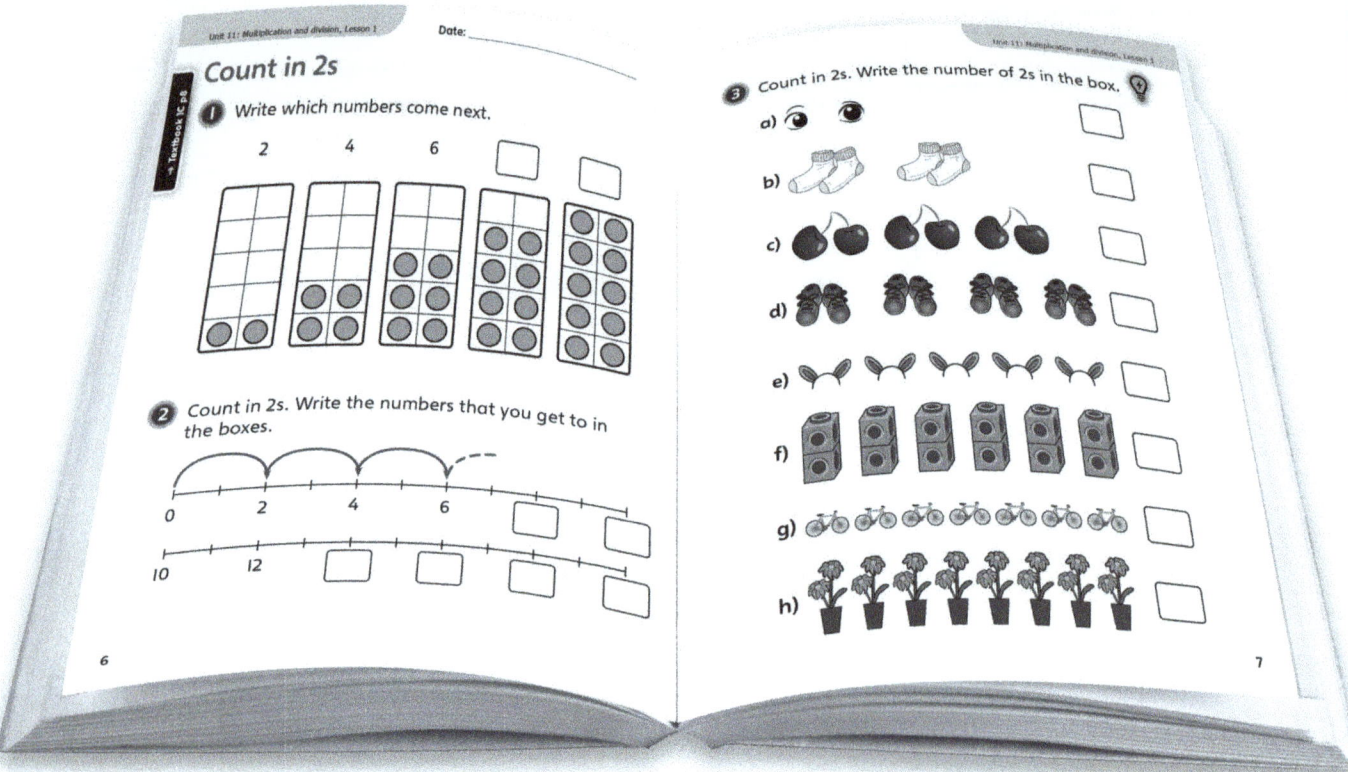

Challenge questions allow children to delve deeper into a concept.

The *Power Maths* characters support and encourage children to think and work in different ways.

Think differently questions encourage children to use reasoning as well as their mathematical knowledge to reach a solution.

Reflect questions reveal the depth of each child's understanding before they move on.

Online subscription

The online subscription will give you access to additional resources and answers from the Textbook and Practice Book.

eTextbooks

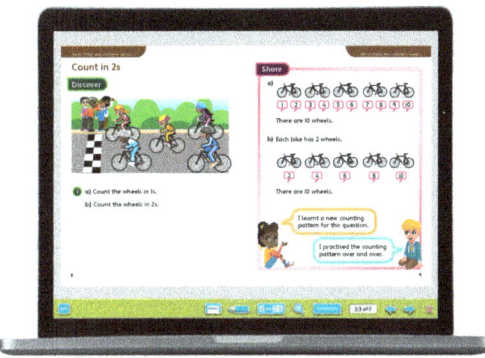

Digital versions of *Power Maths* Textbooks allow class groups to share and discuss questions, solutions and strategies. They allow you to project key structures and representations at the front of the class, to ensure all children are focusing on the same concept.

Teaching tools

Here you will find interactive versions of key *Power Maths* structures and representations.

Power Ups

Use this series of daily activities to promote and check number fluency.

Online versions of Teacher Guide pages

PDF pages give support at both unit and lesson levels. You will also find help with key strategies and templates for tracking progress.

Unit videos

Watch the professional development videos at the start of each unit to help you teach with confidence. The videos explore common misconceptions in the unit, and include intervention suggestions as well as suggestions on what to look out for when assessing mastery in your students.

End of unit Strengthen and Deepen materials

The Strengthen activity at the end of every unit addresses a key misconception and can be used to support children who need it. The Deepen activities are designed to be low ceiling/high threshold and will challenge those children who can understand more deeply. These resources will help you ensure that every child understands and will help you keep the class moving forward together. These printable activities provide an optional resource bank for use after the assessment stage.

Individual Practice Games

These enjoyable games can be used at home or at school to embed key number skills.

Professional Development videos and slides

These slides and videos of *Power Maths* lessons can be used for ongoing training in short bursts or to support new staff.

The *Power Maths* teaching model

At the heart of *Power Maths* is a clearly structured teaching and learning process that helps you make certain that every child masters each maths concept securely and deeply. For each year group, the curriculum is broken down into core concepts, taught in units. A unit divides into smaller learning steps – lessons. Step by step, strong foundations of cumulative knowledge and understanding are built.

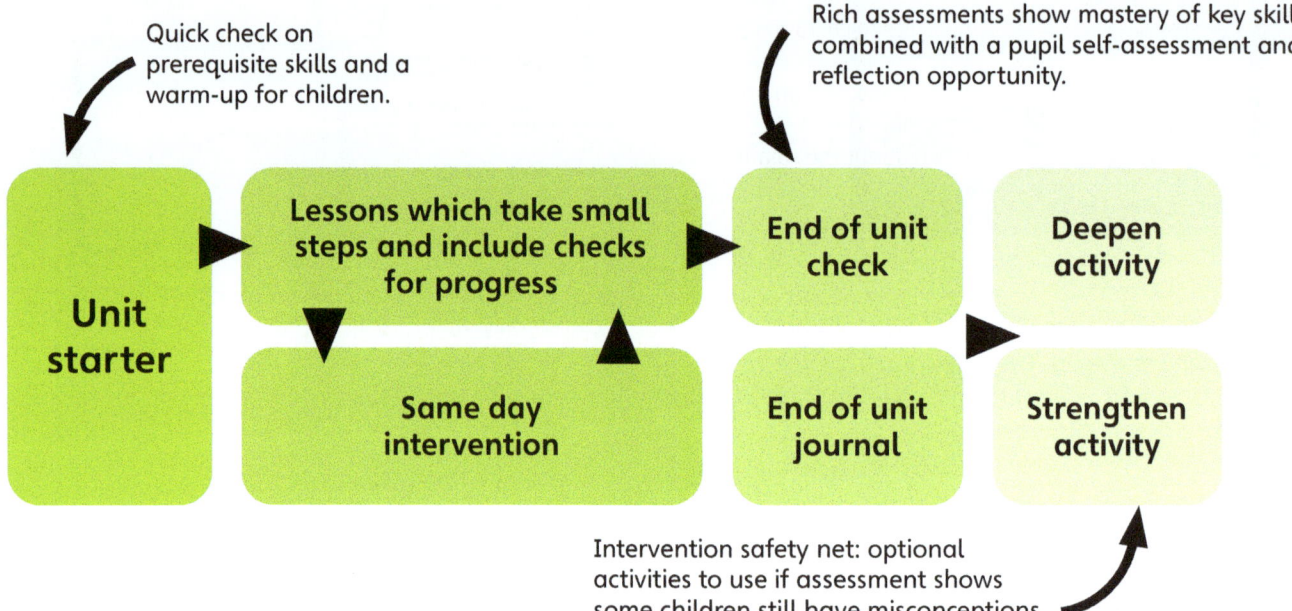

Unit starter

Each unit begins with a unit starter, which introduces the learning context along with key mathematical vocabulary and structures and representations.

- The Textbooks include a check on readiness and a warm-up task for children to complete.
- Your Teacher Guide gives support right from the start on important structures and representations, mathematical language, common misconceptions and intervention strategies.
- Unit-specific videos develop your subject knowledge and insights so you feel confident and fully equipped to teach each new unit. These are available via the online subscription.

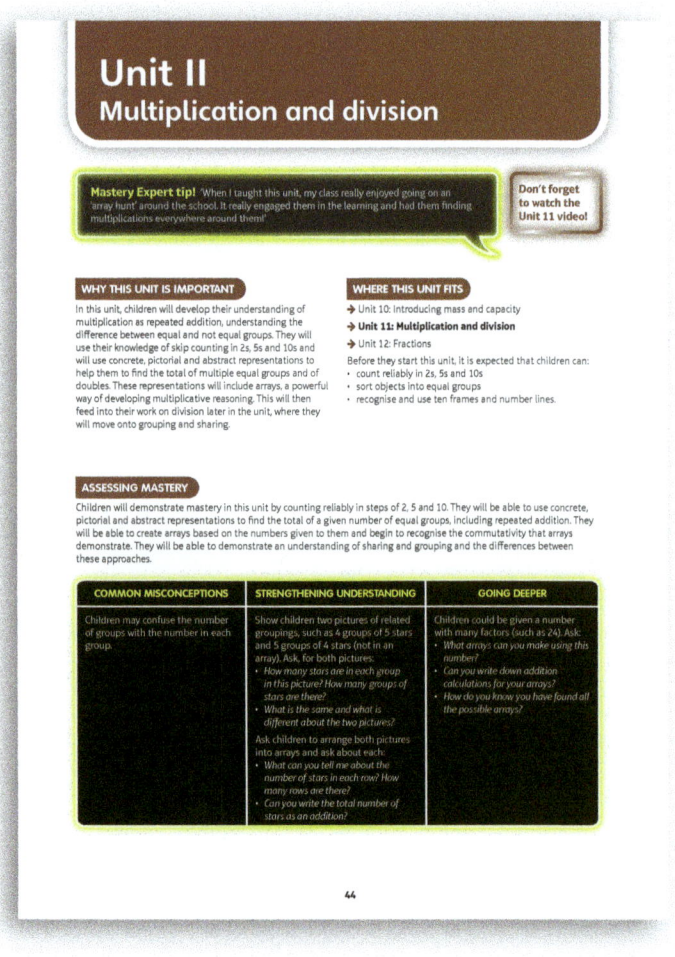

Lesson

Once a unit has been introduced, it is time to start teaching the series of lessons.

- Each lesson is scaffolded with Textbook and Practice Book activities and begins with a Power Up activity (available via online subscription) or the Quick recap activity in the Teacher Guide (see page 15).
- *Power Maths* identifies lesson by lesson what concepts are to be taught.
- Your Teacher Guide offers lots of support for you to get the most from every child in every lesson. As well as highlighting key points, tricky areas and how to handle them, you will also find question prompts to check on understanding and clarification on why particular activities and questions are used.

Same-day intervention

Same-day interventions are vital in order to keep the class progressing together. This can be during the lesson as well as afterwards (see page 27). Therefore, *Power Maths* provides plenty of support throughout the journey.

- Intervention is focused on keeping up now, not catching up later, so interventions should happen as soon as they are needed.
- Practice section questions are designed to bring misconceptions to the surface, allowing you to identify these easily as you circulate during independent practice time.
- Child-friendly assessment questions in the Teacher Guide help you identify easily which children need to strengthen their understanding.

End of unit check and journal

For each unit, the End of unit check in the Textbook lets you see which children have mastered the key concepts, which children have not and where their misconceptions lie. The Practice Books also include an End of unit journal in which children can reflect on what they have learned. Each unit also offers Strengthen and Deepen activities, available via the online subscription.

> The Teacher Guide offers different ways of managing the End of unit assessments as well as giving support with handling misconceptions.

> The End of unit check presents multiple-choice questions. Children think about their answer, decide on a solution and explain their choice.

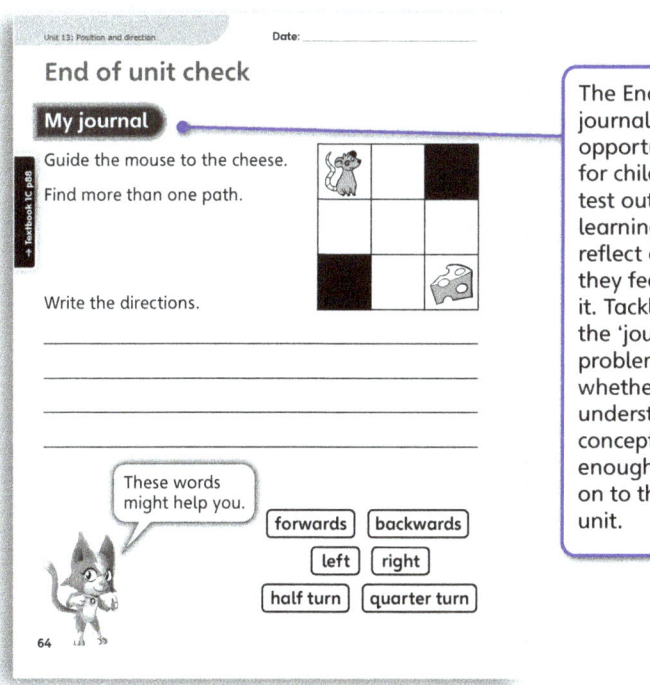

> The End of unit journal is an opportunity for children to test out their learning and reflect on how they feel about it. Tackling the 'journal' problem reveals whether a child understands the concept deeply enough to move on to the next unit.

The *Power Maths* lesson sequence

At the heart of *Power Maths* is a unique lesson sequence designed to empower children to understand core concepts and grow in confidence. Embracing the National Centre for Excellence in the Teaching of Mathematics' (NCETM's) definition of mastery, the sequence guides and shapes every *Power Maths* lesson you teach.

Flexibility is built into the *Power Maths* programme so there is no one-to-one mapping of lessons and concepts and you can pace your teaching according to your class. While some children will need to spend longer on a particular concept (through interventions or additional lessons), others will reach deeper levels of understanding. However, it is important that the class moves forward together through the termly schedules.

Power Up 5 minutes

Each lesson begins with a Power Up activity (available via the online subscription) which supports fluency in key number facts.

The whole-class approach depends on fluency, so the Power Up is a powerful and essential activity.

The Quick recap is an alternative starter, for when you think some or all children would benefit more from revisiting pre-requisite work (see page 15).

TOP TIP
If the class is struggling with the task, revisit it later and check understanding.

Power Ups reinforce the two key things that are essential for success: times-tables and number bonds.

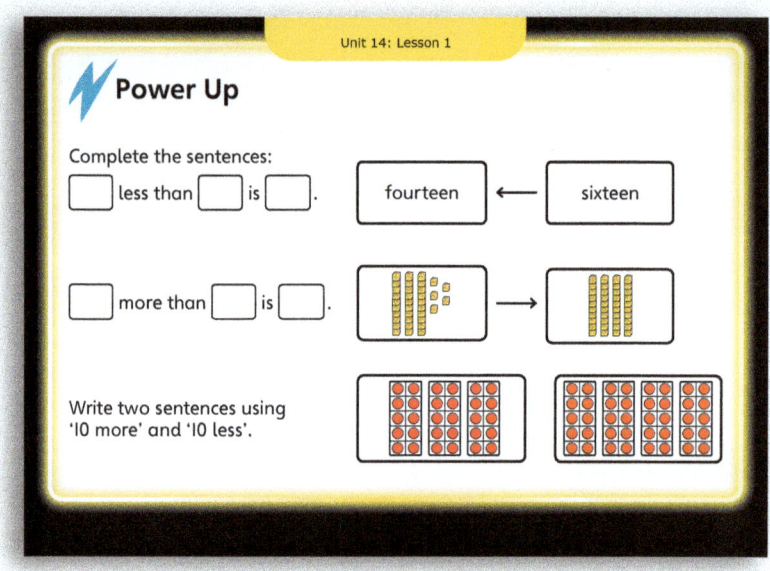

Discover 10 minutes

A practical, real-life problem arouses curiosity. Children find the maths through story telling.

TOP TIP
Discover works best when run at tables, in pairs with concrete objects.

Question ❶ a) tackles the key concept and question ❶ b) digs a little deeper. Children have time to explore, play and discuss possible strategies.

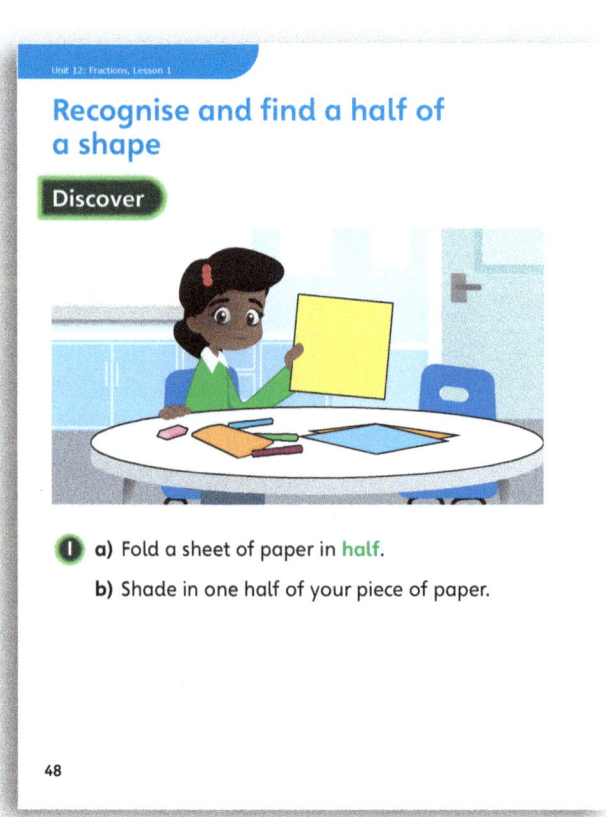

12

Share ⏱ 10 minutes

Teacher-led, this interactive section follows the **Discover** activity and highlights the variety of methods that can be used to solve a single problem.

TOP TIP
You can use the carpet area if you have this. Pairs sharing a textbook is a great format for **Share**!

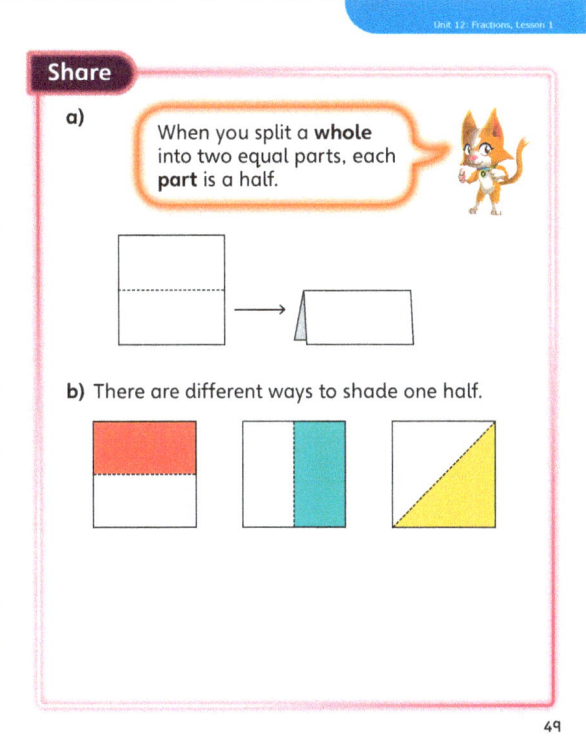

Your Teacher Guide gives target questions for children. The online toolkit provides interactive structures and representations to link concrete and pictorial to abstract concepts.

Bring children to the front to share and celebrate their solutions and strategies.

Think together

⏱ 10 minutes

Children work in groups on the carpet or at tables, using their textbooks or eBooks.

TOP TIP
Make sure children have mini whiteboards or pads to write on if they are not at their tables.

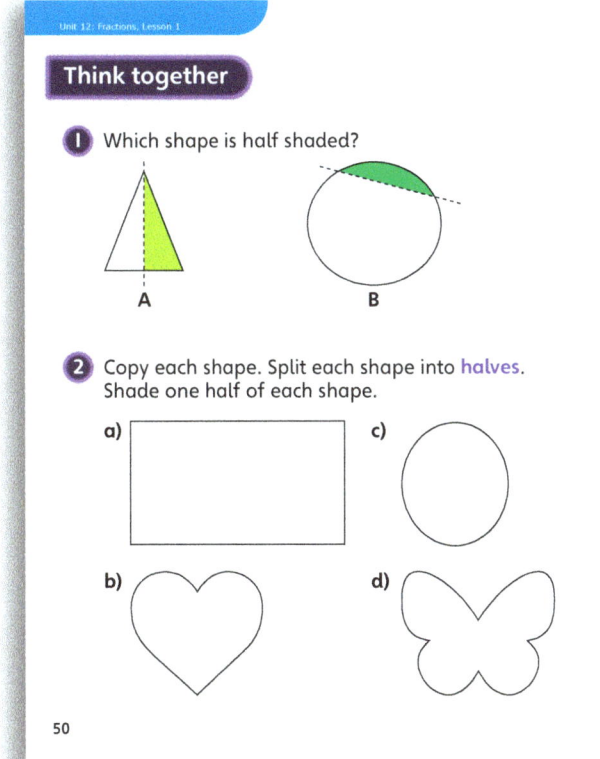

Using the Teacher Guide, model question ❶ for your class.

Question ❷ is less structured. Children will need to think together in their groups, then discuss their methods and solutions as a class.

Question ❸ – the openness of the **Challenge** question helps to check depth of understanding.

Practice ⏱ 15 minutes

Using their Practice Books, children work independently while you circulate and check on progress.

Questions follow small steps of progression to deepen learning.

TOP TIP
Some children could work separately with a teacher or assistant.

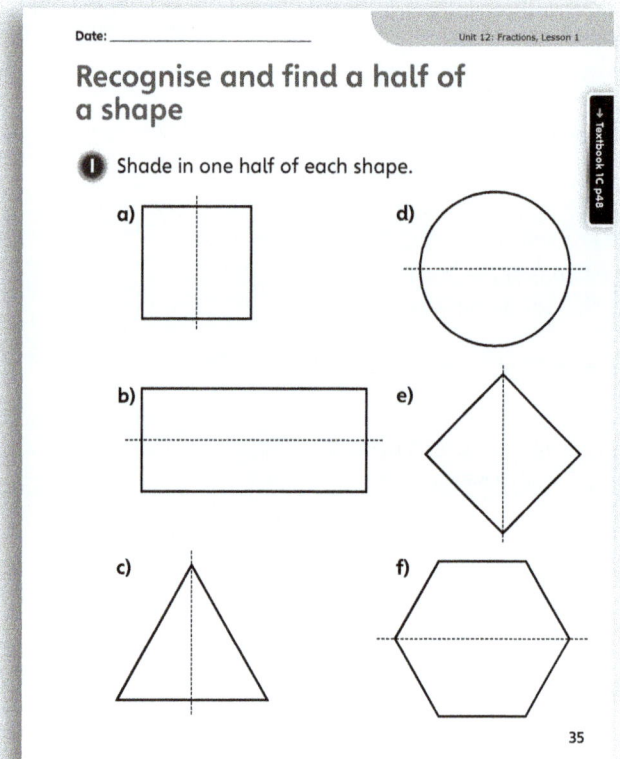

Are some children struggling? If so, work with them as a group, using mathematical structures and representations to support understanding as necessary.

There are no set routines: for real understanding, children need to think about the problem in different ways.

Reflect ⏱ 5 minutes

'Spot the mistake' questions are great for checking misconceptions.

The **Reflect** section is your opportunity to check how deeply children understand the target concept.

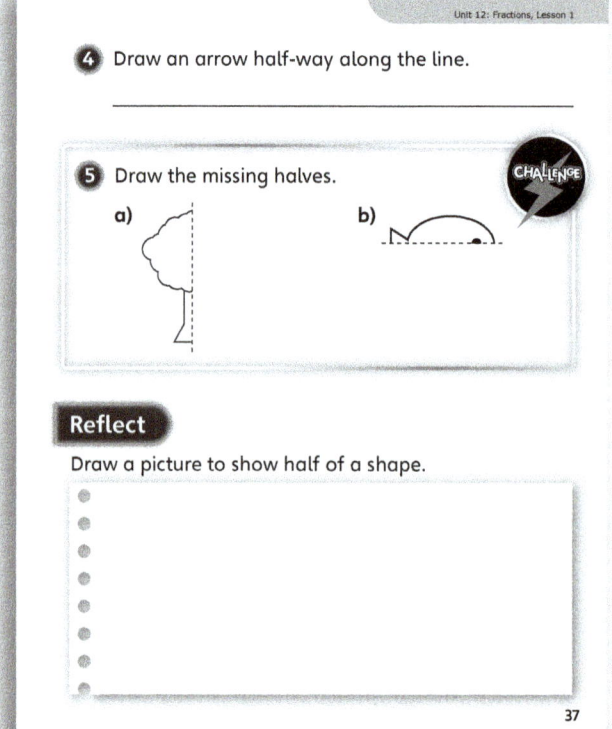

The Practice Books use various approaches to check that children have fully understood each concept.

Looking like they understand is not enough! It is essential that children can show they have grasped the concept.

Using the *Power Maths* Teacher Guide

Think of your Teacher Guides as *Power Maths* handbooks that will guide, support and inspire your day-to-day teaching. Clear and concise, and illustrated with helpful examples, your Teacher Guides will help you make the best possible use of every individual lesson. They also provide wrap-around professional development, enhancing your own subject knowledge and helping you to grow in confidence about moving your children forward together.

There is a Teacher Guide per year group for every term, with unit and lesson level guidance and support.

Never feel stuck! You will find ideas for introducing every unit and lesson and questions to encourage teacher reflection before and after each lesson.

Tips and advice on key elements such as C-P-A approaches, misconceptions, language, modelling growth mindsets and same day intervention.

Annotations for every Textbook and Practice Book page, providing prompts for key questions to ask to expose understanding and explanations as to why key questions have been chosen.

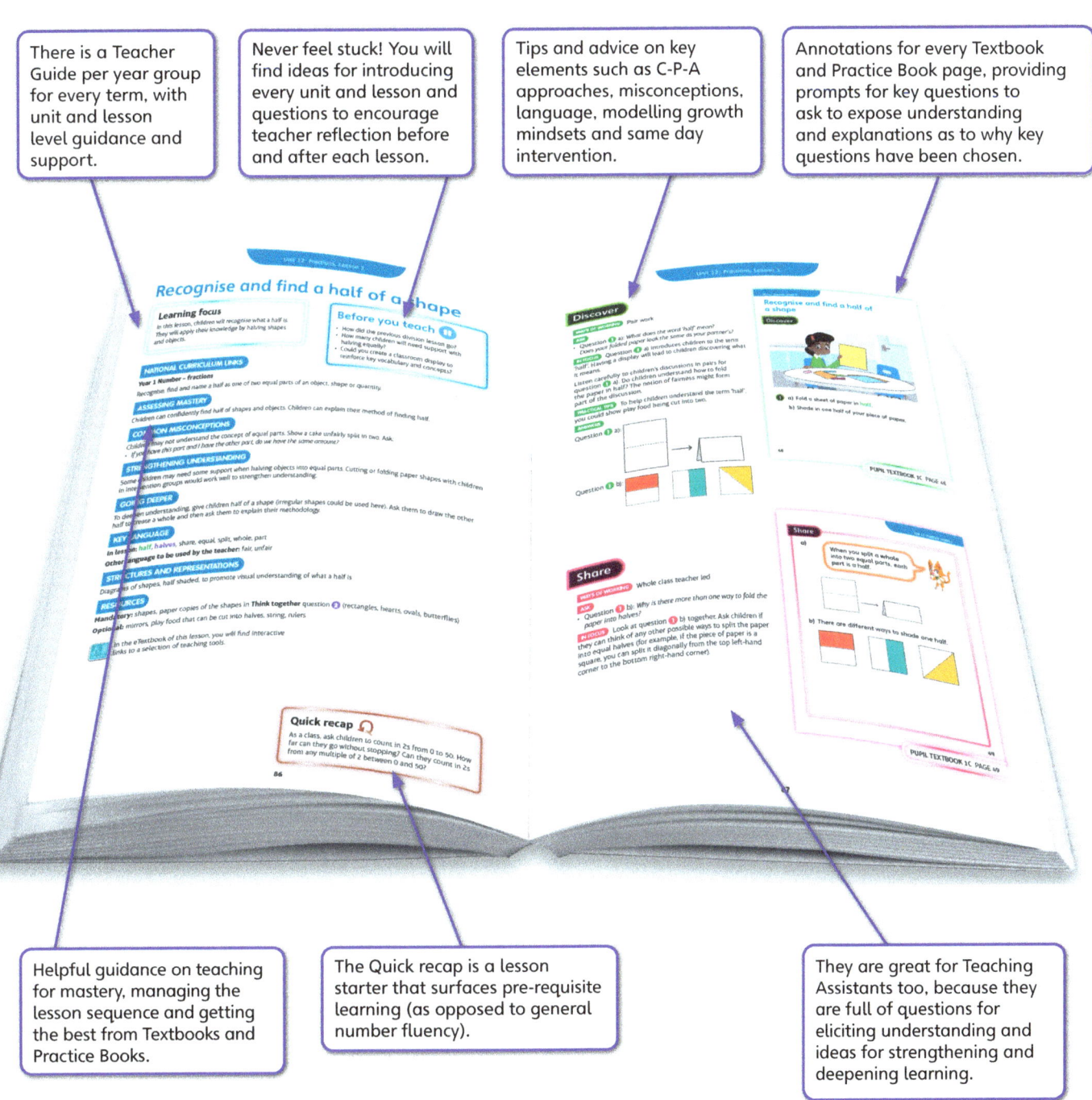

Helpful guidance on teaching for mastery, managing the lesson sequence and getting the best from Textbooks and Practice Books.

The Quick recap is a lesson starter that surfaces pre-requisite learning (as opposed to general number fluency).

They are great for Teaching Assistants too, because they are full of questions for eliciting understanding and ideas for strengthening and deepening learning.

At the end of each unit, your Teacher Guide helps you identify who has fully grasped the concept, who has not and how to move every child forward. This is covered later in the Assessment strategies section.

Power Maths Year 1, yearly overview

Textbook	Strand	Unit		Number of lessons
Textbook A / Practice Book A (Term 1)	Number – number and place value	1	Numbers to 10	14
	Number – addition and subtraction	2	Part-whole within 10	7
	Number – addition and subtraction	3	Addition awithin 10	4
	Number – addition and subtraction	4	Subtraction within 10	8
	Geometry – properties of shape	5	2D and 3D shapes	5
Textbook B / Practice Book B (Term 2)	Number – number and place value	6	Numbers to 20	12
	Number – addition and subtraction	7	Addition and subtraction within 20	11
	Number – number and place value	8	Numbers to 50	7
	Measurement	9	Introducing length and height	4
	Measurement	10	Introducing weight and volume	7
Textbook C / Practice Book C (Term 3)	Number – multiplication and division	11	Multiplication and division	9
	Number – fractions	12	Fractions	4
	Geometry – position and direction	13	Position and direction	5
	Number – number and place value	14	Numbers to 100	6
	Measurement	15	Money	3
	Measurement	16	Time	5

Power Maths Year 1, Textbook 1C (Term 3) overview

Strand	Unit	Unit title	Lesson number	Lesson title	NC Objective 1	NC Objective 2
Number – multiplication and division	11	Multiplication and division	1	Count in 2s	Count, read and write numbers to 100 in numerals; count in multiples of twos, fives and tens	
Number – multiplication and division	11	Multiplication and division	2	Count in 10s	Count, read and write numbers to 100 in numerals; count in multiples of twos, fives and tens	
Number – multiplication and division	11	Multiplication and division	3	Count in 5s	Count, read and write numbers to 100 in numerals; count in multiples of twos, fives and tens	
Number – multiplication and division	11	Multiplication and division	4	Equal groups	Solve one-step problems involving multiplication and division, by calculating the answer using concrete objects, pictorial representations and arrays with the support of the teacher	
Number – multiplication and division	11	Multiplication and division	5	Add equal groups	Solve one-step problems involving multiplication and division, by calculating the answer using concrete objects, pictorial representations and arrays with the support of the teacher	
Number – multiplication and division	11	Multiplication and division	6	Make arrays	Solve one-step problems involving multiplication and division, by calculating the answer using concrete objects, pictorial representations and arrays with the support of the teacher	
Number – multiplication and division	11	Multiplication and division	7	Make doubles	Solve one-step problems involving multiplication and division, by calculating the answer using concrete objects, pictorial representations and arrays with the support of the teacher	Non statutory guidance: through grouping and sharing small quantities, pupils begin to understand: multiplication and division; doubling numbers and quantities; and finding simple fractions of objects, numbers and quantities
Number – multiplication and division	11	Multiplication and division	8	Grouping	Solve one-step problems involving multiplication and division, by calculating the answer using concrete objects, pictorial representations and arrays with the support of the teacher	
Number – multiplication and division	11	Multiplication and division	9	Sharing	Solve one-step problems involving multiplication and division, by calculating the answer using concrete objects, pictorial representations and arrays with the support of the teacher	
Number – fractions	12	Fractions	1	Recognise and find a half of a shape	Recognise, find and name a half as one of two equal parts of an object, shape or quantity	
Number – fractions	12	Fractions	2	Recognise and find a half of a quantity	Recognise, find and name a half as one of two equal parts of an object, shape or quantity	
Number – fractions	12	Fractions	3	Recognise and find a quarter of a shape	Recognise, find and name a quarter as one of four equal parts of an object, shape or quantity.	
Number – fractions	12	Fractions	4	Recognise and find a quarter of a quantity	Recognise, find and name a quarter as one of four equal parts of an object, shape or quantity.	

Strand	Unit	Unit title	Lesson number	Lesson title	NC Objective 1	NC Objective 2
Geometry – position and direction	13	Position and direction	1	Describe turns	Describe position, direction and movement, including whole, half, quarter and three-quarter turns	
Geometry – position and direction	13	Position and direction	2	Describe position – left and right	Non statutory guidance: Pupils use the language of position, direction and motion, including: left and right, top, middle and bottom, on top of, in front of, above, between, around, near, close and far, up and down, forwards and backwards, inside and outside	
Geometry – position and direction	13	Position and direction	3	Describe position – forwards and backwards	Non statutory guidance: Pupils use the language of position, direction and motion, including: left and right, top, middle and bottom, on top of, in front of, above, between, around, near, close and far, up and down, forwards and backwards, inside and outside.	
Geometry – position and direction	13	Position and direction	4	Describe position – above and below	Non statutory guidance: Pupils use the language of position, direction and motion, including: left and right, top, middle and bottom, on top of, in front of, above, between, around, near, close and far, up and down, forwards and backwards, inside and outside.	
Geometry – position and direction	13	Position and direction	5	Ordinal numbers	Non-statutory guidance: Pupils practise counting (1, 2, 3…), ordering (for example, first, second, third…), and to indicate a quantity (for example, 3 apples, 2 centimetres), including solving simple concrete problems, until they are fluent.	
Number – number and place value	14	Numbers to 100	1	Count from 50 to 100	Count, read and write numbers to 100 in numerals; count in multiples of twos, fives and tens	
Number – number and place value	14	Numbers to 100	2	10s to 100	Count, read and write numbers to 100 in numerals; count in multiples of twos, fives and tens	
Number – number and place value	14	Numbers to 100	3	Partition into 10s and 1s	Identify and represent numbers using objects and pictorial representations including the number line, and use the language of: equal to, more than, less than (fewer), most, least	Recognise the place value of each digit in a two-digit number (tens, ones) (year 2)
Number – number and place value	14	Numbers to 100	4	Number line to 100	Identify and represent numbers using objects and pictorial representations including the number line, and use the language of: equal to, more than, less than (fewer), most, least	
Number – number and place value	14	Numbers to 100	5	One more and one less	Given a number, identify one more and one less	
Number – number and place value	14	Numbers to 100	6	Compare numbers	Identify and represent numbers using objects and pictorial representations including the number line, and use the language of: equal to, more than, less than (fewer), most, least	
Measurement	15	Money	1	Recognise coins	Recognise and know the value of different denominations of coins and notes	

Strand	Unit	Unit title	Lesson number	Lesson title	NC Objective 1	NC Objective 2
Measurement	15	Money	2	Recognise notes	Recognise and know the value of different denominations of coins and notes	
Measurement	15	Money	3	Count in coins	Recognise and know the value of different denominations of coins and notes	
Measurement	16	Time	1	Before and after	Sequence events in chronological order using language [for example, before and after, next, first, today, yesterday, tomorrow, morning, afternoon and evening]	
Measurement	16	Time	2	Days of the week	Recognise and use language relating to dates, including days of the week, weeks, months and years	
Measurement	16	Time	3	Months of the year	Recognise and use language relating to dates, including days of the week, weeks, months and years	
Measurement	16	Time	4	Tell the time to the hour	Tell the time to the hour and half past the hour and draw the hands on a clock face to show these times	
Measurement	16	Time	5	Tell the time to the half hour	Tell the time to the hour and half past the hour and draw the hands on a clock face to show these times	

Mindset: an introduction

Global research and best practice deliver the same message: learning is greatly affected by what learners perceive they can or cannot do. What is more, it is also shaped by what their parents, carers and teachers perceive they can do. Mindset – the thinking that determines our beliefs and behaviours – therefore has a fundamental impact on teaching and learning.

Everyone can!

Power Maths and mastery methods focus on the distinction between 'fixed' and 'growth' mindsets (Dweck, 2007).[1] Those with a fixed mindset believe that their basic qualities (for example, intelligence, talent and ability to learn) are pre-wired or fixed: 'If you have a talent for maths, you will succeed at it. If not, too bad!' By contrast, those with a growth mindset believe that hard work, effort and commitment drive success and that 'smart' is not something you are or are not, but something you become. In short, everyone can do maths!

Key mindset strategies

A growth mindset needs to be actively nurtured and developed. *Power Maths* offers some key strategies for fostering healthy growth mindsets in your classroom.

It is okay to get it wrong

Mistakes are valuable opportunities to re-think and understand more deeply. Learning is richer when children and teachers alike focus on spotting and sharing mistakes as well as solutions.

Praise hard work

Praise is a great motivator, and by focusing on praising effort and learning rather than success, children will be more willing to try harder, take risks and persist for longer.

Mind your language!

The language we use around learners has a profound effect on their mindsets. Make a habit of using growth phrases, such as, 'Everyone can!', 'Mistakes can help you learn' and 'Just try for a little longer'. The king of them all is one little word, 'yet'… I can't solve this…yet!' Encourage parents and carers to use the right language too.

Build in opportunities for success

The step-by-small-step approach enables children to enjoy the experience of success. In addition, avoid ability grouping and encourage every child to answer questions and explain or demonstrate their methods to others.

[1]Dweck, C (2007) *The New Psychology of Success*, Ballantine Books: New York

The *Power Maths* characters

The *Power Maths* characters model the traits of growth mindset learners and encourage resilience by prompting and questioning children as they work. Appearing frequently in the Textbooks and Practice Books, they are your allies in teaching and discussion, helping to model methods, alternatives and misconceptions, and to pose questions. They encourage and support your children, too: they are all hardworking, enthusiastic and unafraid of making and talking about mistakes.

Meet the team!

Creative Flo is open-minded and sometimes indecisive. She likes to think differently and come up with a variety of methods or ideas.

Determined Dexter is resolute, resilient and systematic. He concentrates hard, always tries his best and he'll never give up – even though he doesn't always choose the most efficient methods!

'Let's try again.'
'Mistakes are cool!'
'Have I found all of the solutions?'

'Let's try it this way…'
'Can we do it differently?'
'I've got another way of doing this!'

'I'm going to try this!'
'I know how to do that!'
'Want to share my ideas?'

Curious Ash is eager, interested and inquisitive, and he loves solving puzzles and problems. Ash asks lots of questions but sometimes gets distracted.

'What if we tried this…?'
'I wonder…'
'Is there a pattern here?'

Sparks the Cat

Miaow!

Brave Astrid is confident, willing to take risks and unafraid of failure. She's never scared to jump straight into a problem or question, and although she often makes simple mistakes, she's happy to talk them through with others.

Mathematical language

Traditionally, we in the UK have tended to try simplifying mathematical language to make it easier for young children to understand. By contrast, evidence and experience show that by diluting the correct language, we actually mask concepts and meanings for children. We then wonder why they are confused by new and different terminology later down the line! *Power Maths* is not afraid of 'hard' words and avoids placing any barriers between children and their understanding of mathematical concepts. As a result, we need to be deliberate, precise and thorough in building every child's understanding of the language of maths. Throughout the Teacher Guides you will find support and guidance on how to deliver this, as well as individual explanations throughout the pupil Textbooks.

Use the following key strategies to build children's mathematical vocabulary, understanding and confidence.

Precise and consistent

Everyone in the classroom should use the correct mathematical terms in full, every time. For example, refer to 'equal parts', not 'parts'. Used consistently, precise maths language will be a familiar and non-threatening part of children's everyday experience.

Full sentences

Teachers and children alike need to use full sentences to explain or respond. When children use complete sentences, it both reveals their understanding and embeds their knowledge.

Stem sentences

These important sentences help children express mathematical concepts accurately, and are used throughout the *Power Maths* books. Encourage children to repeat them frequently, whether working independently or with others. Examples of stem sentences are:

'4 is a part, 5 is a part, 9 is the whole.'

'There are …. groups. There are …. in each group.'

Key vocabulary

The unit starters highlight essential vocabulary for every lesson. In the pupil books, characters flag new terminology and the Teacher Guide lists important mathematical language for every unit and lesson. New terms are never introduced without a clear explanation.

Symbolic language

Symbols are used early on so that children quickly become familiar with them and their meaning. Often, the *Power Maths* characters will highlight the connection between language and particular symbols.

The role of talk and discussion

When children learn to talk purposefully together about maths, barriers of fear and anxiety are broken down and they grow in confidence, skills and understanding. Building a healthy culture of 'maths talk' empowers their learning from day one.

Explanation and discussion are integral to the *Power Maths* structure, so by simply following the books, your lessons will stimulate structured talk. The following key 'maths talk' strategies will help you strengthen that culture and ensure that every child is included.

Sentences, not words

Encourage children to use full sentences when reasoning, explaining or discussing maths. This helps both speaker and listeners to clarify their own understanding. It also reveals whether or not the speaker truly understands, enabling you to address misconceptions as they arise.

Working together

Working with others in pairs, groups or as a whole class is a great way to support maths talk and discussion. Use different group structures to add variety and challenge. For example, children could take timed turns for talking, work independently alongside a 'discussion buddy', or perhaps play different *Power Maths* character roles within their group.

Think first – then talk

Provide clear opportunities within each lesson for children to think and reflect, so that their talk is purposeful, relevant and focused.

Give every child a voice

Where the 'hands up' model allows only the more confident child to shine, *Power Maths* involves everyone. Make sure that no child dominates and that even the shyest child is encouraged to contribute – and praised when they do.

Assessment strategies

Teaching for mastery demands that you are confident about what each child knows and where their misconceptions lie; therefore, practical and effective assessment is vitally important.

Formative assessment within lessons

The **Think together** section will often reveal any confusions or insecurities; try ironing these out by doing the first **Think together** question as a class. For children who continue to struggle, you or your Teaching Assistant should provide support and enable them to move on.

▶ Performance in practice can be very revealing: check Practice Books and listen out both during and after practice to identify misconceptions.

▶ The **Reflect** section is designed to check on the all-important depth of understanding. Be sure to review how the children performed in this final stage before you teach the next lesson.

End of unit check – Textbook

Each unit concludes with a summative check to help you assess quickly and clearly each child's understanding, fluency, reasoning and problem solving skills. Your Teacher Guide will suggest ideal ways of organising a given activity and offer advice and commentary on what children's responses mean. For example, 'What misconception does this reveal?'; 'How can you reinforce this particular concept?'

For Year 1 and Year 2 children, assess in small, teacher-led groups, giving each child time to think and respond while also consolidating correct mathematical language. Assessment with young children should always be an enjoyable activity, so avoid one-to-one individual assessments, which they may find threatening or scary. If you prefer, the End of unit check can be carried out as a whole-class group using whiteboards and Practice Books.

End of unit check – Practice Book

The Practice Book contains further opportunities for assessment, and can be completed by children independently whilst you are carrying out diagnostic assessment with small groups. Your Teacher Guide will advise you on what to do if children struggle to articulate an explanation – or perhaps encourage you to write down something they have explained well. It will also offer insights into children's answers and their implications for next learning steps. It is split into three main sections, outlined below.

My journal is designed to allow children to show their depth of understanding of the unit. It can also serve as a way of checking that children have grasped key mathematical vocabulary. The question children should answer is first presented in the Textbook in the Think! section. This provides an opportunity for you to discuss the question first as a class to ensure children have understood their task. Children should have some time to think about how they want to answer the question, and you could ask them to talk to a partner about their ideas. Then children should write their answer in their Practice Book, using the word bank provided to help them with vocabulary.

The **Power check** allows pupils to self-assess their level of confidence on the topic by colouring in different smiley faces. You may want to introduce the faces as follows:

Each unit ends with either a Power play or a Power puzzle. This is an activity, puzzle or game that allows children to use their new knowledge in a fun, informal way.

Progress Tests

There are *Power Maths* Progress Tests for each half term and at the end of the year, including an Arithmetic test and Reasoning test in each case. You can enter results in the online markbook to track and analyse results and see the average for all schools' results. The tests use a 6-step scale to show results against age-related expectation.

How to ask diagnostic questions

The diagnostic questions provided in children's Practice Books are carefully structured to identify both understanding and misconceptions (if children answer in a particular way, you will know why). The simple procedure below may be helpful:

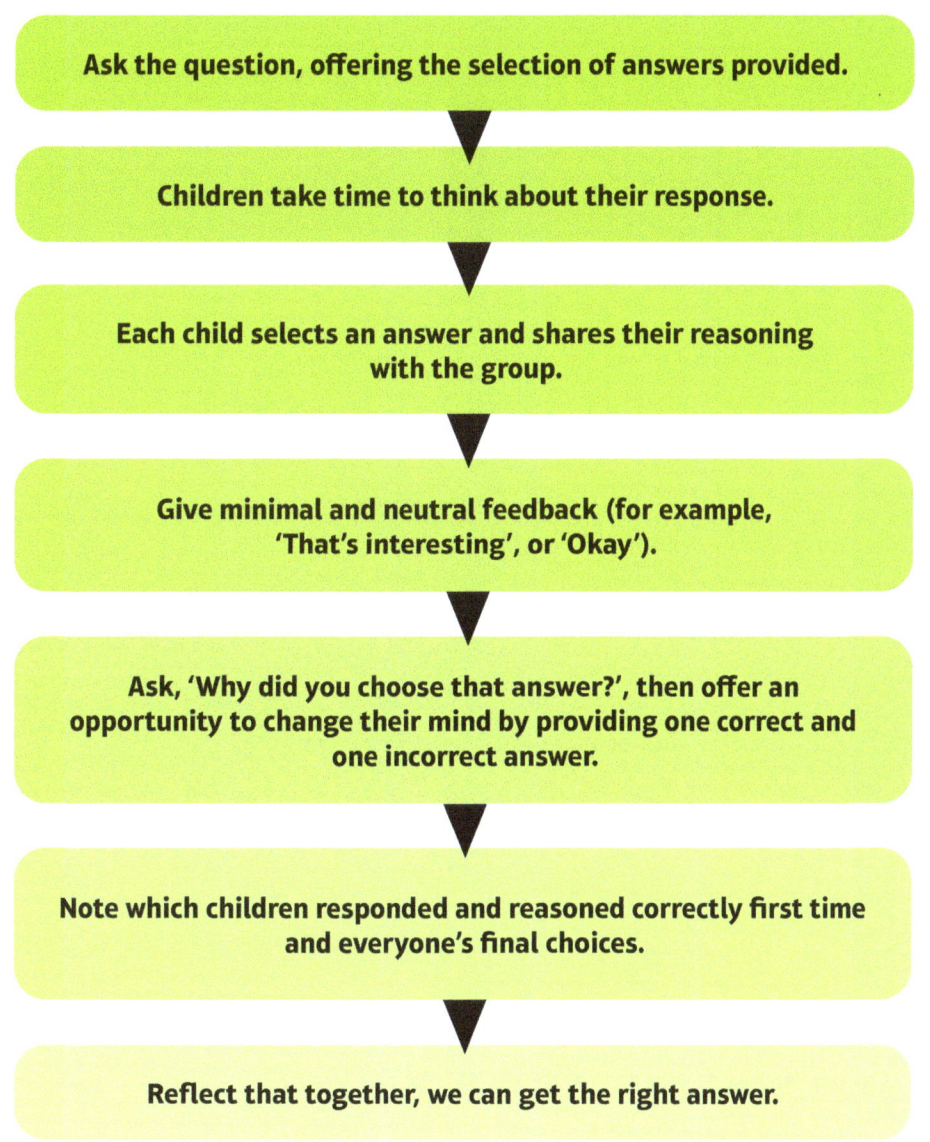

Keeping the class together

Traditionally, children who learn quickly have been accelerated through the curriculum. As a consequence, their learning may be superficial and will lack the many benefits of enabling children to learn with and from each other.

By contrast, *Power Maths'* mastery approach values real understanding and richer, deeper learning above speed. It sees all children learning the same concept in small, cumulative steps, each finding and mastering challenge at their own level. Remember that when you teach for mastery, EVERYONE can do maths! Those who grasp a concept easily have time to explore and understand that concept at a deeper level. The whole class therefore moves through the curriculum at broadly the same pace via individual learning journeys.

For some teachers, the idea that a whole class can move forward together is revolutionary and challenging. However, the evidence of global good practice clearly shows that this approach drives engagement, confidence, motivation and success for all learners, and not just the high flyers. The strategies below will help you keep your class together on their maths journey.

Mix it up

Do not stick to set groups at each table. Every child should be working on the same concept, and mixing up the groupings widens children's opportunities for exploring, discussing and sharing their understanding with others.

Recycling questions

Reuse the Textbook and Practice Book questions with concrete materials to allow children to explore concepts and relationships and deepen their understanding. This strategy is especially useful for reinforcing learning in same-day interventions.

Strengthen at every opportunity

The next lesson in a *Power Maths* sequence always revises and builds on the previous step to help embed learning. These activities provide golden opportunities for individual children to strengthen their learning with the support of Teaching Assistants.

Prepare to be surprised!

Children may grasp a concept quickly or more slowly. The 'fast graspers' won't always be the same individuals, nor does the speed at which a child understands a concept predict their success in maths. Are they struggling or just working more slowly?

Same-day intervention

Since maths competence depends on mastering concepts one by one in a logical progression, it is important that no gaps in understanding are ever left unfilled. Same-day interventions – either within or after a lesson – are a crucial safety net for any child who has not fully made the small step covered that day. In other words, intervention is always about keeping up, not catching up, so that every child has the skills and understanding they need to tackle the next lesson. That means presenting the same problems used in the lesson, with a variety of concrete materials to help children model their solutions.

We offer two intervention strategies below, but you should feel free to choose others if they work better for your class.

Within-lesson intervention

The **Think together** activity will reveal those who are struggling, so when it is time for practice, bring these children together to work with you on the first practice questions. Observe these children carefully, ask questions, encourage them to use concrete models and check that they reach and can demonstrate their understanding.

After-lesson intervention

You might like to use the **Think together** questions to recap the lesson with children who are working behind expectations during assembly time. Teaching Assistants could also work with these children at other convenient points in the school day. Some children may benefit from revisiting work from the same topic in the previous year group. Note also the suggestion for recycling questions from the Textbook and Practice Book with concrete materials on page 26.

The role of practice

Practice plays a pivotal role in the *Power Maths* approach. It takes place in class groups, smaller groups, pairs, and independently, so that children always have the opportunities for thinking as well as the models and support they need to practise meaningfully and with understanding.

Intelligent practice

In *Power Maths*, practice never equates to the simple repetition of a process. Instead we embrace the concept of intelligent practice, in which all children become fluent in maths through varied, frequent and thoughtful practice that deepens and embeds conceptual understanding in a logical, planned sequence. To see the difference, take a look at the following examples.

Traditional practice
- Repetition can be rote – no need for a child to think hard about what they are doing
- Praise may be misplaced
- Does this prove understanding?

Intelligent practice
- Varied methods – concrete, pictorial and abstract
- Equation expressed in different ways, requiring thought and understanding
- Constructive feedback

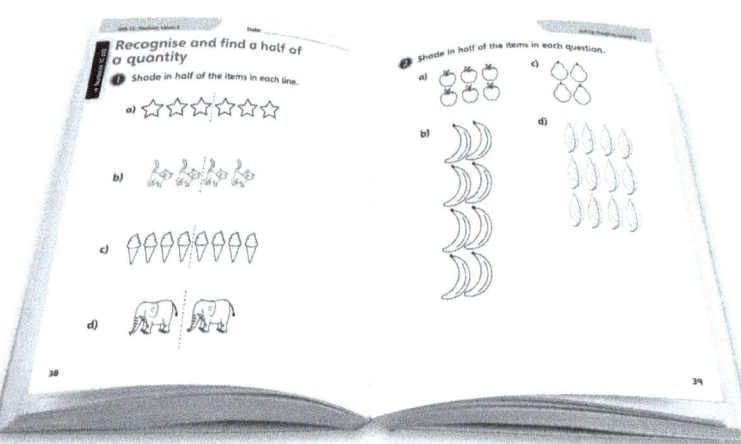

- All practice questions are designed to move children on and reveal misconceptions.
- Simple, logical steps build onto earlier learning.
- C-P-A runs throughout – different ways of modelling and understanding the same concept.
- Conceptual variation – children work on different representations of the same maths concept.
- Friendly characters offer support and encourage children to try different approaches.

A carefully designed progression

The Practice Books provide just the right amount of intelligent practice for children to complete independently in the final sections of each lesson. It is really important that all children are exposed to the practice questions, and that children are not directed to complete different sections. That is because each question is different and has been designed to challenge children to think about the maths they are doing. The questions become more challenging so children grasping concepts more quickly will start to slow down as they progress. Meanwhile, you have the chance to circulate and spot any misconceptions before they become barriers to further learning.

Homework and the role of parents and carers

While *Power Maths* does not prescribe any particular homework structure, we acknowledge the potential value of practice at home. For example, practising fluency in key facts, such as number bonds and times-tables, is an ideal homework task. You can share the Individual Practice Games for homework (see pages 6 and 9), or parents and carers could work through uncompleted Practice Book questions with children at either primary stage.

However, it is important to recognise that many parents and carers may themselves lack confidence in maths, and few, if any, will be familiar with mastery methods. A Parents' and Carers' evening that helps them understand the basics of mindsets, mastery and mathematical language is a great way to ensure that children benefit from their homework. It could be a fun opportunity for children to teach their families that everyone can do maths!

Structures and representations

Unlike most other subjects, maths comprises a wide array of abstract concepts – and that is why children and adults so often find it difficult. By taking a concrete-pictorial-abstract (C-P-A) approach, *Power Maths* allows children to tackle concepts in a tangible and more comfortable way.

Non-linear stages

Concrete

Replacing the traditional approach of a teacher working through a problem in front of the class, the concrete stage introduces real objects that children can use to 'do' the maths – any familiar object that a child can manipulate and move to help bring the maths to life. It is important to appreciate, however, that children must always understand the link between models and the objects they represent. For example, children need to first understand that three cakes could be represented by three pretend cakes, and then by three counters or bricks. Frequent practice helps consolidate this essential insight. Although they can be used at any time, good concrete models are an essential first step in understanding.

Pictorial

This stage uses pictorial representations of objects to let children 'see' what particular maths problems look like. It helps them make connections between the concrete and pictorial representations and the abstract maths concept. Children can also create or view a pictorial representation together, enabling discussion and comparisons. The *Power Maths* teaching tools are fantastic for this learning stage, and bar modelling is invaluable for problem solving throughout the primary curriculum.

Abstract

Our ultimate goal is for children to understand abstract mathematical concepts, symbols and notation and of course, some children will reach this stage far more quickly than others. To work with abstract concepts, a child must be comfortable with the meaning of and relationships between concrete, pictorial and abstract models and representations. The C-P-A approach is not linear, and children may need different types of models at different times. However, when a child demonstrates with concrete models and pictorial representations that they have grasped a concept, we can be confident that they are ready to explore or model it with abstract symbols such as numbers and notation.

Use at any time and with any age to support understanding

Variation helps visualisation

Children find it much easier to visualise and grasp concepts if they see them presented in a number of ways, so be prepared to offer and encourage many different representations.

For example, the number six could be represented in various ways:

Practical aspects of *Power Maths*

One of the key underlying elements of *Power Maths* is its practical approach, allowing you to make maths real and relevant to your children, no matter their age.

Manipulatives are essential resources for both key stages and *Power Maths* encourages teachers to use these at every opportunity, and to continue the Concrete-Pictorial-Abstract approach right through to Year 6.

The Textbooks and Teacher Guides include lots of opportunities for teaching in a practical way to show children what maths means in real life.

Discover and Share

The **Discover** and **Share** sections of the Textbook give you scope to turn a real-life scenario into a practical and hands-on section of the lesson. Use these sections as inspiration to get active in the classroom. Where appropriate, use the **Discover** contexts as a springboard for your own examples that have particular resonance for your children – and allow them to get their hands dirty trying out the mathematics for themselves.

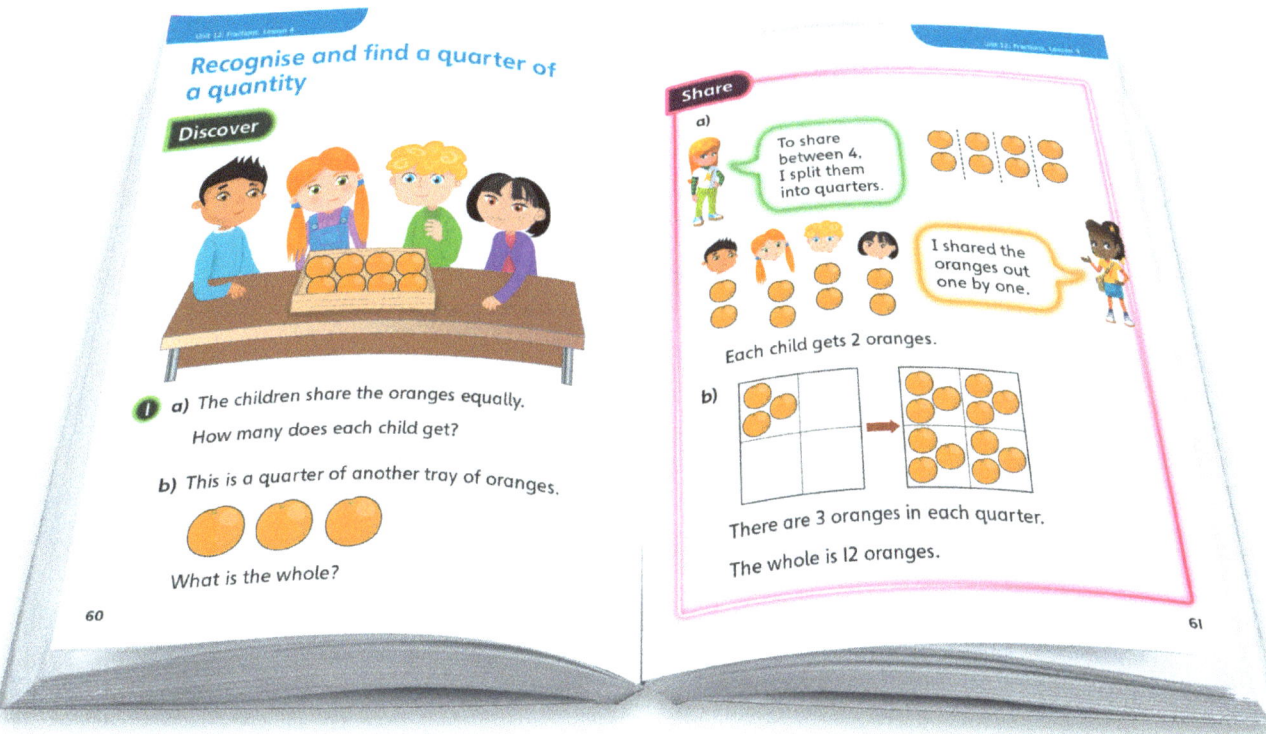

Unit videos

Every term has one unit video which incorporates real-life classroom sequences.

These videos show you how the reasoning behind mathematics can be carried out in a practical manner by showing real children using various concrete and pictorial methods to come to the solution. You can see how using these practical models, such as part-whole and bar models, helps them to find and articulate their answer.

Mastery tips

Mastery Experts give anecdotal advice on where they have used hands-on and real-life elements to inspire their children.

Mastery Expert tip! 'When I taught this unit, my class really enjoyed going on an 'array hunt' around the school. It really engaged them in the learning and had them finding multiplications everywhere around them!'

Don't forget to watch the Unit 11 video!

Concrete-Pictorial-Abstract (C-P-A) approach

Each **Share** section uses various methods to explain an answer, helping children to access abstract concepts by using concrete tools, such as counters. Remember, this isn't a linear process, so even children who appear confident using the more abstract method can deepen their knowledge by exploring the concrete representations. Encourage children to use all three methods to really solidify their understanding of a concept.

Pictorial representation – drawing the problem in a logical way that helps children visualise the maths

Concrete representation – using manipulatives to represent the problem. Encourage children to physically use resources to explore the maths.

Abstract representation – using words and calculations to represent the problem.

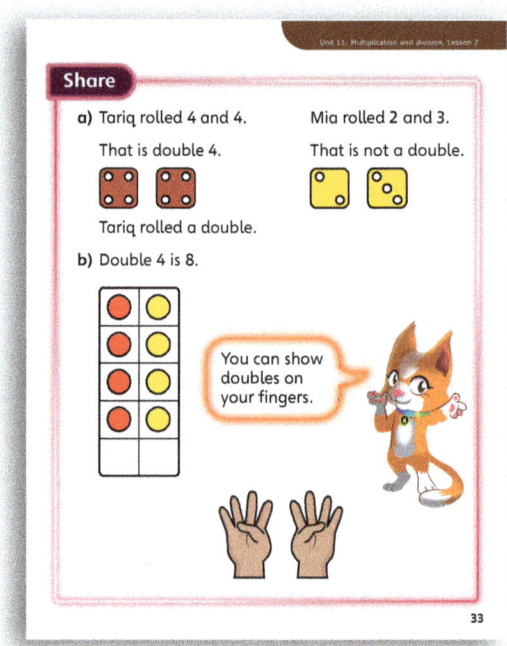

Practical tips

Every lesson suggests how to draw out the practical side of the **Discover** context.

You'll find these in the **Discover** section of the Teacher Guide for each lesson.

PRACTICAL TIPS Provide children with counters and ten frames to represent the question. Encourage them to count aloud.

Resources

Every lesson lists the practical resources you will need or might want to use. There is also a summary of all of the resources used throughout the term on page 40 to help you be prepared.

RESOURCES

Mandatory: counters

Optional: ten frames, number cards with numbers 1–20

Using *Power Maths* flexibly in Key Stage I

Power Maths lessons have a coherent, regular structure that supports you in building up children's understanding in a series of small steps. This is something most classes will need to build up to, rather than running in from a standing start at the beginning of Year 1.

Start by using the Practice Books in small groups

In most Year 1 classes, it won't be realistic for the whole class to complete the Practice Book pages independently at the start of the year, but they will learn to do this gradually. For the Textbooks, children will need to get used to direct teaching and recording answers in their own books. And, of course, this will set them up well for the rest of Primary school.

Small teacher-led groups are likely to be the best approach for independent practice. This format allows you to talk children through the question, discuss their ideas using manipulatives (often there will be manipulatives on the page as a hint), and guide them in representing their answer. (For instance, they can tell you the answer is 5, but they may need help writing 5 or knowing that they should colour in 5 apples.)

Go through the questions one-by-one with the group. You can mark their work/give feedback there and then. As children get used to the materials, the next stage could be for the small group to work through the questions at their own pace. The style of questions in *Power Maths* is quite regular, so children will get better at knowing what they need to do.

To facilitate small group work, you are likely to need some other activities as a carousel. A good way to do this is by turning a question from the Textbook into a game (usually **Think together** question 3 will work well) and teaching this to children before you break into groups. For instance, look at the example below (pages 78–79 in Textbook 1A). You could teach children a game with a part-whole model where one child puts in the whole using counters and the other children have to put in the parts. Or they could try this with beanbags and hoops. Base the practice on the key learning from the lesson.

Are there any other ways to use the resources flexibly?

Don't be afraid to bring the **Discover** activity to life! Perhaps you could turn it into a game, or a role play. For instance, if the context is a teddy bear's picnic, you could share out fruit between teddies in the class. Or could you find a toy rocket to launch for the lesson below? (Textbook 1A page 32).

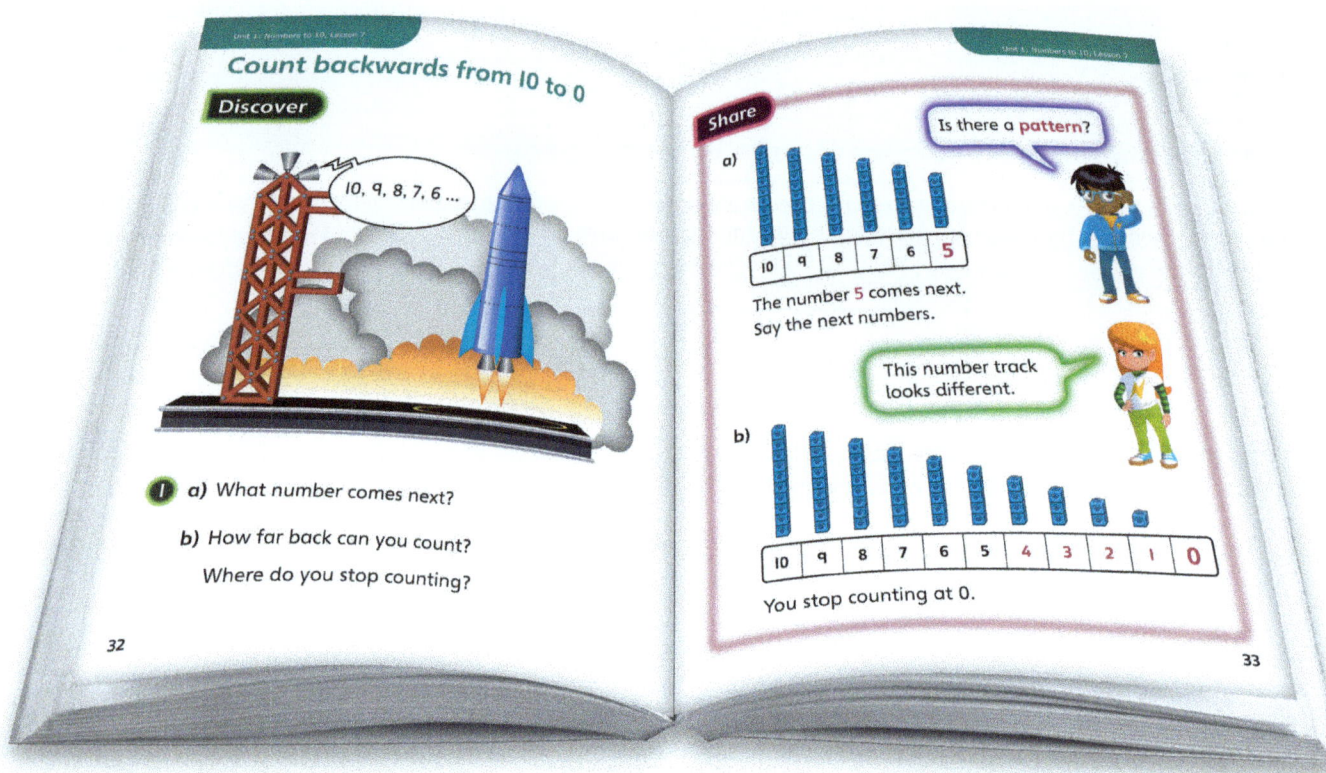

For some lessons you could consider a slightly different approach where you move backwards and forwards between the Textbook and Practice Book. If **Think together** question 1 links well with Practice Book question 1, you could do the **Think together** question together and then let children complete the Practice Book question, then the same for question 2, etc. This works better for some lessons than others, but it is one way of making practice more independent in short bursts, as a way of building up independence.

Don't forget, there isn't a *Power Maths* lesson for every lesson in the year. You can take more time where you need to, so that children's understanding is secure. In Key Stage 1, it will be all the more important to take your time, because children need to get used to the format as well as master the key learning. If using some of the ideas above means that a *Power Maths* lesson actually takes two lessons, e.g. for the first part of the year, then that's fine!

There are some further ideas for using the materials flexibly in the next section.

Working with children below age-related expectation

This section offers advice on using *Power Maths* with children who are significantly behind age-related expectation. Teacher judgement will be crucial in terms of where and why children are struggling, and in choosing the right approach. The suggestions can of course be adapted for children with special educational needs, depending on the specific details of those needs.

General approaches to support children who are struggling

Keeping the pace manageable
Remember, you have more teaching days than *Power Maths* lessons so you can cover a lesson over more than one day, and revisit key learning, to ensure all children are ready to move on. You can use the + and – buttons to adjust the time for each unit in the online planning. The NCETM's Ready-to-Progress criteria can be used to help determine what should be highest priority.

Same-day intervention
You could go over the Textbook pages or revisit the previous year's work if necessary. Remember that same-day intervention can be within the lesson, as well as afterwards (see page 29). As children start their independent practice, you can work with those who found the first part of the lesson difficult, checking understanding using manipulatives.

Fluency sessions
Fit in as much practice as you can for number bonds and times-tables, etc., at other times of the day. If you can, plan a short 'maths meeting' for this in the afternoon. You might choose to use a Power Up you haven't used already.

Pre-teaching
Find a 5- to 10-minute slot before the lesson to work with the children you feel would benefit. The afternoon before the lesson can work well, because it gives children time to think in between. Recap previous work on the topic (addressing any gaps you're aware of) and do some fluency practice, targeting number facts etc. that will help children access the learning.

Focusing on the key concepts
If children are a long way behind, it can be helpful to take a step back and think about the key concepts for children to engage with, not just the fine detail of the objective for that year group (e.g. addition with a specific number of columns). Bearing that in mind, how could children advance their understanding of the topic?

Providing extra support within the lesson

Support in the Teacher Guide
First of all, use the Strengthen support in the Teacher Guide for guided and independent work in each lesson, and share this with Teaching Assistants, where relevant. As you read through the lesson content and corresponding Teacher Guide pages before the lesson, ask yourself what key idea or nugget of understanding is at the heart of the lesson. If children are struggling, this should help you decide what's essential for all children before they move on.

Annotating pages
You can annotate questions to provide extra scaffolding or hints if you need to, but aim to build up children's ability to access questions independently wherever you can. Children tend to get used to the style of the *Power Maths* questions over time.

Quick recap as lesson starter
The Quick recap for each lesson in the Teacher Guide is an alternative starter activity to the Power Up. You might choose to use this with some or all children if you feel they will need support accessing the main lesson.

Consolidation questions
If you think some children would benefit from additional questions at the same level before moving on, write one or two similar questions on the board. (This shouldn't be at the expense of reasoning and problem-solving opportunities: take longer over the lesson if you need to.)

Hard copy Textbooks
The Textbooks help children focus in more easily on the mathematical representations, read the text more comfortably, and revisit work from a previous lesson that you are building on, as well as giving children ownership of their learning journey. In main lessons, it can work well to use the e-Textbook for **Discover** and give out the books when discussing the methods in the **Share** section.

Reading support
It's important that all children are exposed to problem solving and reasoning questions, which often involve reading. For whole-class work you can read questions together. For independent practice you could consider annotating pages to help children see what the question is asking, and stem sentences to help structure their answer. A general focus on specific mathematical language and vocabulary will help children access the questions. You could consider pairing weaker readers with stronger readers, or read questions as a group if those who need support are on the same table.

Providing extra depth and challenge with *Power Maths*

Just as prescribed in the National Curriculum, the goal of *Power Maths* is never to accelerate through a topic but rather to gain a clear, deep and broad understanding. Here are some suggestions to help ensure all children are appropriately challenged as you work with the resources.

Overall approaches

First of all, remember that the materials are designed to help you keep the class together, allowing all children to master a concept while those who grasp it quickly have time to explore it in more depth. Use the Deepen support in the Teacher Guide (see below) to challenge children who work through the questions quickly. Here are some questions and ideas to encourage breadth and depth during specific parts of the lesson, or at any time (where no part of the lesson sequence is specified):

- **Discover**: 'Can you demonstrate your solution another way?'
- **Share**: Make sure every child is encouraged to give answers and engage with the discussion, not just the most confident.
- **Think together**: 'Can you model your answers using concrete materials? Can you explain your solution to a partner?'
- Practice: Allow all children to work through the full set of questions, so that they benefit from the logical sequence.
- **Reflect**: 'Is there another way of working out the answer? And another way?'
  'Have you found all the solutions?'
  'Is that always true?'
  'What's different between this question and that question? And what's the same?'

Note that the **Challenge** questions are designed so that all children can access and attempt them, if they have worked through the steps leading up to them. There may be some children in a given lesson who don't manage to do the **Challenge**, but it is not supposed to be a distinct task for a subset of the class. When you look through the lesson materials before teaching, think about what each question is specifically asking, and compare this with the key learning point for the lesson. This will help you decide which questions you feel it's essential for all children to answer, before moving on. You can at least aim for all children to try the **Challenge**!

Deepen activities and support

The Teacher Guide provides valuable support for each stage of the lesson. This includes Deepen tips for the guided and independent practice sections, which will help you provide extra stretch and challenge within your lesson, without having to organise additional tasks. If you have a Teaching Assistant, they can also make use of this advice. There are also suggestions for the lesson as a whole in the 'Going Deeper' section on the first page of the Teacher Guide section for that lesson. Every class is different, so you can always go a bit further in the direction indicated, if appropriate, and build on the suggestions given.

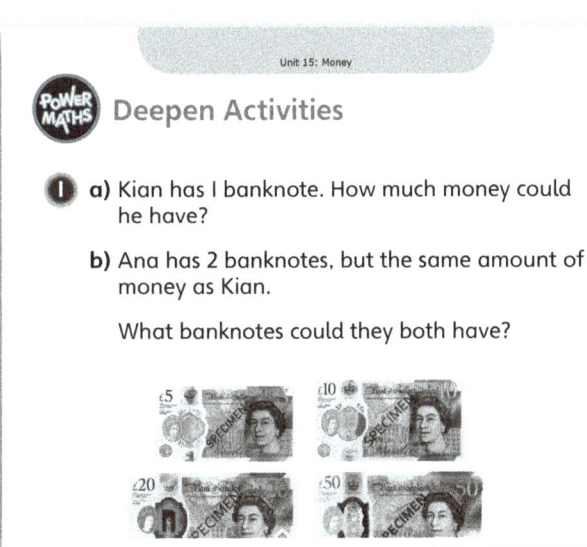

There is a Deepen activity for each unit. These are designed to follow on from the End of unit check, stretching children who have a firm understanding of the key learning from the unit. Children can work on them independently, which makes it easier for the teacher to facilitate the Strengthen activity for children who need extra support. Deepen activities could also be introduced earlier in the unit if the necessary work has been covered. The Deepen activities are on *ActiveLearn* on the Planning page for each unit, and also on the Resources page).

Using the questions flexibly to provide extra challenge

Sometimes you may want to write an extra question on the board or provide this on paper. You can usually do this by tweaking the lesson materials. The questions are designed to form a carefully structured sequence that builds understanding step by step, but, with careful thought about the purpose of each question, you can use the materials flexibly where you need to. Sometimes you might feel that children would benefit from another similar question for consolidation before moving on to the next one, or you might feel that they would benefit from a harder example in the same style. It should be quick and easy to generate 'more of the same' type questions where this is the case.

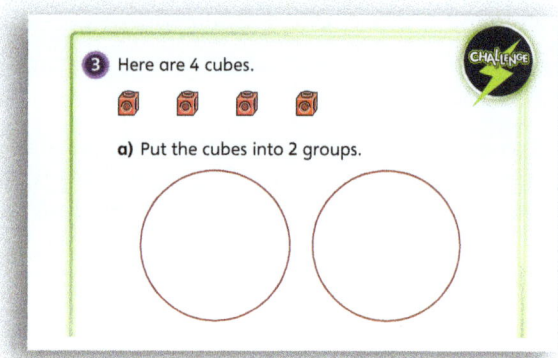

When you see a question like this one (from Unit 2, Lesson 1), it's easy to make harder examples to do afterwards if you need them. What if there were 9 cubes? Can children write the parts and wholes and find lots of different ways?

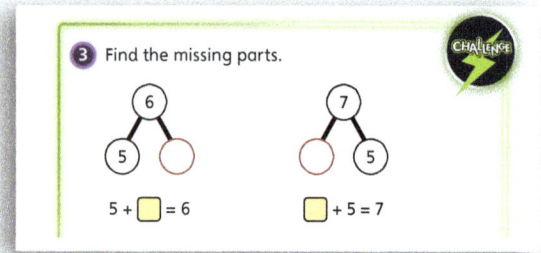

For this example (from Unit 3, Lesson 4), you could ask children to make up their own question(s) for a partner to solve. (In fact, for any of these examples you could ask early finishers to create their own question for a partner.)

Here's an example (from Unit 8, Lesson 2) where you could use the original context to provide extra challenge at the end of the lesson. For example, you could ask how far the frogs have to go to reach the next frog, or to be the winner.

Besides creating additional questions, you should be able to find a question in the lesson that you can adapt into a game or open-ended investigation, if this helps to keep everyone engaged. It could simply be that, instead of answering 5 + 6 etc. on the page, they could build a robot with 5 cubes and 6 cylinders.

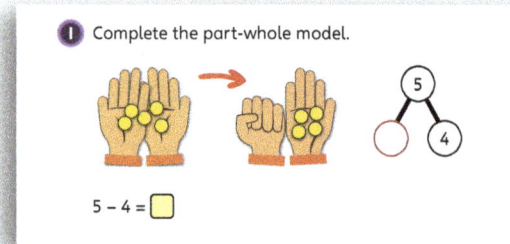

With a question like this one (from Unit 4, Lesson 4), children could play the game in pairs, taking an agreed number of counters and then showing what's in one hand. They could write a subtraction sentence each time, varying the whole and the parts.

See the bullets on the previous page for some general ideas that will help with 'opening out' questions in the books, e.g. 'Can you find all the solutions?' type questions.

Other suggestions

Another way of stretching children is through mixed ability pairs, or via other opportunities for children to explain their understanding in their own way. This is a good way of encouraging children to go deeper into the learning, rather than, for instance, tackling questions that are computationally more challenging but conceptually equivalent in level.

Using *Power Maths* with mixed age classes

Overall approaches

There are many variables between schools that would make it inadvisable to recommend a one-size-fits-all approach to mixed age teaching with *Power Maths*. These include how year groups are merged, availability of Teaching Assistants, experience and preference of teaching staff, range in pupil attainment across years, classroom space and layout, level of flexibility around timetables, and overall organisational structure (whether the school is part of a trust).

Some schools will find it best to timetable separate maths lessons for the different year groups. Others will aim to teach the class together as much as possible using the mixed age planning support on *ActiveLearn* (see the lesson exemplars for ways of organising lessons with strong/medium/weak correlation between year groups). There will also be ways of adapting these general approaches. For example, offset lessons where Year A start their lesson with the teacher, while Year B work independently on the practice from the previous lesson, and then start the next lesson with the teacher while Year A work independently; or teachers may choose to base their provision around the lesson from one year group and tweak the content up/down for the other group.

Key strategies for mixed age teaching

The mixed age teaching webinar on *ActiveLearn* provides advice on all aspects of mixed age teaching, including more detail on the ideas below.

Developing independence over time
Investing time in building up children's independence will pay off in the medium term.

Clear rationale
If someone asked, 'Why did you teach both Unit 3 and 4 in the same lesson/separate lessons?', what would your answer be?

Designing a lesson
1. Identify the core learning for each group
2. Identify any number skills necessary to access the core
3. Consider the flow of concepts and how one core leads to the other

Challenging all children
The questions are designed to build understanding step by step, but with careful thought about the purpose of each question you can tweak them to increase the challenge.

Multiple years combined
With more than two years together, teachers will inevitably need to use the resources flexibly if delivering a single lesson.

Enjoy the positives!

Comparison deepens understanding and there will be lots of opportunities for children, as well as misconceptions to explore. There is also in-built pre-teaching and the chance to build up a concept from its foundations. For teachers there is double the material to draw on! Mixed age teachers require a strong understanding of the progression of ideas across year groups, which is highly valuable for all teachers. Also, it is necessary to engage deeply with the lesson to see how to use the materials flexibly – this is recommended for all teachers and will help you bring your lesson to life!

List of practical resources

Year 1C Mandatory resources

Resource	Lesson
100 squares	**Unit 11** Lesson 3
100 squares (blank)	**Unit 14** Lesson 3
100 squares (large, printed)	**Unit 14** Lesson 1
Banknotes (images of, real and pretend)	**Unit 15** Lesson 2
Calendars (range, printed and digital, showing 1 month/page and also 1 year/page	**Unit 16** Lesson 3
Calendars (showing a week to a page)	**Unit 16** Lesson 2
Card (blank pieces of)	**Unit 11** Lesson 8
Coins (real or realistic plastic)	**Unit 15** Lesson 3
Coins (real, with numbers not just words, or realistic plastic versions)	**Unit 15** Lesson 1
Counters	**Unit 11** Lessons 1, 2, 3, 4, 5, 6 **Unit 12** Lessons 2, 4 **Unit 14** Lessons 1, 3, 5
Cubes	**Unit 11** Lesson 4
Multilink cubes	**Unit 11** Lesson 8
Paper shapes (copies of shapes in **Think together** question 2: rectangles, hearts, ovals, butterflies)	**Unit 12** Lesson 1
Paper squares, rectangles and circles	**Unit 12** Lesson 3
Shapes	**Unit 12** Lesson 1
Ten frames	**Unit 11** Lesson 2 **Unit 14** Lesson 3
Ten frames (blank)	**Unit 11** Lessons 1, 3
Ten frames (large, printed)	**Unit 14** Lesson 5

Year 1C Optional resources

Resource	Lesson
100 squares	**Unit 14** Lessons 1, 3, 5 **Unit 15** Lesson 3
100 squares (large, printed)	**Unit 14** Lessons 2, 6
Apples (8; or other items that can be shared between children)	**Unit 12** Lesson 2
Bead strings	**Unit 11** Lesson 1 **Unit 14** Lessons 2, 3, 5
Boards (as bases for the arrays)	**Unit 11** Lesson 6
Calendars (range, printed and digital)	**Unit 16** Lesson 2
Cards	**Unit 13** Lesson 5
Cards (to represent events pictorially)	**Unit 16** Lesson 1
Cards (with arrows labelled Left and Right)	**Unit 13** Lesson 2
Cards (with times written on)	**Unit 16** Lesson 4
Clock tool	**Unit 16** Lesson 4
Clock tool (marked with the numbers 1 to 12)	**Unit 16** Lesson 5
Clocks (real)	**Unit 16** Lesson 4
Construction equipment	**Unit 13** Lesson 4
Countable objects (36; such as shells or other items found in the classroom or outside)	**Unit 14** Lesson 5
Countable objects (collections)	**Unit 11** Lessons 4, 5, 6
Countable objects (selection, orderable)	**Unit 13** Lesson 5
Counters	**Unit 14** Lessons 2, 6
Counters (different colours)	**Unit 11** Lesson 7
Counters (translucent)	**Unit 11** Lesson 3
Counters (or countable objects, such as playing cards, marbles, printed circles to represent groups; or pictures of these)	**Unit 11** Lesson 9
Counters (or countable objects, such as toy people, toy animals; or pictures of these)	**Unit 11** Lesson 8
Cubes	**Unit 11** Lesson 2 **Unit 12** Lesson 2 **Unit 14** Lessons 2, 5
Cubes (or counters)	**Unit 13** Lesson 4
Dice	**Unit 11** Lesson 3
Dice (6-sided)	**Unit 11** Lesson 7
Digit cards	**Unit 11** Lesson 9
Grids (blank; 2 × 3)	**Unit 13** Lesson 4
Grids (blank; 5 × 5)	**Unit 13** Lesson 3
Grids (showing the group sizes suitable for the Textbook questions)	**Unit 11** Lesson 4
Interlocking cubes	**Unit 14** Lesson 3
Leaves	**Unit 14** Lesson 6
Mirrors	**Unit 12** Lesson 1
Model clocks	**Unit 16** Lesson 4
Multilink cubes	**Unit 12** Lesson 3 **Unit 13** Lesson 4
Number cards (displaying ordinal numbers)	**Unit 13** Lesson 5
Number lines	**Unit 11** Lessons 1, 3 **Unit 14** Lesson 5 **Unit 15** Lesson 3
Number lines (dry wipe)	**Unit 14** Lesson 4
Number lines (laminated)	**Unit 11** Lesson 2
Number lines (large, for display in class)	**Unit 14** Lesson 4
Number lines (large, printed)	**Unit 11** Lessons 5, 6
Number tracks	**Unit 14** Lesson 5
Number tracks (1–20)	**Unit 11** Lesson 7

Year 1C Optional resources – *continued*

Resource	Lesson
Number tracks (blank, printable)	**Unit 14** Lesson 6
Number tracks (listing ordinal numbers, in order)	**Unit 13** Lesson 5
Objects (classroom, such as pens and pencils)	**Unit 13** Lesson 4
Objects (to rotate and move)	**Unit 13** Lesson 3
Objects (to rotate, such as toys, programmable toys or toy cars)	**Unit 13** Lesson 1
Objects (used in this lesson: bicycle wheels or bicycles, shoes, socks – or pictorial representations of them)	**Unit 11** Lesson 1
Paper circles	**Unit 13** Lesson 1
Paper clips	**Unit 11** Lesson 8
PE cones	**Unit 13** Lesson 3
PE equipment (4 different types, such as: hoola-hoops, bean bags, small benches, small hurdles for children to jump over)	**Unit 13** Lesson 5
Pencils (colouring)	**Unit 13** Lesson 1
Play food that can be cut into halves	**Unit 12** Lesson 1
Printed circles (to represent groups)	**Unit 11** Lesson 8
Printed clock faces	**Unit 16** Lesson 4
Printed clock faces (for children to complete)	**Unit 16** Lesson 5
Printed rectangles (to represent tables)	**Unit 11** Lesson 8
Rekenreks	**Unit 14** Lesson 2
Rulers	**Unit 12** Lesson 1
Skipping ropes	**Unit 13** Lesson 3
Sorting hoops	**Unit 12** Lesson 2
String	**Unit 12** Lesson 1
Ten frames	**Unit 14** Lesson 2
Ten frames (large, printed)	**Unit 11** Lesson 7 **Unit 14** Lesson 6
Tennis balls (or pictures of tennis balls)	**Unit 11** Lesson 3
Vocabulary prompts (showing key words with pictures to illustrate them)	**Unit 13** Lessons 3, 4

Getting started with Power Maths

As you prepare to put *Power Maths* into action, you might find the tips and advice below helpful.

STEP 1: Train up!

A practical, up-front full day professional development course will give you and your team a brilliant head-start as you begin your *Power Maths* journey. You will learn more about the ethos, how it works and why.

STEP 2: Check out the progression

Take a look at the yearly and termly overviews. Next take a look at the unit overview for the unit you are about to teach in your Teacher Guide, remembering that you can match your lessons and pacing to your class.

STEP 3: Explore the context

Take a little time to look at the context for this unit: what are the implications for the unit ahead? (Think about key language, common misunderstandings and intervention strategies, for example.) If you have the online subscription, don't forget to watch the corresponding unit video.

STEP 4: Prepare for your first lesson

Familiarise yourself with the objectives, essential questions to ask and the resources you will need. The Teacher Guide offers tips, ideas and guidance on individual lessons to help you anticipate children's misconceptions and challenge those who are ready to think more deeply.

STEP 5: Teach and reflect

Deliver your lesson – and enjoy!

Afterwards, reflect on how it went … Did you cover all five stages? Does the lesson need more time? How could you improve it?

Unit 11
Multiplication and division

Mastery Expert tip! 'When I taught this unit, my class really enjoyed going on an 'array hunt' around the school. It really engaged them in the learning and had them finding multiplications everywhere around them!'

Don't forget to watch the Unit 11 video!

WHY THIS UNIT IS IMPORTANT

In this unit, children will develop their understanding of multiplication as repeated addition, understanding the difference between equal and not equal groups. They will use their knowledge of skip counting in 2s, 5s and 10s and will use concrete, pictorial and abstract representations to help them to find the total of multiple equal groups and of doubles. These representations will include arrays, a powerful way of developing multiplicative reasoning. This will then feed into their work on division later in the unit, where they will move onto grouping and sharing.

WHERE THIS UNIT FITS

→ Unit 10: Introducing mass and capacity
→ **Unit 11: Multiplication and division**
→ Unit 12: Fractions

Before they start this unit, it is expected that children can:
- count reliably in 2s, 5s and 10s
- sort objects into equal groups
- recognise and use ten frames and number lines.

ASSESSING MASTERY

Children will demonstrate mastery in this unit by counting reliably in steps of 2, 5 and 10. They will be able to use concrete, pictorial and abstract representations to find the total of a given number of equal groups, including repeated addition. They will be able to create arrays based on the numbers given to them and begin to recognise the commutativity that arrays demonstrate. They will be able to demonstrate an understanding of sharing and grouping and the differences between these approaches.

COMMON MISCONCEPTIONS	STRENGTHENING UNDERSTANDING	GOING DEEPER
Children may confuse the number of groups with the number in each group.	Show children two pictures of related groupings, such as 4 groups of 5 stars and 5 groups of 4 stars (not in an array). Ask, for both pictures: • *How many stars are in each group in this picture? How many groups of stars are there?* • *What is the same and what is different about the two pictures?* Ask children to arrange both pictures into arrays and ask about each: • *What can you tell me about the number of stars in each row? How many rows are there?* • *Can you write the total number of stars as an addition?*	Children could be given a number with many factors (such as 24). Ask: • *What arrays can you make using this number?* • *Can you write down addition calculations for your arrays?* • *How do you know you have found all the possible arrays?*

Unit 11: Multiplication and division

UNIT STARTER PAGES

Use these pages to introduce the focus to children. You can use the characters to explore different ways of working too.

STRUCTURES AND REPRESENTATIONS

Array: Arrays are a visual representation of multiplication and division. They are an excellent tool for showing what *x groups of y are equivalent to*. They also clearly show the commutativity of multiplication (how '*x groups of y*' has the same total as '*y groups of x*').

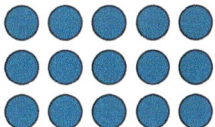

Number line: Number lines help children to represent their skip counting. They will help children count on and back from a given starting point and help them identify patterns and groups within the count.

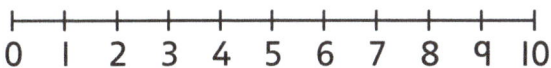

Ten frame: A ten frame will help children visualise 10. In this unit, it is used to demonstrate and cement children's understanding of doubles.

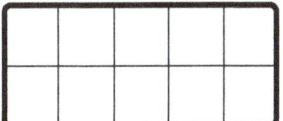

KEY LANGUAGE

There is some key language that children will need to know as part of the learning in this unit:

→ equal group
→ array
→ add, addition, adding, altogether, total
→ row, column
→ double, twice
→ share

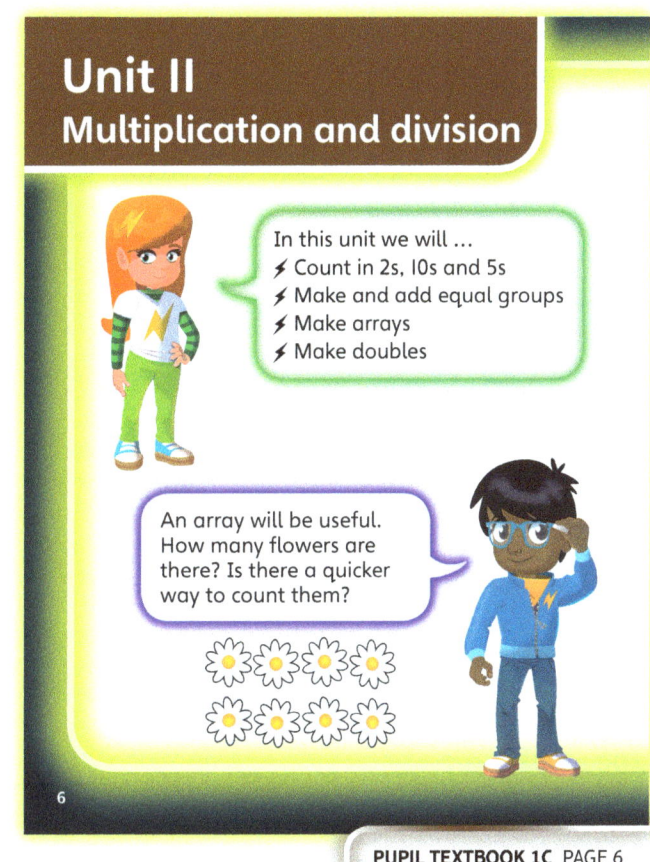

PUPIL TEXTBOOK 1C PAGE 6

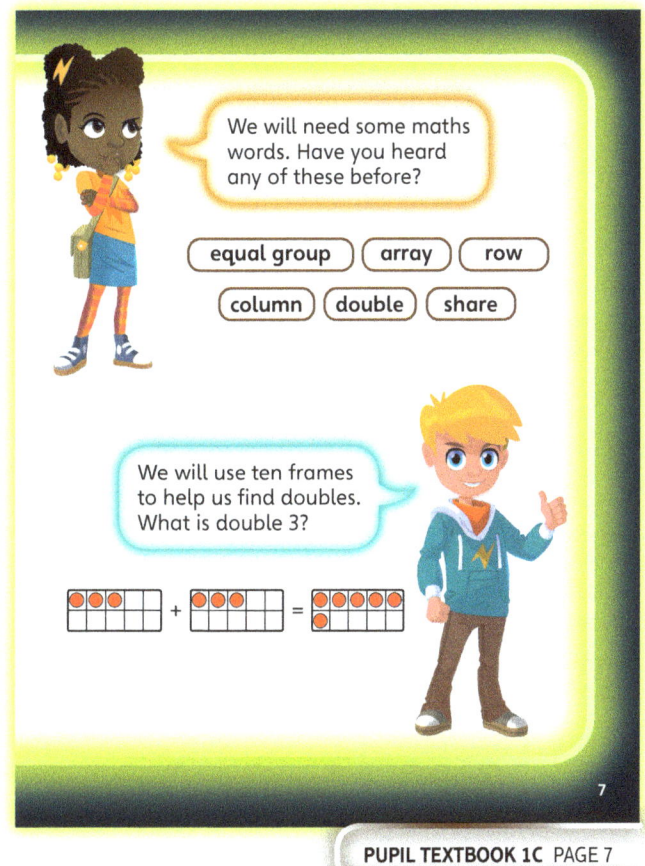

PUPIL TEXTBOOK 1C PAGE 7

Unit 11: Multiplication and division, Lesson 1

Count in 2s

Learning focus

In this lesson, children will explore counting on and back in 2s.

Before you teach
- Are children confident counting on and back?
- Are children secure with one-to-one correspondence?

NATIONAL CURRICULUM LINKS

Year 1 Number – multiplication and division

Count, read and write numbers to 100 in numerals; count in multiples of twos, fives and tens.

ASSESSING MASTERY

Children can count on and back in 2s from an even starting point. Children realise that, when counting in 2s from an even starting point, all numbers end in 0, 2, 4, 6, or 8 (note that children will be introduced to the terms 'even' and 'odd' in Year 2).

COMMON MISCONCEPTIONS

Children may give odd numbers when counting in 2s from an even starting point. Use a ten frame to represent counting in 2s. Ask:

• *What do you notice about the 1s digits when you count in 2s?*

STRENGTHENING UNDERSTANDING

Allow children to physically manipulate the objects (or pictorial representations of them) used in the questions in this lesson. Children could practise counting them in groups of 2 and grouping them into 2s.

Use ten frames to help children spot that, when counting in 2s, you only ever say numbers ending in 0, 2, 4, 6 or 8.

Provide practical activities to strengthen understanding as well. Ask children to line up in 2s for assembly and hand out fruit or snacks in groups of 2.

GOING DEEPER

Challenge children to explain if certain numbers would be said when counting on or back in 2s from an even starting point. For example, ask: *Would you say 37 when counting in 2s from 10? How do you know?*

KEY LANGUAGE

In lesson: count, 2s, pairs, in total

Other language to be used by the teacher: 1s, multiples

STRUCTURES AND REPRESENTATIONS

Ten frame, number line, 100 square

RESOURCES

Mandatory: blank ten frames, counters

Optional: objects (or pictorial representations of the objects) used in this lesson (bicycle wheels or bicycles, shoes, socks), number lines, bead strings

 In the eTextbook of this lesson, you will find interactive links to a selection of teaching tools.

Quick recap

Check that children can confidently count objects in 1s to 20. For example, as a class, children may count the number of shapes/beanbags in a box or try and clap 20 times as a class, counting in 1s as they go.

46

Unit 11: Multiplication and division, Lesson 1

Discover

WAYS OF WORKING Pair work

ASK

- Question 1 a): *What does the picture show?*
- Questions 1 a) and b): *What is the question asking you? How many wheels are there on each bike? What representations could you use?*

IN FOCUS Question 1 a) asks children to count the wheels in 1s and then question 1 b) introduces counting the wheels in 2s. Children should use counters and ten frames to help them count in 2s. Children may reflect that counting in 2s is quicker and more efficient than counting in 1s.

This scenario could be recreated in the classroom with five pairs of shoes representing the five bikes.

ANSWERS

Question 1 a): 1, 2, 3, 4, 5, 6, 7, 8, 9, 10
There are 10 wheels.

Question 1 b): 2, 4, 6, 8, 10
There are 10 wheels.

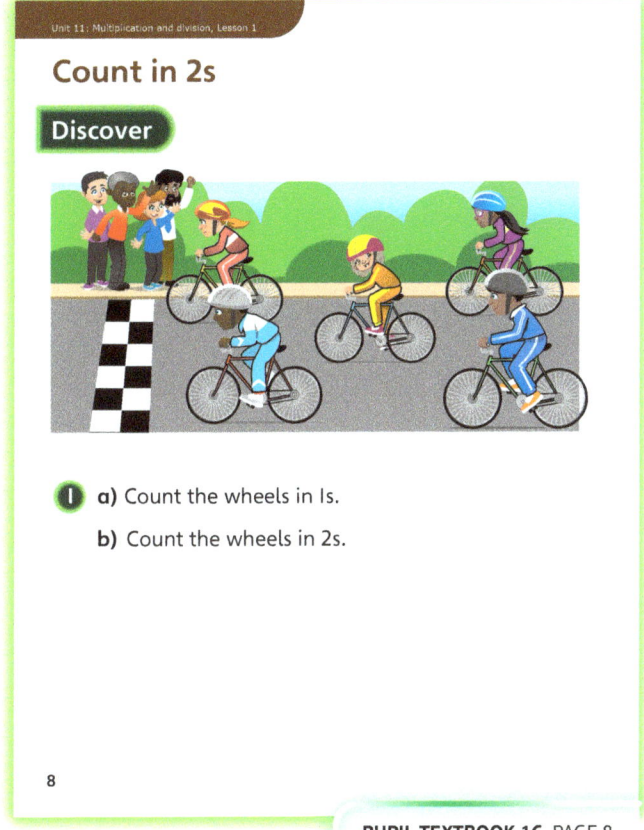

PUPIL TEXTBOOK 1C PAGE 8

Share

WAYS OF WORKING

ASK

- Question 1 a): *How did you count in 1s? How can you make sure you did not miss any wheels?*
- Question 1 b): *How did you solve this problem? How could you count in 2s? What representations could you use to help you count in 2s?*

IN FOCUS In question 1 b), use Flo and Dexter's comments to encourage children to explore the differences between counting in 1s and counting in 2s. Children can also use a ten frame and counters to help. They should be encouraged to make the representations on their ten frames with counters in pairs next to each other. This makes counting in 2s easier.

PUPIL TEXTBOOK 1C PAGE 9

Think together

WAYS OF WORKING Whole class teacher led (I do, We do, You do)

ASK
- Question ❶: *Where have you seen this before? Can you practise counting in 2s and pointing with a partner? What do you think comes after 10?*
- Question ❷: *How does this number line show counting in 2s and not 1s? What number do you think comes next? Why?*
- Question ❸: *How many of each object are there? How can you count them quickly? How can you make sure you do not miss any?*

IN FOCUS In question ❶, children practise further counting in 2s to understand the 2, 4, 6, 8 pattern. They should count out loud as a class and then in pairs. Spend some time counting in 2s to make sure they understand the pattern. Question ❷ introduces children to counting in 2s on a number line. They should see that they miss out one number on the number line each time when they count in 2s. This is the first time they are likely to have missed a number when counting on a number line, so it could be beneficial to reinforce this a couple of times. Question ❸ applies this understanding to counting groups of different objects.

STRENGTHEN Use a variety of models and images to support children. Show them the links between counting in 2s using a ten frame or bead string and also how this can be recorded in jumps of 2 on a number line.

DEEPEN Encourage children to identify the patterns they notice when counting in 2s from 0 or any multiple of 2. They should be able to identify that all the numbers they say end in 0, 2, 4, 6 or 8, and they should use this knowledge to identify if a given number will be said when counting in 2s. It is important not to use the term 'even' yet as this word is not introduced until Year 2.

ASSESSMENT CHECKPOINT Use questions ❶ and ❷ to determine whether children are comfortable counting in 2s from 0, using a combination of ten frames and number lines to help them. Use question ❸ to assess children's ability to confidently count the total number of objects in 2s.

ANSWERS

Question ❶ a): Children count 2, 4, 6, 8, 10.

Question ❷ a): Children count 2, 4, 6, 8, 10, 12, 14, 16, 18, 20.

Question ❸ a): 2

Question ❸ b): 4

Question ❸ c): 6

Question ❸ d): 8

Question ❸ e): 10

Question ❸ f): 12

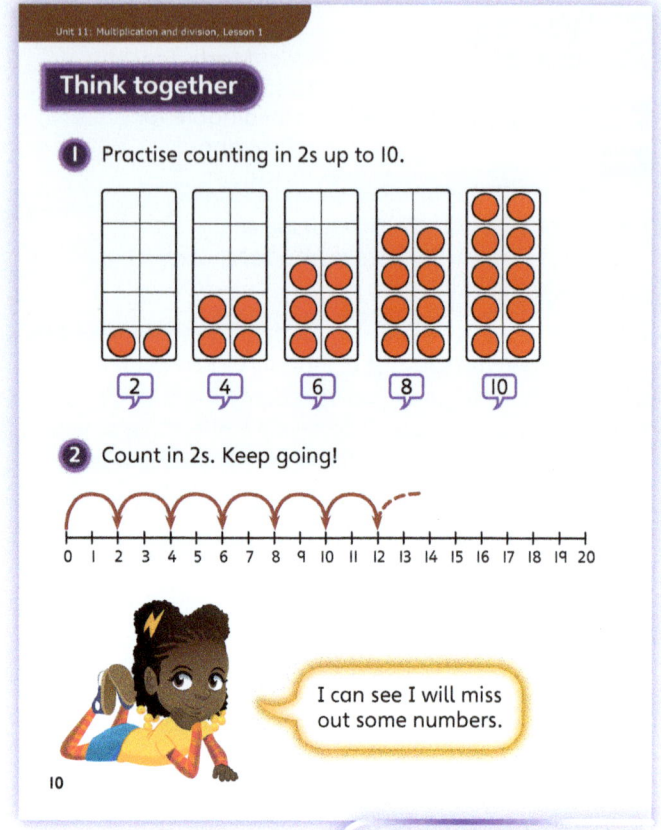

PUPIL TEXTBOOK 1C PAGE 10

PUPIL TEXTBOOK 1C PAGE 11

Practice

WAYS OF WORKING Independent thinking

IN FOCUS Question ❶ helps children to notice patterns when counting in 2s using ten frames. They can point as they count to help them if needed. Question ❷ checks their understanding of using a number line to count in 2s. Questions ❸ and ❹ then provide children with plenty of practice in counting in 2s, counting objects first and then joining the dots by knowing the counting sequence for counting in 2s.

STRENGTHEN For question ❸, children could group and count pictures of bicycles or bicycle wheels and socks, or actual socks and other objects. Listen to children counting. If children say an odd number, ask: *What do you notice about the 1s digits when you count in 2s?*

DEEPEN Encourage children to identify the patterns they notice when counting in 2s from 0 or any multiple of 2. Children should be able to identify that all the numbers they say end in 0, 2, 4, 6 or 8, and they should use this knowledge to identify if a given number will be said when counting in 2s. You could encourage children to use a 100 square to shade in the numbers they say when counting in 2s. This will help them to describe the patterns they notice.

THINK DIFFERENTLY Question ❸ requires children to apply their knowledge of counting in 2s to counting pairs of objects.

ASSESSMENT CHECKPOINT Use question ❸ to assess children's understanding of counting in 2s. Can children confidently count the number of each object?

ANSWERS Answers for the **Practice** part of the lesson can be found in the *Power Maths* online subscription.

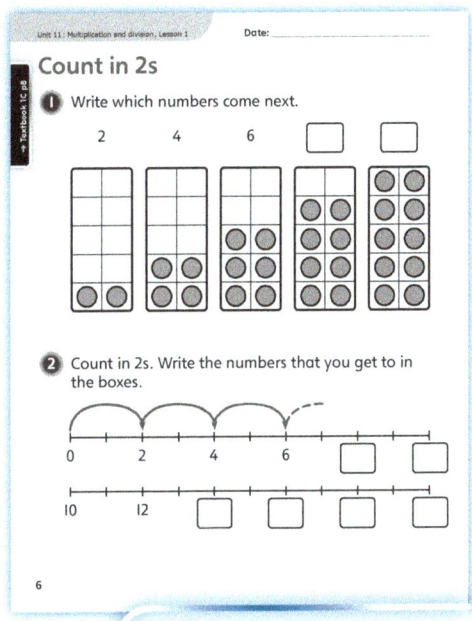

PUPIL PRACTICE BOOK 1C PAGE 6

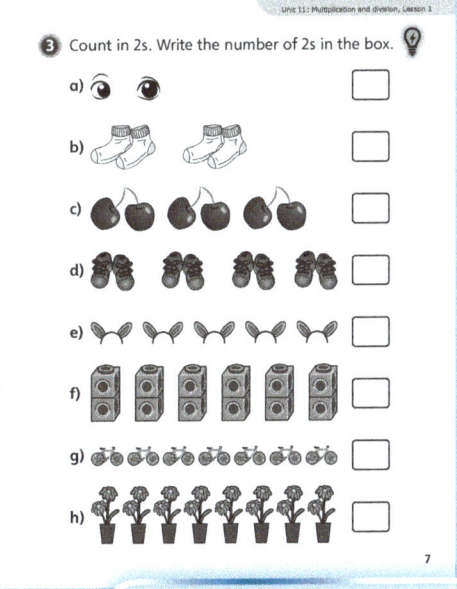

PUPIL PRACTICE BOOK 1C PAGE 7

Reflect

WAYS OF WORKING Independent thinking

IN FOCUS This question allows children to practise the skills they have developed in this lesson, as well as reinforcing the pattern or rule when counting in 2s.

ASSESSMENT CHECKPOINT Check what numbers children have shaded in. Check they have not shaded in any numbers ending in 1, 3, 5, 7 or 9 (odd numbers, although children will not know this term yet). Ask them what they notice about all the numbers they have shaded in.

ANSWERS Answers for the **Reflect** part of the lesson can be found in the *Power Maths* online subscription.

After the lesson

- How can you reinforce counting in 2s and provide practice within your classroom routine?
- Have children made the connection between counting in 2s and the pattern in the 1s digit in each number?

PUPIL PRACTICE BOOK 1C PAGE 8

Unit 11: Multiplication and division, Lesson 2

Count in 10s

Learning focus
In this lesson, children count on and back in 10s. They will investigate the patterns this count creates using different concrete, pictorial and abstract representations.

Before you teach
- Can children recognise different representations of 10, such as ten frames, egg boxes and bead strings?
- Can children count confidently to 50?

NATIONAL CURRICULUM LINKS

Year 1 Number – multiplication and division

Count, read and write numbers to 100 in numerals; count in multiples of twos, fives and tens.

ASSESSING MASTERY

Children can count on and back in 10s from 0 to 50. They notice that numbers in the count lie in the last column of a 100 square and that all the numbers end in 0.

COMMON MISCONCEPTIONS

Children may try and work out the total number of objects (that are given in groups of 10) by counting in 1s or even in 2s or 5s, rather than counting in 10s. Ask:

• *Can you see the groups of 10? Do you think it will be quicker to count in 10s instead of 1s? How can you check that you have got the correct answer?*

Children may keep counting beyond where they need to count to. Ask:

• *Can you show me the count on a number line? How many jumps will you need to make? Can you show me where to stop counting? How is that different from your first count?*

STRENGTHENING UNDERSTANDING

Use ten frames and counters to help children count in 10s. Count aloud and point with children as they count. They will have already heard the numbers when they worked earlier in the year on counting objects to 50. You may also want to use a number line to support their understanding.

GOING DEEPER

This may be an opportunity for more confident children to continue to count beyond 50 to 100. Use a 100 square to help them count.

KEY LANGUAGE

In lesson: count, 10s

Other language to be used by the teacher: how many?

STRUCTURES AND REPRESENTATIONS

Number lines, ten frames

RESOURCES

Mandatory: counters, ten frames

Optional: laminated number lines, cubes

 In the eTextbook of this lesson, you will find interactive links to a selection of teaching tools.

Quick recap

Ask pairs of children to make 20 using counters on ten frames. Ask: *How do you know there are 20?* Repeat, this time making 30 using counters on ten frames.

Unit 11: Multiplication and division, Lesson 2

Discover

WAYS OF WORKING Pair work

ASK
- Question 1 a): *How can you show 10 on your fingers or using cubes?*
- Question 1 b): *How many are you showing? How many is your partner showing? How many are you showing altogether?*

IN FOCUS Children use their fingers or cubes to help them count in 10s. Encourage children to represent the numbers using counters on ten frames, so they can see a different representation of 10.

PRACTICAL TIPS This activity could be done as a whole class together.

ANSWERS
Question 1 a): Children show all 10 fingers on two hands or 10 cubes.
Question 1 b): 2 tens is 20 altogether.

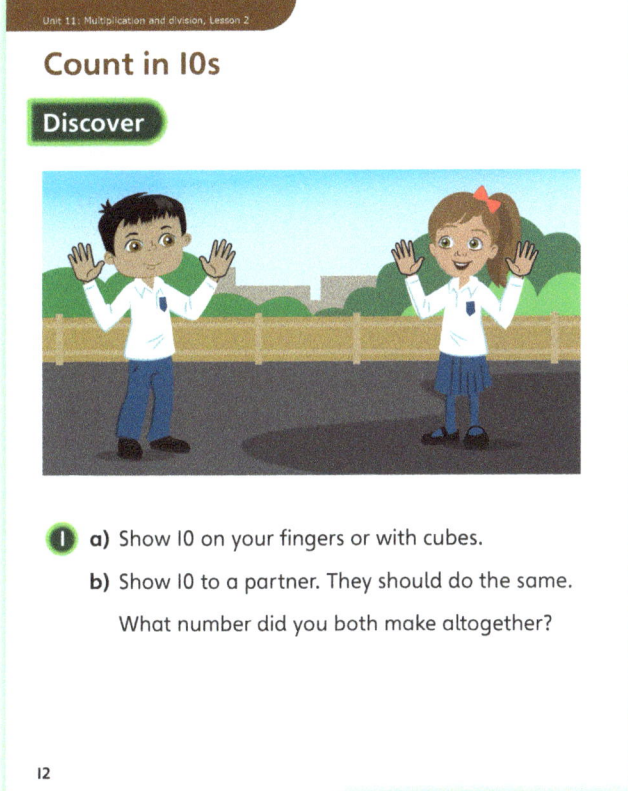

Share

WAYS OF WORKING Whole class teacher led

ASK
- Question 1 a): *Can you show me 10 with your hands or with cubes?*
- Question 1 b): *How can you represent 2 tens on ten frames? What number are you showing now?*

IN FOCUS For question 1 a), ask children to show you 10 on their hands or using cubes. Ask them to represent this on a ten frame. For question 1 b), ask children to copy what Dexter has done and look at their ten frame along with their partner's. Ask them to count in 10s. Count aloud and point to the hands or cubes first and then count aloud and point to the ten frames to reinforce counting in 10s. You may want to discuss at this stage that you notice all the numbers end with the digit 0.

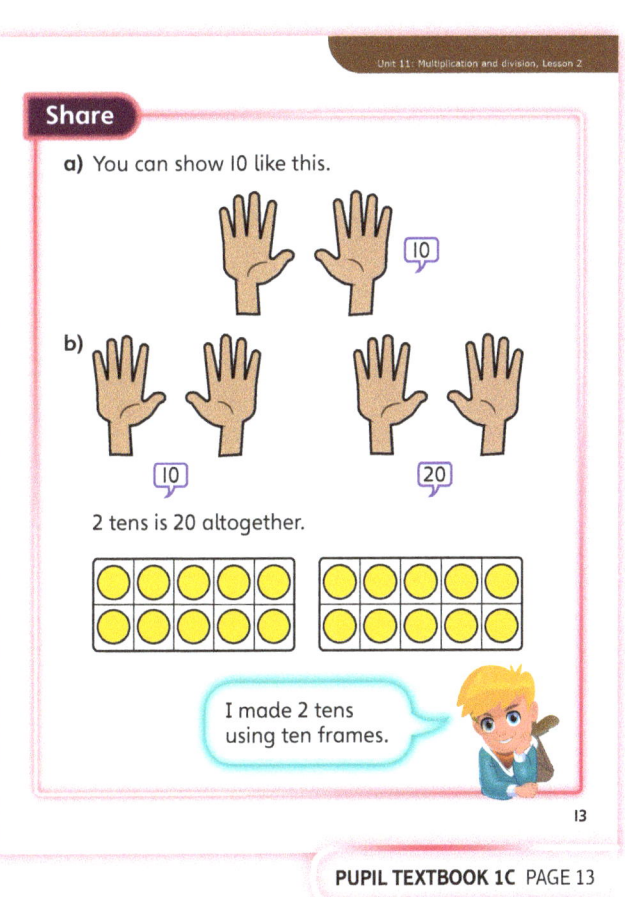

Think together

WAYS OF WORKING Whole class teacher led (I do, We do, You do)

ASK

- Question ❶: *Can you see 10? Can you see 20? Can you see 30?*
- Question ❷: *How many does each ten frame show? How many 10s can you see?*
- Question ❸ a): *Have you seen this grid before? Where?*
- Question ❸ a) and b): *What do you notice about all the numbers that you have pointed at and counted to? What is the same and what is different?*

IN FOCUS In question ❶, children further practise counting in 10s. Question ❷ shows standard ten frames and counters to represent 10s. Use the representations to get children to point and count aloud as they go from 0 to 40. Question ❸ moves to more of an abstract representation of 10s on half a 100 square, where children start to see that the numbers are all in the last column and that they all end in 0. Use Ash's comment to discuss the common misconception that children can make when they say 30.

STRENGTHEN Use ten frames and counters to help children count in 10s. Count aloud and point with children as they count. They will have already heard the numbers when they worked earlier in the year on counting objects to 50. You may also want to use a number line to support their understanding.

DEEPEN Challenge children to identify whether a certain number would or would not be said when counting in 10s from 0. For example, ask: *Would you say 75 when counting in 10s from 0? Can you explain why? What do you know about all numbers that are in the 10s count?*

ASSESSMENT CHECKPOINT Use questions ❶ and ❷ to check children can confidently count in 10s. Use question ❸ to check that children have started to understand that all numbers in the 10s count end in a 0.

ANSWERS

Question ❶: Children count 10, 20, 30.

Question ❷: Children count 10, 20, 30, 40.

Question ❸ a): Children point to and count 10, 20, 30, 40, 50.

Question ❸ b): Children point to and count 10, 20, 30, 40, 50.

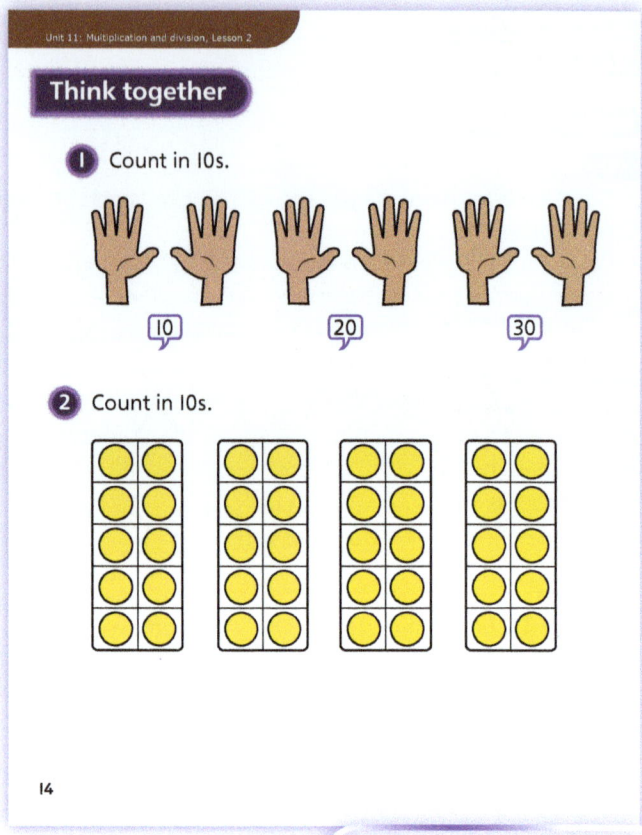

PUPIL TEXTBOOK 1C PAGE 14

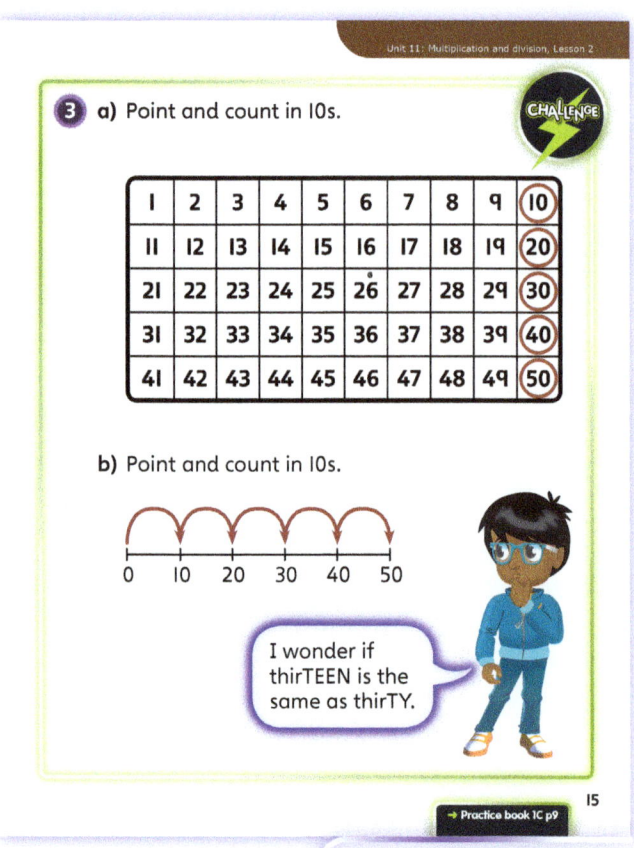

PUPIL TEXTBOOK 1C PAGE 15

Unit 11: Multiplication and division, Lesson 2

Practice

WAYS OF WORKING Independent thinking

IN FOCUS These questions are designed to encourage children to understand that the same numbers are always said when counting in 10s from 0. Questions ❶ and ❷ provide visual representations to support counting. These questions mirror exactly what children have just done as a whole class. Question ❸ asks children to identify where the 10s are on half of a 100 square. Questions ❹ and ❺ show the 10s count on a number line, providing another representation that they can use. Question ❻ challenges children to notice which numbers are in both the 10s count and the 2s count.

STRENGTHEN Use ten frames and counters to help children count in 10s. Count aloud and point with children as they count. They will have already heard the numbers when they worked on counting objects to 50 earlier in the year. You may also want to use a number line to support their understanding.

DEEPEN Ask children to explain if a given number will be within the numerical pattern they have noticed. Children could be asked to apply their knowledge of this numerical pattern to numbers greater than 50. For example, show children the number 65. Ask: *Would you say this number when counting in 10s from 0? How do you know?*

THINK DIFFERENTLY In question ❹, children need to use their knowledge of counting in 10s to label the missing numbers on the number line.

ASSESSMENT CHECKPOINT Use questions ❶ and ❷ to check children can count in 10s using images and use questions ❸, ❹ and ❺ to check that children have a secure understanding of counting in 10s.

ANSWERS Answers for the **Practice** part of the lesson can be found in the *Power Maths* online subscription.

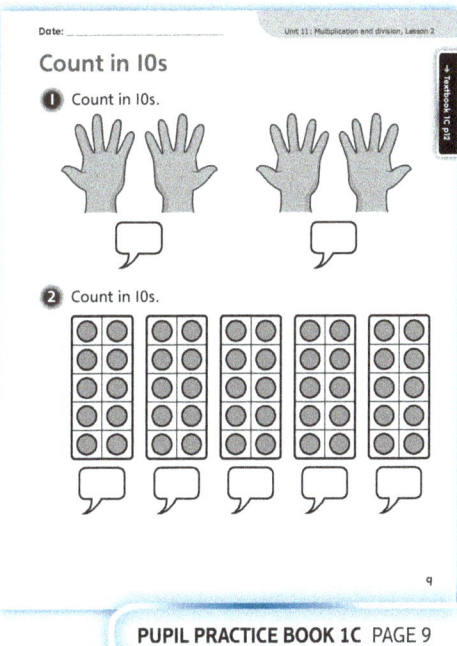

PUPIL PRACTICE BOOK 1C PAGE 9

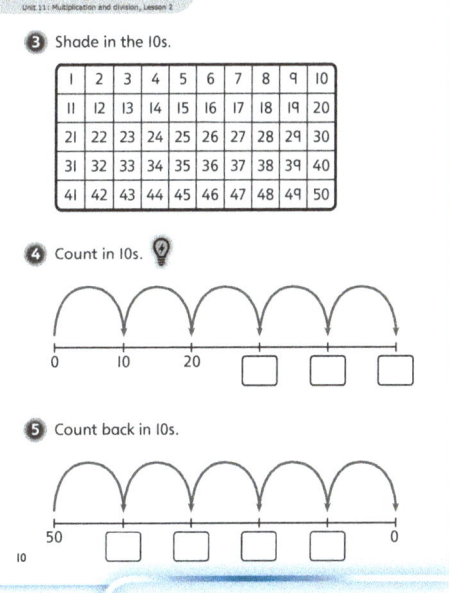

PUPIL PRACTICE BOOK 1C PAGE 10

Reflect

WAYS OF WORKING Pair work

IN FOCUS Children practise counting in 10s with a partner. Encourage them to first count together and then ask them to count alternately. For example, child 1 says 10, child 2 says 20, and so on.

ASSESSMENT CHECKPOINT Check that children can count in 10s confidently to at least 50. Encourage them to go beyond 50 if they are confident with counting in 10s.

ANSWERS Answers for the **Reflect** part of the lesson can be found in the *Power Maths* online subscription.

After the lesson

- Can children count in 10s from 0 to 50?
- Do children know that numbers in the 10s count end in a 0?

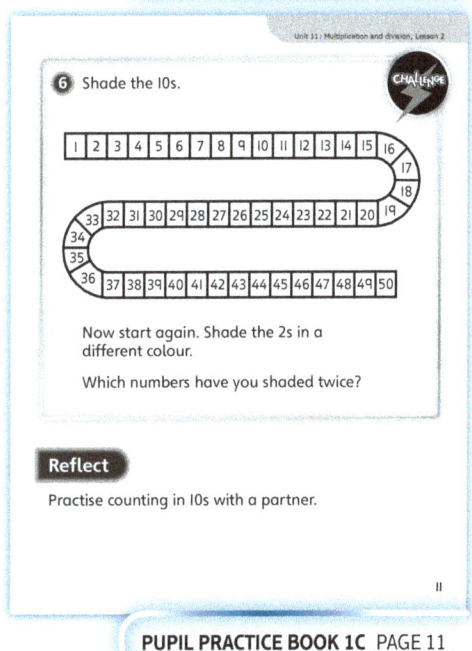

PUPIL PRACTICE BOOK 1C PAGE 11

53

Unit 11: Multiplication and division, Lesson 3

Count in 5s

Learning focus

In this lesson, children will explore counting on and back in 5s, to 50. They will explore the patterns that exist when counting in 5s.

Before you teach

- Are there any children who still need support with numbers to 50?
- How will you support children in identifying patterns and making connections within this lesson?

NATIONAL CURRICULUM LINKS

Number – multiplication and division

Count, read and write numbers to 100 in numerals; count in multiples of twos, fives and tens.

ASSESSING MASTERY

Children can count on and back in 5s from 0 and other starting points that are multiples of 5. Children can identify the numerical patterns that exist when counting in 5s.

COMMON MISCONCEPTIONS

When counting in 5s, children may say numbers that do not end in a 0 or 5. Draw their attention to the patterns that exist when counting in 5s using a number line and 100 square. Ask:
- What do you notice about the digits that all the numbers end in?

Children may identify the spatial pattern when counting in 5s on the 100 square (that numbers are in two columns) but not link this to the numerical pattern. Ask:
- What do all the numbers in this column end in?

STRENGTHENING UNDERSTANDING

Encourage children to mark on a 100 square the numbers they say when counting in 5s from 0, and to identify the spatial pattern (all numbers are in two columns) and link this to the numerical pattern (all numbers alternate between ending in 0 and 5).

GOING DEEPER

Challenge children to identify whether a certain number would or would not be said when counting in 5s from 0. For example, ask: *Would you say 43 when counting in 5s from 0? Can you explain why?*

KEY LANGUAGE

In lesson: count, 5s

Other language to be used by the teacher: digit, multiples

STRUCTURES AND REPRESENTATIONS

Number line, ten frame, 100 square, number track

RESOURCES

Mandatory: 100 square, counters, blank ten frames

Optional: number lines, tennis balls (or pictures of tennis balls), translucent counters, dice

 In the eTextbook of this lesson, you will find interactive links to a selection of teaching tools.

Quick recap

Ask children to pick between 1 and 5 counters and place the counters in front of them. They should show a partner who then tells them how many counters they have got. Children should try and do this by subitising (without counting). You can do this as a whole class by displaying the counters on the board in different arrangements.

Unit 11: Multiplication and division, Lesson 3

Discover

WAYS OF WORKING Pair work

ASK

• Questions ① a) and b): *What does the picture show? What is the question asking you? What representations could you use?*

IN FOCUS Question ① a) asks children to count the number of balls in each tube. They may count the number of balls in one tube by subitising rather than counting. Question ① b) introduces children to counting in 5s. Children should use counters and ten frames or dice arrangements of 5 to help them do this. Children may reflect that counting in 5s is quicker and more efficient than counting in 1s.

PRACTICAL TIPS This activity could be replicated using easily available classroom objects that are in groups of 5, such as bundles of 5 pencils.

ANSWERS

Question ① a): Each tube has 5 tennis balls.

Question ① b): There are 25 tennis balls in total.

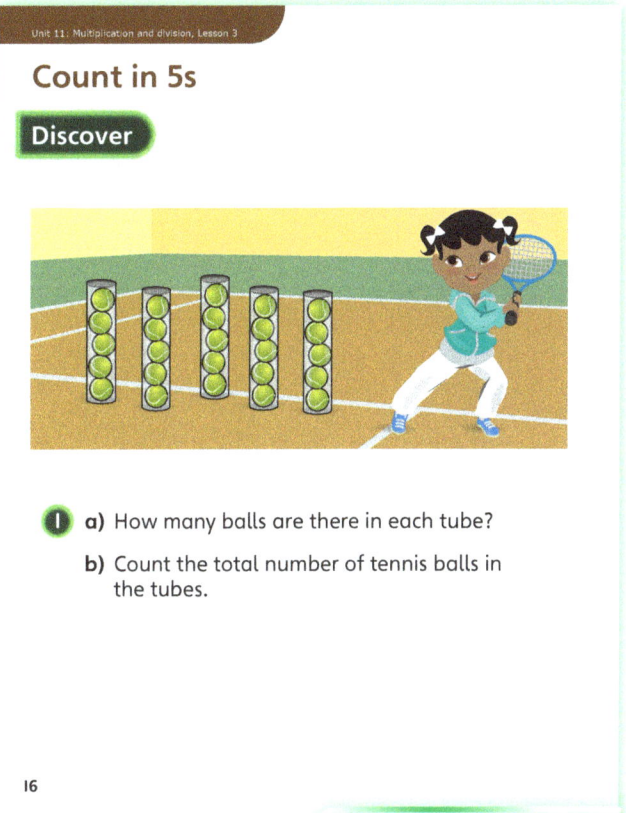

PUPIL TEXTBOOK 1C PAGE 16

Share

WAYS OF WORKING Whole class teacher led

ASK

• Question ① a): *How did you count the tennis balls? Did you know how many tennis balls were in one tube without counting them?*
• Question ① b): *How did you solve this problem? How could you count in 5s? What representations could you use to help you count in 5s?*

IN FOCUS Use Dexter and Flo's comments to help children count in 5s. Use dice patterns of 5 or counters on ten frames to support them. Children may also use their fingers to help them count in 5s. They need to understand the counting pattern in 5s.

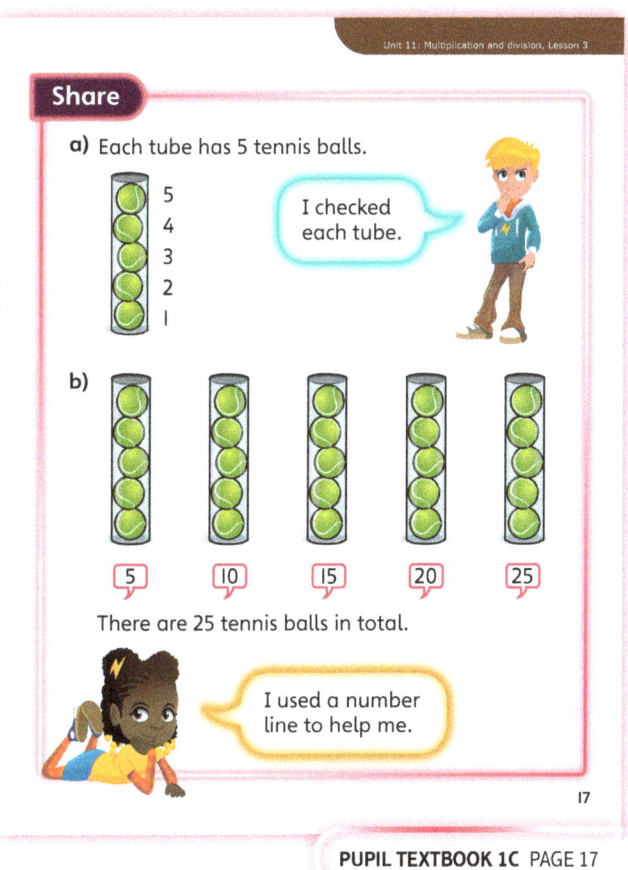

PUPIL TEXTBOOK 1C PAGE 17

Think together

WAYS OF WORKING Whole class teacher led (I do, We do, You do)

ASK
- Question ❶: *Can you practise counting in 5s with a partner? Do you notice anything?*
- Question ❷: *What number is on each dice? How do you know? If they are all the same, how many dots will there be? How many dots on two dice, three dice, four dice …?*
- Question ❸: *Where on the 100 square are all the numbers that you have said? What do you notice about them? What do they all end in?*

IN FOCUS In question ❶, children practise further counting in 5s to understand the 0, 5 pattern. They should count out loud as a class and then in pairs. Spend some time counting in 5s to make sure they understand the pattern. Question ❷ shows a typical dice representation of 5. Use the dice representations to get children to point and count as they go. Question ❸ moves to a more abstract representation on half a 100 square, where children start to see that the numbers are in two columns, and that they all end in a 0 or a 5. This will be useful later in their maths journey. In question ❸ b), children use the number line to count in 5s, as they did with counting in 2s.

STRENGTHEN Children may benefit from having concrete objects available for them to manipulate. Encourage children to mark the numbers in the 5s count on their own 100 squares, either by circling or by placing translucent counters over them.

DEEPEN Challenge children to identify whether a certain number would or would not be said when counting in 5s from 0. For example, ask: *Would you say 43 when counting in 5s from 0? Can you explain why? What do you know about all numbers that are in the 5s count?*

ASSESSMENT CHECKPOINT Use questions ❶ and ❷ to check children can confidently count in 5s. Use question ❸ to check that children have started to understand that numbers that are in the 5s count end in a 0 or a 5.

ANSWERS

Question ❶: Children count 5, 10, 15, 20, 25, 30, 35, 40, 45, 50.

Question ❷: Children count 5, 10, 15, 20, 25, 30, 35, 40.

Question ❸ a): Children point to and count 5, 10, 15, 20, 25, 30, 35, 40, 45, 50.

Question ❸ b): Children point to and count 5, 10, 15, 20, 25, 30, 35, 40.

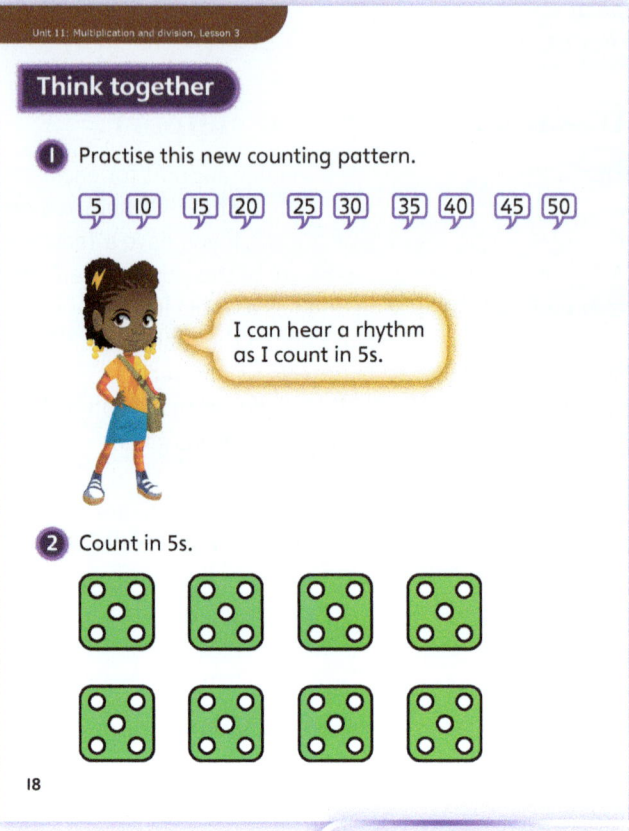

PUPIL TEXTBOOK 1C PAGE 18

PUPIL TEXTBOOK 1C PAGE 19

Unit 11: Multiplication and division, Lesson 3

Practice

WAYS OF WORKING Independent thinking

IN FOCUS These questions are designed to encourage children to understand that the same numbers are always said when counting in 5s from 0. Encourage children to use their answers to previous questions to help them. For example, can they use their answers to the earlier questions to count back in 5s in question ❺? Questions ❶ and ❷ provide visual representations to support children counting in 5s.

Questions ❸, ❹ and ❺ provide more abstract examples that rely on children knowing the 5s count. Question ❻ challenges children to notice which numbers are in both the 5s count and the 2s count.

STRENGTHEN Listen to children counting. If they say a number that does not end in a 0 or 5, ask: *What patterns have you noticed when you count in 5s?* Encourage children to continue to use a 100 square to support their counting in 5s, using the spatial pattern (two columns) to link to the numerical pattern (numbers alternating between ending in 0 and 5).

DEEPEN Ask children to explain if a given number will be in the numerical pattern they have noticed. Children could be asked to apply their knowledge of this numerical pattern to numbers greater than 50. For example, show children the number 65. Ask: *Would you say this number when counting in 5s from 0? How do you know?*

THINK DIFFERENTLY In question ❹, children need to use their knowledge of counting in 5s to label the missing numbers on the number line.

ASSESSMENT CHECKPOINT Use questions ❶ and ❷ to check children can count in 5s using images; and use questions ❸, ❹ and ❺ to check that children have a secure understanding of counting in 5s.

ANSWERS Answers for the **Practice** part of the lesson can be found in the *Power Maths* online subscription.

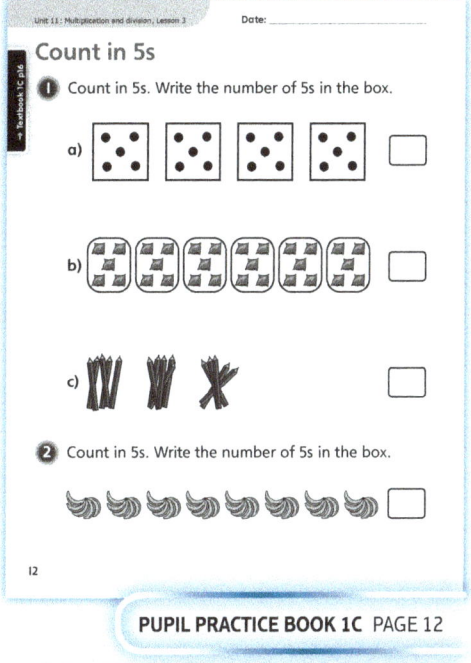

PUPIL PRACTICE BOOK 1C PAGE 12

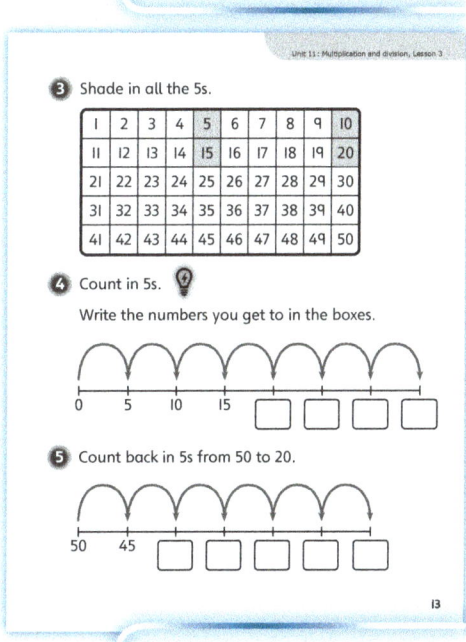

PUPIL PRACTICE BOOK 1C PAGE 13

Reflect

WAYS OF WORKING Pair work

IN FOCUS This activity allows children to apply their knowledge of counting in 5s by taking it in turns with a partner to count in 5s.

ASSESSMENT CHECKPOINT Assess whether children can count in 5s. Have they securely identified the numerical pattern?

ANSWERS Answers for the **Reflect** part of the lesson can be found in the *Power Maths* online subscription.

After the lesson

- Are children secure in counting in 5s and have they identified the numerical pattern?
- What opportunities can you create for children to practise counting in 5s in a range of different situations?

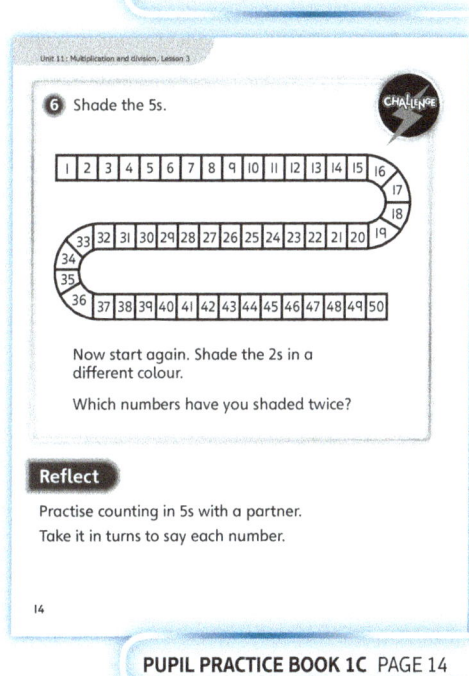

PUPIL PRACTICE BOOK 1C PAGE 14

Unit 11: Multiplication and division, Lesson 4

Equal groups

Learning focus

In this lesson, children will explore counting on and back in 5s, to 50. They will explore the patterns that exist when counting in 5s.

Before you teach

- How confident are children in their understanding of 'equal'?
- How could you use children's real-life experiences to broaden and add depth to this lesson?

NATIONAL CURRICULUM LINKS

Year 1 Number – multiplication and division

Solve one-step problems involving multiplication and division, by calculating the answer using concrete objects, pictorial representations and arrays with the support of the teacher.

ASSESSING MASTERY

Children can recognise and explain how they know when groups are equal. They can recognise and explain how this can help them count totals more efficiently.

COMMON MISCONCEPTIONS

Children may confuse the number of groups with the number of objects in each group. Ask:
- *How many objects are in each group? How can you check? How many groups of these objects are there? Can you say in a sentence how many groups of objects there are?*

STRENGTHENING UNDERSTANDING

Introduce the topic of this lesson through contexts easily recognisable to children: for example, when handing out resources. How do they know when something has been grouped equally? Ask: *I am going to give each person on this table two pencils. How many groups of 2 will that be?*

GOING DEEPER

Children could investigate numbers from 1 to 20. How many ways can each number be put into equal groups? For example, 6 can be put into 3 groups of two, or 2 groups of three, or 6 groups of one. Ask: *Can you find out how many ways 10 can be put into equal groups? Can you tell me how many groups of 2 make 8? Can 9 be put into equal groups of 2? Which is the smallest number that can be put into equal groups of 2? Or 5? Or 10? Which numbers cannot be put into any equal groups of 2, 5 or 10? With a set of six objects, can you count 1 group of 6 and 6 groups of 1 as equal groups?*

KEY LANGUAGE

In lesson: equal groups

Other language to be used by the teacher: counting, groups, number, pattern, unequal, amount

STRUCTURES AND REPRESENTATIONS

Cubes, counters

RESOURCES

Mandatory: cubes, counters

Optional: grids showing the group sizes appropriate to the questions in the Textbook, collections of countable objects

 In the eTextbook of this lesson, you will find interactive links to a selection of teaching tools.

Quick recap

Ask 5 children to come to the front of the class. Give another child 4 pencils. Ask: *Are there enough pencils for the children? Does this mean there are more pencils or more children?* Play similar activities, working out whether there are more or fewer objects or children.

Unit 11: Multiplication and division, Lesson 4

Discover

WAYS OF WORKING Pair work

ASK

- Questions 1 a) and b): *What is the same and what is different about the boats? In which type of boat will the people be easier to count? What is different about the size of the groups in the rowing boats and sailing boats?*

IN FOCUS Use this picture to focus on why it is easier and more efficient to count the people in the rowing boats. See whether children can recognise that, as there are unequal numbers of people in the sailing boats, it is not possible to count them using a repeating pattern.

ANSWERS

Question 1 a): There are 4 rowing boats.
There are 2 people in each rowing boat.

Question 1 b): There are 3 sailing boats.
There is not the same number of people in each sailing boat.

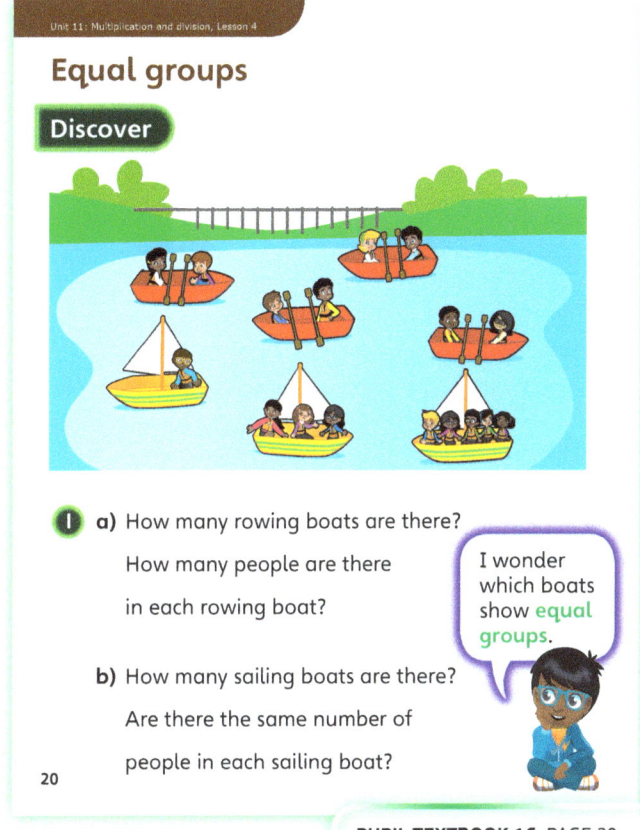

PUPIL TEXTBOOK 1C PAGE 20

Share

WAYS OF WORKING Whole class teacher led

ASK

- Question 1 a): *How did you count the total number of people in the rowing boats? Did you count them one by one, or is there another way? How could you use jumps on a number line to count the people in the rowing boats?*
- Question 1 b): *Could the people in the sailing boats be grouped equally? How do you know?*
- Question 1 b): *Is there more than one way to group the people in the sailing boats equally?*

IN FOCUS At this point in the lesson, it would be beneficial to let children explore putting different numbers of objects into equal groups. Start with the children in the sailing boats. Children should discover that there is only one way to make equal groups: 3 groups of 3. As an additional point to explore with children, 12 would be a useful number to investigate as it can be used to make equal groups of 2, 3, 4 and 6.

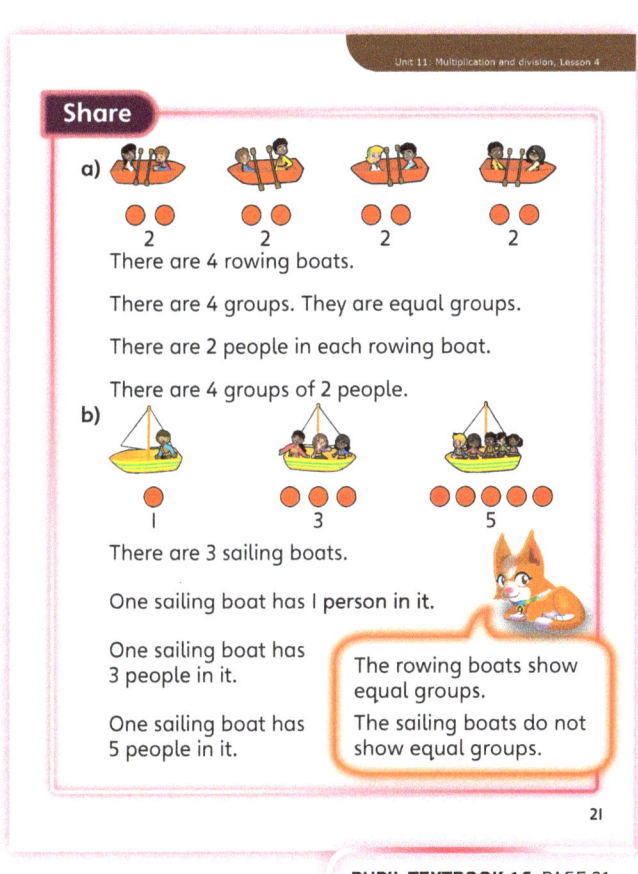

PUPIL TEXTBOOK 1C PAGE 21

Think together

WAYS OF WORKING Whole class teacher led (I do, We do, You do)

ASK
- Question ❷: *How will you know if the counters are in equal groups? What two pieces of information will you need?*
- Question ❸: *Does it matter if the objects are arranged differently?*

IN FOCUS All questions in this section will develop children's ability to recognise and describe equal groups. Question ❸ requires children to begin differentiating between equal and unequal groups.

STRENGTHEN If children struggle to recognise equal groups in the pictures, give them an opportunity to use concrete manipulatives to represent the groups shown. This could be supported further by giving children grids with a number of boxes matching the group size (for groups of 2, have a grid with 2 boxes). Ask children to place the objects into the grid. Do they have enough to have an equal quantity in each box?

DEEPEN Once children have solved question ❸, challenge them to answer Ash's question. Can they arrange the other shapes into equal groups? Explain that they do not have to be put into 3 equal groups as in the diagrams, but could be a different number of equal groups – even equal groups with one shape in each. Ask: *Is there only one way of grouping the (triangles, for example) into equal groups? Could you put them into 4 equal groups? How many triangles would be in each group then? Are groups where there is one triangle in each group still equal groups? How many groups would you have then?* Discuss the fact that groups of 1 are still equal groups.

ASSESSMENT CHECKPOINT All questions will show whether children understand and recognise equal groups, with question ❸ identifying children who do or do not recognise equal groups when the objects are arranged differently. Question ❸ will also identify which children can rearrange objects so that they are in equal groups. Some children may just see that moving one triangle from the first group to the last group will make 3 groups of 4. Other children, having been told that there do not need to be only 3 groups, may explore further to discover more or all solutions.

ANSWERS

Question ❶: There are 5 groups of 2 ice cubes.

Question ❷: Answers will vary depending on how many counters are put in each circle. For example, 3 equal groups of 2; 3 equal groups of 3.

Question ❸: Set A shows equal groups.

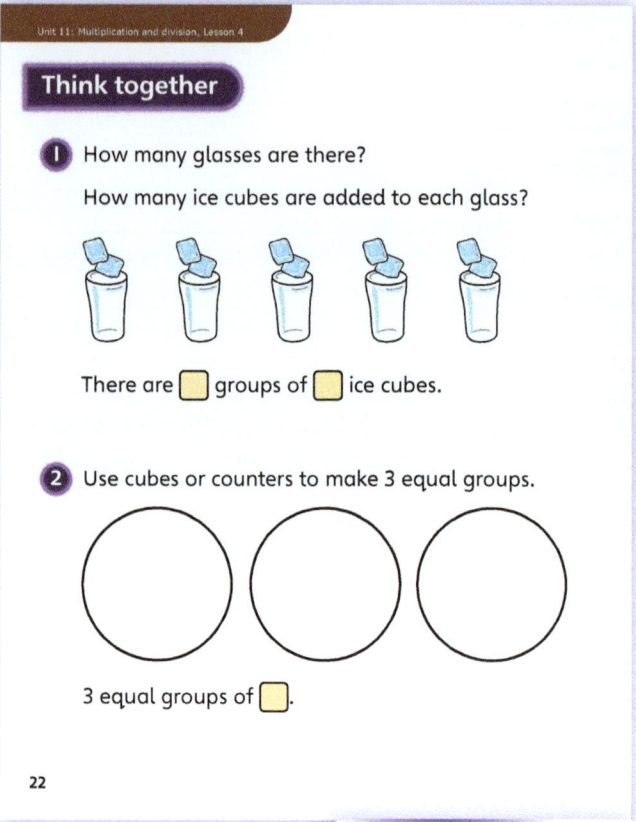

PUPIL TEXTBOOK 1C PAGE 22

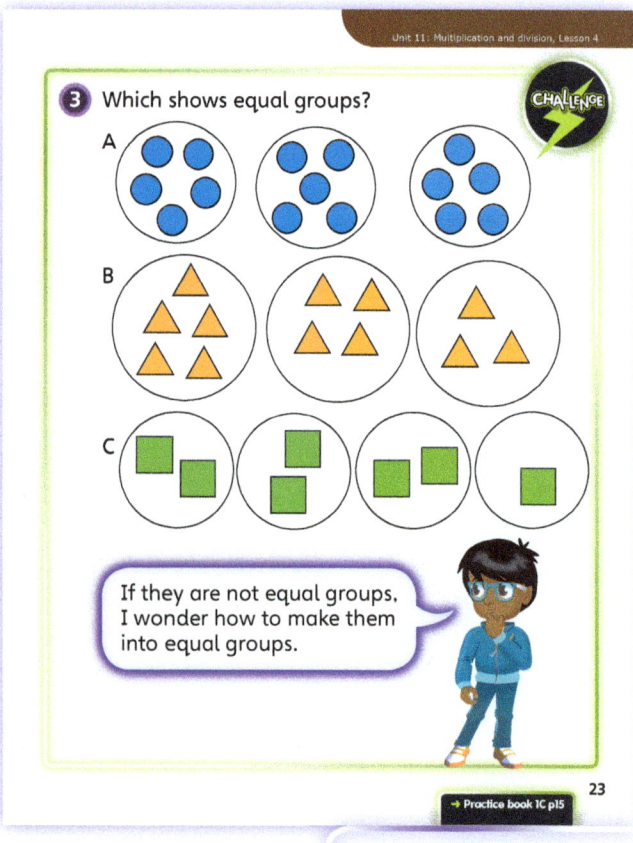

PUPIL TEXTBOOK 1C PAGE 23

Unit 11: Multiplication and division, Lesson 4

Practice

WAYS OF WORKING Independent thinking

IN FOCUS These questions build on each other in making it possible to assess whether children can recognise equal groups and if they can make their own equal groups. In question ①, children are given one group and are then asked to draw objects to make equal groups. In question ②, children make their own equal groups by drawing their own objects. Question ③ scaffolds children's ability to describe equal groups, focusing on how many groups there are and how many objects are in each group. In question ④, children have to complete sentences that show equal groups.

STRENGTHEN If children are finding it tricky to identify equal groups, ask them to explain what the word 'equal' means. Ask: *What could be the same about each group? Can you spot where the groups have the same number of objects or pictures?*

DEEPEN To deepen children's understanding of how the *arrangement* of objects in a group does not impact on the *number* it represents, ask: *Do groups need to look equal?* Challenge children to explain why mistakes could be made. Ask: *Which sets of objects are clearly equal? Which set of objects look unequal but are actually equal?*

ASSESSMENT CHECKPOINT These questions assess whether children can differentiate between the number of groups and the number *in* each group, identify equal groups, and show understanding that equal groups are equal even if the objects in the groups are arranged differently.

ANSWERS Answers for the **Practice** part of the lesson can be found in the *Power Maths* online subscription.

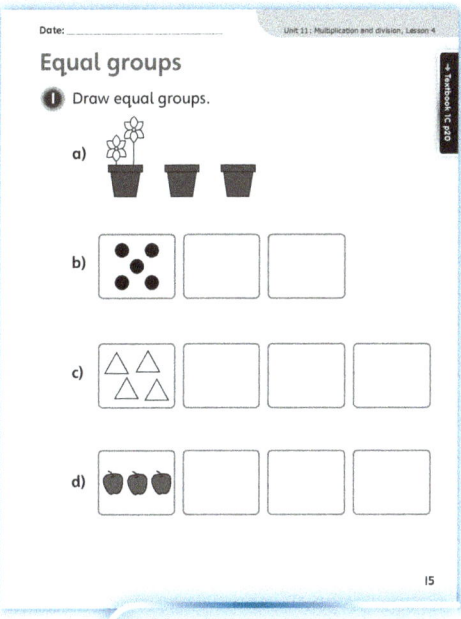

PUPIL PRACTICE BOOK 1C PAGE 15

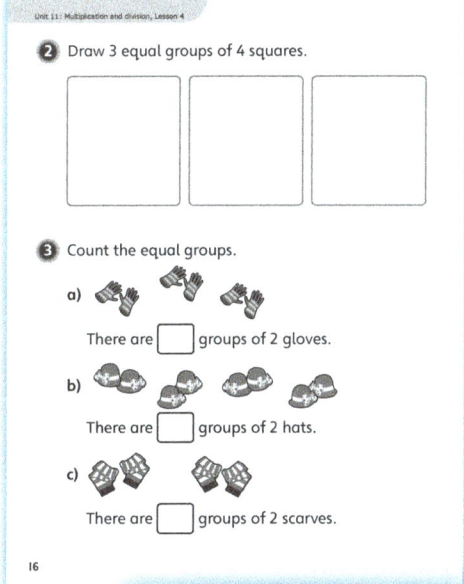

PUPIL PRACTICE BOOK 1C PAGE 16

Reflect

WAYS OF WORKING Pair work

IN FOCUS Ask children to focus on just one set of groups. Listen to words they use such as 'equal' and why they are equal.

ASSESSMENT CHECKPOINT Check that children are able to explain how many groups there are, how many objects are in each group, and whether the groups are equal or not. This question provides the opportunity to discuss if groups that do not look equal are actually equal.

ANSWERS Answers for the **Reflect** part of the lesson can be found in the *Power Maths* online subscription.

After the lesson ⏸

- Can children recognise equal groups?
- Can they draw their own equal groups?
- Do they know that equal groups do not have to be arranged in the same way to be equal?

Unit 11: Multiplication and division, Lesson 5

Add equal groups

Learning focus

In this lesson, children will build on their understanding of equal groups and begin adding equal groups together to find a total.

Before you teach

- How confident are children at recognising when groups are equal and when they are not equal?
- What support will you offer children who count on in 1s rather than using the count number?

NATIONAL CURRICULUM LINKS

Year 1 Number – multiplication and division

Solve one-step problems involving multiplication and division, by calculating the answer using concrete objects, pictorial representations and arrays with the support of the teacher.

ASSESSING MASTERY

Children can recognise where groups are equal and add these groups together to find a total. They can use a number line to reliably and fluently represent their additions.

COMMON MISCONCEPTIONS

Children may not be able to translate the problem into an addition. Ask:
- *How many counts of (5) do you need to make? How many jumps on the number line are there? How many (5s) should there be in the addition? How can you write this problem as an addition?*

STRENGTHENING UNDERSTANDING

Ask children to arrange counters or cubes into a set number of equal groups, for example 4 groups of 2. Children should then practise recording what the counters show in two ways: 4 groups of 2 and as an addition 2 + 2 + 2 + 2. Repeat by asking children to add one counter to each group making 4 groups of 3: 3 + 3 + 3 + 3. Continue to show 4 groups of 4 and 4 groups of 5. Ask: *Which number is changing?* [Number of counters.] *Which number stays the same?* [Number of groups.] *How many numbers are in each addition?* [4, the same as the number of groups.]

GOING DEEPER

Encourage children to investigate a skip count sequence and its related additions. For example, ask children to look into the patterns found if the skip count is in groups of 4. Ask: *How far can you skip count in 4s? What would the addition look like for each new step of the sequence?* Ask children to describe any patterns they notice in the numbers and in the addition calculations. Could they use the patterns to help them count further? Children could also use number lines to draw skip counts, recording each as an addition.

KEY LANGUAGE

In lesson: pair, skip counting, addition, in total, adding

Other language to be used by the teacher: equal groups, counts, add, plus, sum, equals, number line, problem, sequence, pattern, represent

STRUCTURES AND REPRESENTATIONS

Number line

RESOURCES

Mandatory: counters

Optional: collections of countable objects, large printed number line

 In the eTextbook of this lesson, you will find interactive links to a selection of teaching tools.

Quick recap

Ask children to make their own equal groups of counters. Ask: *How many groups did you make? How many counters are in each group?* Ask children to write or say a sentence about their groups. Now ask children to make 4 groups of 2 counters. Ask them how they can work out how many counters there are in total.

Unit 11: Multiplication and division, Lesson 5

Discover

WAYS OF WORKING Pair work

ASK

- Question 1 a): *How many shoes are in each box? How many boxes are there? How can you count the shoes?*
- Question 1 b): *How many vases are there? How many flowers are in each vase? How can you count the flowers?*

IN FOCUS In questions 1 a) and b), children need to apply their knowledge of counting in 2s and 5s and equal groups to work out the total number of each object. They may represent the objects using counters and then count out loud. They may also choose to use a number line to help them. Ask children if they can write their counting as a number sentence.

PRACTICAL TIPS Count other groups of concrete objects that are in equal groups of 2s, 5s and 10s.

ANSWERS

Question 1 a): There are 6 shoes.

Question 1 b): There are 15 flowers.

PUPIL TEXTBOOK 1C PAGE 24

Share

WAYS OF WORKING Whole class teacher led

ASK

- Question 1 a): *Are the groups equal? How do you know? Do equal groups make counting easier? Can you write this as an addition sentence?*
- Question 1 b): *What is the same about this question and what is different?*

IN FOCUS In question 1 a), use the language of equal groups from the previous lesson to discuss why the groups are equal. This then means children can count in 2s to find the total. Count aloud with children and then use the number of groups to show how they find the total. Explain that they can write down an addition number sentence to represent this. 2 + 2 + 2 = 6, which shows that there are 3 groups and there are 2 shoes in each group; and there are 6 shoes in total. Question 1 b) goes through the same process for counting in 5s and working out the total number of flowers in 3 vases.

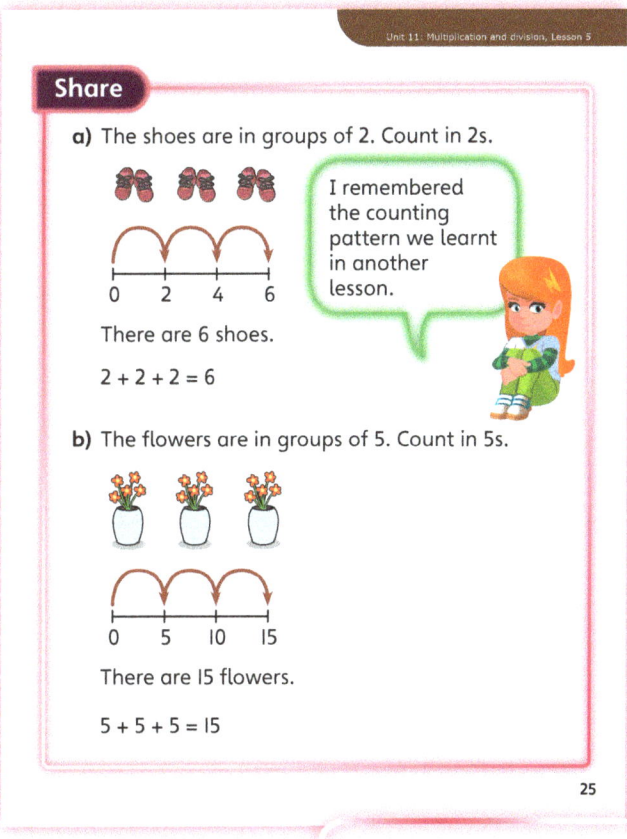

PUPIL TEXTBOOK 1C PAGE 25

Unit 11: Multiplication and division, Lesson 5

Think together

WAYS OF WORKING Whole class teacher led (I do, We do, You do)

ASK

- Question ❶: *How many eggs are there in each box? How many boxes are there? How many eggs are there in total? What addition sentence shows the total number of eggs?*
- Question ❷: *What addition sentences can you write for each score?*
- Question ❸: *What is the same and what is different about each number sentence? How does the answer to each calculation match the picture?*

IN FOCUS Throughout these questions, children write down addition sentences to show the totals. They should start to see the connection between the number of groups and number of objects in each group and the numbers they need to add together.

STRENGTHEN For children who are struggling to relate the pictures of the groups to the number, use concrete resources to represent the objects pictured. These could be arranged along a large printed number line to help children link the concrete with the abstract. Ask: *How many are in each group? What number do you need to skip count in? How many skip counts do there need to be? What numbers will be recorded along the number line?*

DEEPEN For children who have completed question ❸, deepen their understanding by drawing on the previous lesson. Ask: *Can you use the cubes to make different equal groups?* Challenge children to record their new equal groups as an addition. Ask: *How many different ways can you find to do this? How will you know when you have found all the possibilities?*

ASSESSMENT CHECKPOINT Use questions ❶ and ❷ to assess whether children can use pictorial representations to add equal groups. Use question ❸ to assess whether children can link pictorial and abstract representations of equal groups together.

ANSWERS

Question ❶: 10 + 10 + 10 = 30

Question ❷ a): 5 + 5 + 5 + 5 + 5 = 25

Question ❷ b): 2 + 2 + 2 + 2 = 8

Question ❸:

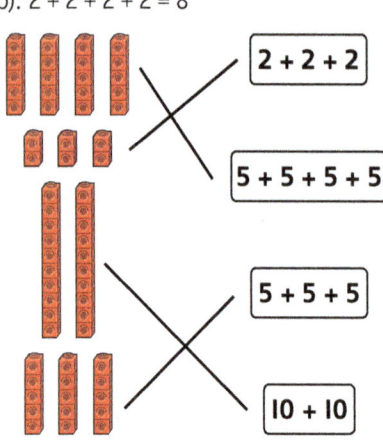

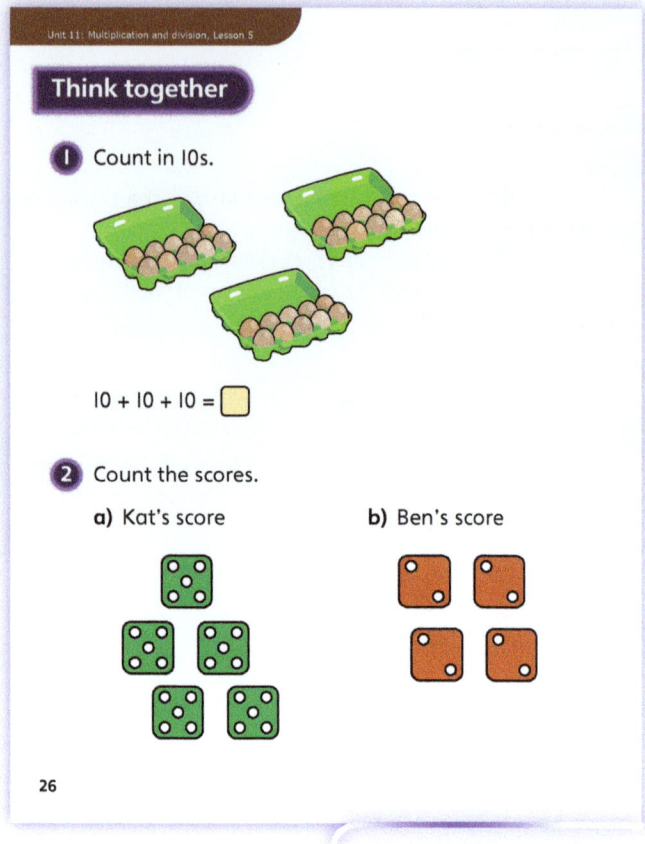

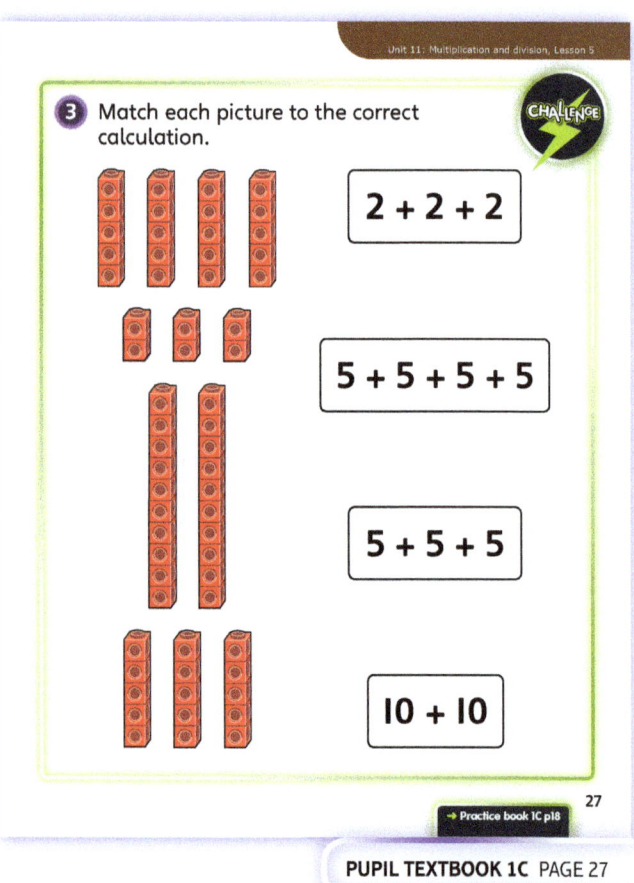

Unit 11: Multiplication and division, Lesson 5

Practice

WAYS OF WORKING Independent thinking

IN FOCUS At this point in the lesson, where children are working independently, it is very important to look out for any children who confuse the number of groups with the number of objects in each group. Question ① a) helps to scaffold children's understanding of this concept, with the elements of the count partially completed for them.
In question ②, children write down calculations to work out the total scores. Children use their knowledge of counting in 2s and 5s to work out the answers.

STRENGTHEN To support children in answering question ②, ask: *How many dots are there on each dice? How many dice do you need to count? Are there any patterns in your counting that can help you find the total?*

DEEPEN Question ④ offers an excellent opportunity to develop children's reasoning. As the two sticker sheets both have the same number of stickers, the question will also challenge children's assumptions about group size and number of groups, and how these affect the total amounts. Ask: *Can you explain how the two sticker sheets are similar and how they are different? Do you know what links the two sheets? Can you describe how to count the stickers more efficiently?*

ASSESSMENT CHECKPOINT At this point in the lesson, children should be able to explain clearly how skip counting is an efficient method for finding the total of a number of equal groups. They will be able to link this to a number line and the written addition calculation. Using these methods, children should be fluent when finding the total of a given number of equal groups.

ANSWERS Answers for the **Practice** part of the lesson can be found in the *Power Maths* online subscription.

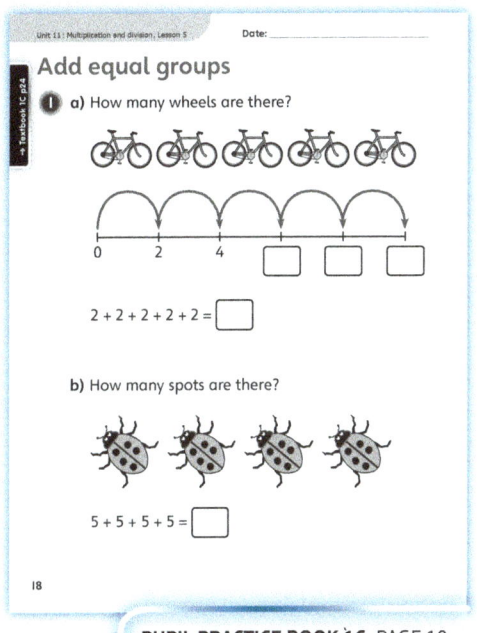

PUPIL PRACTICE BOOK 1C PAGE 18

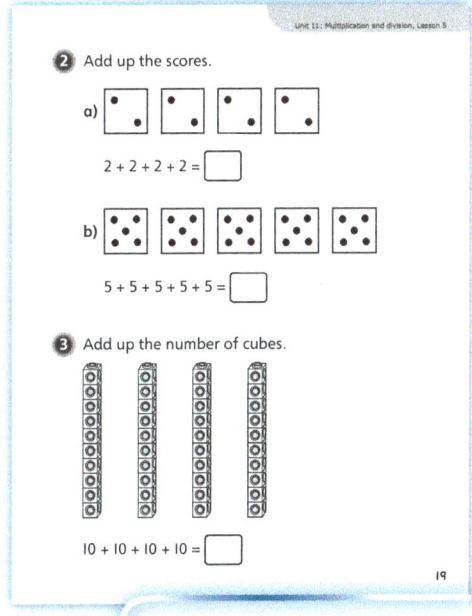

PUPIL PRACTICE BOOK 1C PAGE 19

Reflect

WAYS OF WORKING Whole class

IN FOCUS Discuss children's different approaches to the question and encourage them to share their ideas with the class. Ask: *What size groups did you use? Could you have used different-sized groups? How many of these groups were there?*

ASSESSMENT CHECKPOINT Look for children recognising the equal groups within the picture and using these to skip count. Children may have seen different groups (6 groups of 5, 3 groups of 10 or possibly 2 groups of 15). Use this as an opportunity to assess children's fluency and flexibility with the idea of finding and using equal groups to assist counting.

ANSWERS Answers for the **Reflect** part of the lesson can be found in the *Power Maths* online subscription.

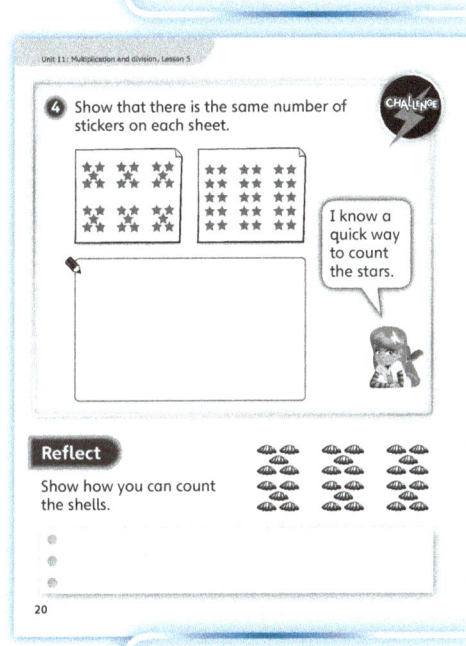

PUPIL PRACTICE BOOK 1C PAGE 20

After the lesson

- Could children confidently explain the link between skip counting and addition?
- Are there any opportunities you could offer, outside of the mathematics lesson, for children to practise skip counting?

Unit 11: Multiplication and division, Lesson 6

Make arrays

Learning focus
In this lesson, children will learn to recognise, understand and create simple arrays. They will link this representation to their learning about repeated addition.

Before you teach
- How could you introduce the lesson's vocabulary of 'rows' and 'columns' using children's real-life experiences?
- How will you make this lesson as hands-on as possible? Could you begin or end with an array hunt around the school environment?

NATIONAL CURRICULUM LINKS

Year 1 Number – multiplication and division

Solve one-step problems involving multiplication and division, by calculating the answer using concrete objects, pictorial representations and arrays with the support of the teacher.

ASSESSING MASTERY

Children can recognise an array and explain what it represents fluently using the vocabulary of 'columns' and 'rows'. They can identify and explain how an array can be read in two ways and can show the related repeated additions along a number line.

COMMON MISCONCEPTIONS

When counting or creating an array, children may mistakenly count too many or too few objects in a row or column. Ask:
- *How many should there be in all the rows? How many should there be in all the columns? Can you prove that all the rows are equal and all the columns are equal? Where did you make a mistake?*

STRENGTHENING UNDERSTANDING

Encourage children to look for arrays in the world around them. Examples include egg cartons, window panes, bookshelves, some chocolate box arrangements, cake trays and chess boards. Discuss with children the patterns they can see in the arrays. Ask: *Which equal groups can you see?* Discuss and practise counting using first the rows and then the columns, establishing that the total of all the rows is the same as the total of all the columns.

GOING DEEPER

Encourage children to look for more arrays in the world around them. Ask them to take photos or draw pictures of the arrays that they have found. Challenge children to order their arrays from largest to smallest and to draw them a different way. Ask them to choose one of the arrays. Ask: *Can you create a different array that represents the same amount? Could you turn or change this array so that the rows become the columns?*

KEY LANGUAGE

In lesson: row, array, column, arrangement, total, count

Other language to be used by the teacher: addition, largest, smallest, groups

STRUCTURES AND REPRESENTATIONS

Array, number line

RESOURCES

Mandatory: counters

Optional: collections of countable objects, large printed number lines, boards as a base for the arrays

 In the eTextbook of this lesson, you will find interactive links to a selection of teaching tools.

Quick recap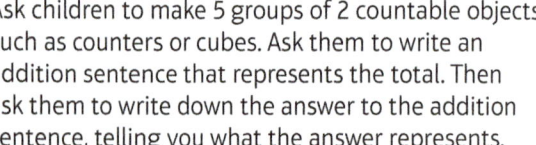

Ask children to make 5 groups of 2 countable objects, such as counters or cubes. Ask them to write an addition sentence that represents the total. Then ask them to write down the answer to the addition sentence, telling you what the answer represents. Repeat for other groups of 2, 5 and 10.

Unit 11: Multiplication and division, Lesson 6

Discover

WAYS OF WORKING Pair work

ASK

- Question 1 a): *How have the seeds been arranged? What equal groups can you see in the picture?*

IN FOCUS To help reinforce the structure of arrays, in question 1 a) focus on how both the columns and rows represent equal groups. In question 1 b), children are asked to use concrete materials to replicate the pictorial array. This will help secure their understanding of what the array shows.

PRACTICAL TIPS Using cubes or counters in question 1 b) will give children hands-on experience of creating their own arrays.

ANSWERS

Question 1 a): There are 10 seeds in each row.
There are 2 rows.
10 + 10 = 20
There are 20 seeds in total.

Question 1 b): Children show 2 equal rows of ten counters or cubes.

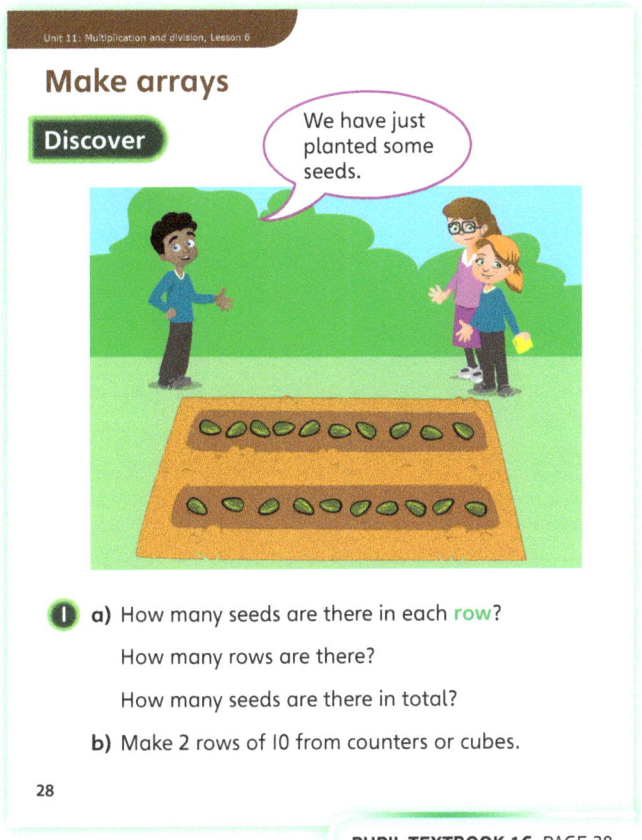

PUPIL TEXTBOOK 1C PAGE 28

Share

WAYS OF WORKING Whole class teacher led

ASK

- Question 1 a): *What is the same about all the columns? What is the same about all the rows?*

IN FOCUS At this point in the lesson, it is important for children to recognise how an array is made up of rows and columns. In question 1 b), they should be encouraged to line their counters or cubes up side by side, so they are aligned. Encourage children to investigate the different arrays they can make with a given number of objects. Can they find a number where the number of objects in both the array's rows and columns are the same?

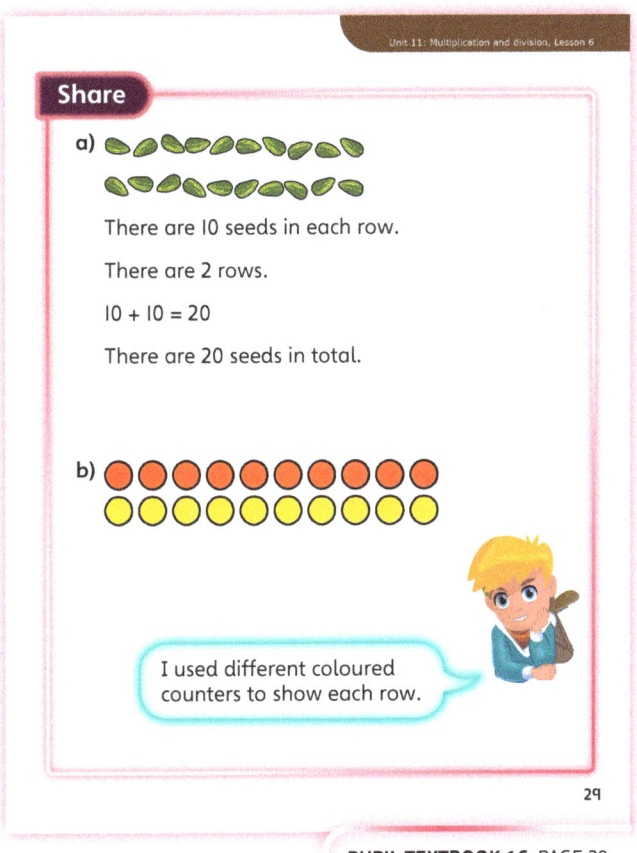

PUPIL TEXTBOOK 1C PAGE 29

Think together

WAYS OF WORKING Whole class teacher led (I do, We do, You do)

ASK

- Question ①: *How many rows are there? How many columns are there? How many counters are in each row? How can you work out how many counters there are in the array by counting quickly?*
- Question ②: *Is it easier to count the columns or count the rows? Do you get the same answer if you count the columns and rows?*

IN FOCUS In question ①, children look at how arrays are made up of rows and columns. They start to grow confident in finding out how many counters are in the array by adding the counters in the rows and/or columns.

In question ② a), explore finding the total number of counters by adding the rows and then adding the columns. Show children that the answer is the same when counting in rows or columns. Remind them that they should count efficiently by counting in 2s, 5s and 10s. In question ③ b), children discuss why the set of counters is not an array.

STRENGTHEN During questions ① and ②, encourage children to build the arrays using concrete resources to strengthen their understanding. Ask them to make the arrays they see on the page. Then ask them to explain how many counters there are in total. Is there another way of looking at the array?

DEEPEN In question ③ a), ask children to see if they can work out the number of counters in each array using different methods. For example, can they count in 2s? Can they work out the next answer from the previous one?

ASSESSMENT CHECKPOINT Children should be able to recognise an array, observing the rows and columns and explaining how many objects are in each one. Children should be able to explain that the total should be the same whether they count the objects in rows or in columns.

ANSWERS

Question ① a): Children show 3 equal rows of 5.

Question ① b): There are 3 rows of 5.

Question ① c): There are 15 in total.

Question ② a): 5 + 5 = 10

Question ② b): 5 + 5 + 5 + 5 = 20

Question ③ a): 2 + 2 + 2 = 6 or 3 + 3 = 6
2 + 2 + 2 + 2 = 8 or 4 + 4 = 8
2 + 2 + 2 + 2 + 2 = 10 or 5 + 5 = 10

Question ③ b): The rows and columns are not equal. The rows do not have an equal number of counters in them.

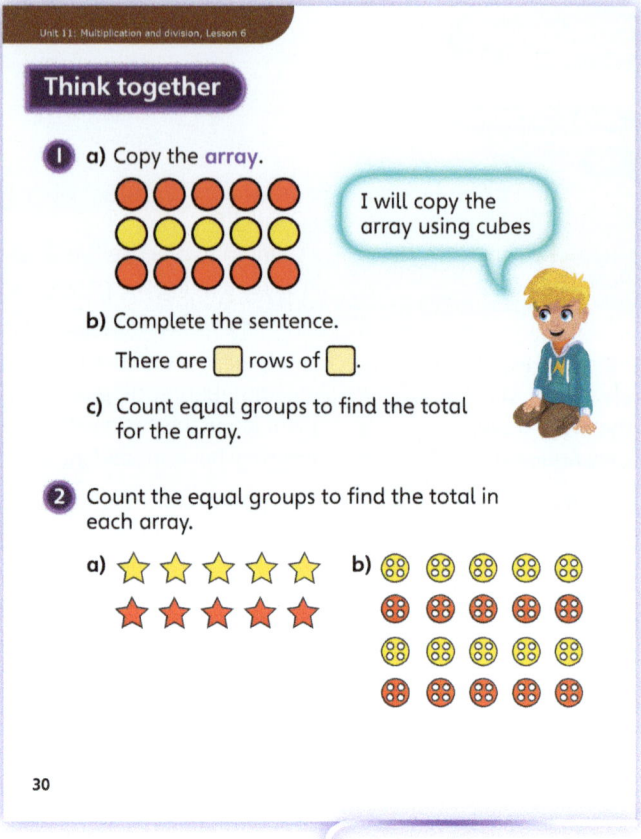

PUPIL TEXTBOOK 1C PAGE 30

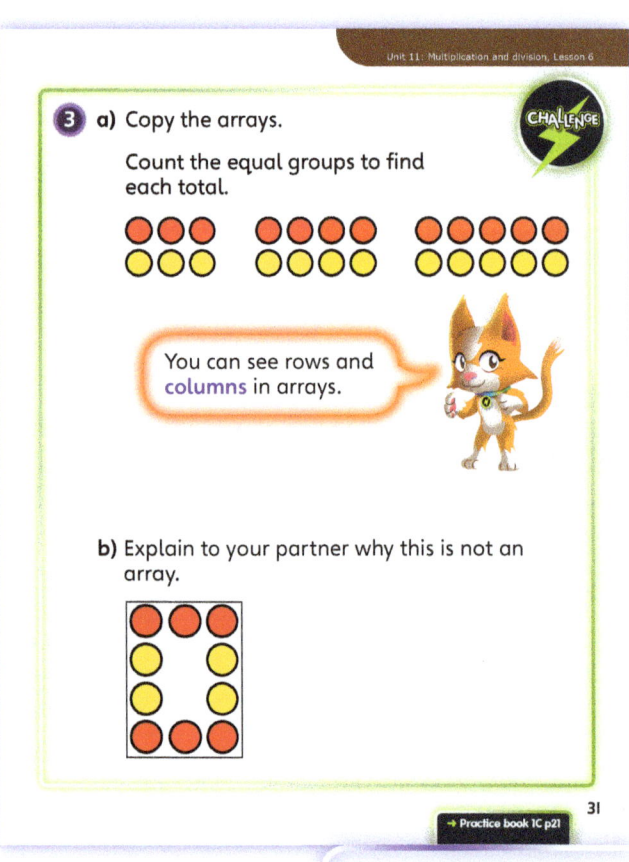

PUPIL TEXTBOOK 1C PAGE 31

Unit 11: Multiplication and division, Lesson 6

Practice

WAYS OF WORKING Independent thinking

IN FOCUS Questions ❶ and ❷ provide scaffolding for children to find out the total from arrays by asking children to shade in rows or columns. They then use addition sentences and counting in 2s and 5s to find the answers. In question ❸, children match arrays to the correct descriptions.

STRENGTHEN If children are struggling to link the arrays with repeated addition calculations, ask them to find equal groups within the array. Did they look at the rows or columns? Encourage the use of a number line to skip count, linking this to addition. It may be beneficial to allow children to make the array using counters or other physical resources, then place these along a printed number line, to help reinforce the link between the models.

DEEPEN Ask children to draw or make their own 10 × 2 array. Ask them to explain how they can work out how many counters there are in the array. Discuss why it is easier to count in 10s than 2s. Show children that whichever way they add, they get the same number of counters. Now give children 30 counters and ask them to make different arrays.

THINK DIFFERENTLY In question ❹, children are shown equal groups within an array, and have to complete an addition sentence for each array, with reduced scaffolding.

ASSESSMENT CHECKPOINT At this point in the lesson, children should be confidently identifying and describing arrays using the correct vocabulary of rows and columns. They should be able to explain which two repeated additions an array shows, and represent these using number lines and addition calculations.

ANSWERS Answers for the **Practice** part of the lesson can be found in the *Power Maths* online subscription.

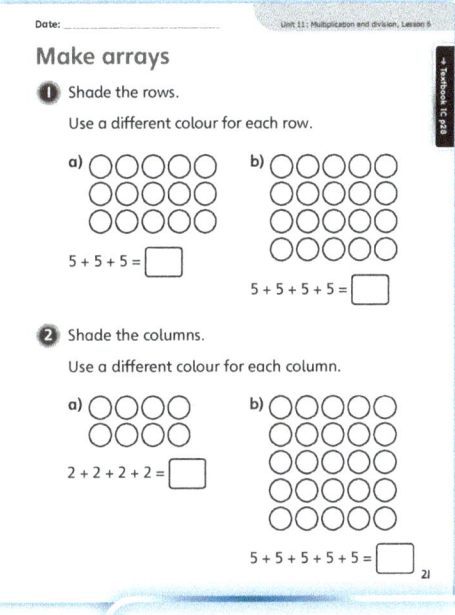

PUPIL PRACTICE BOOK 1C PAGE 21

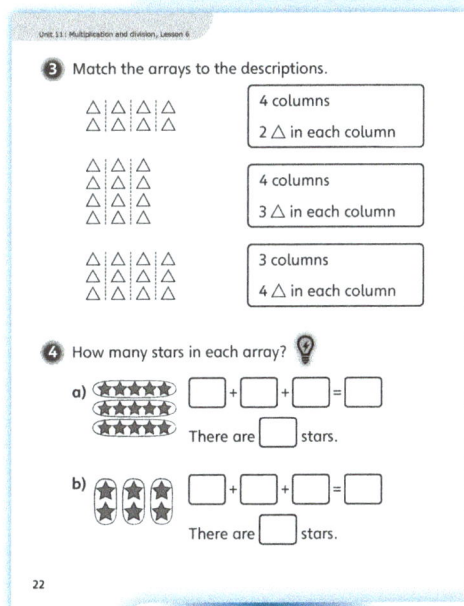

PUPIL PRACTICE BOOK 1C PAGE 22

Reflect

WAYS OF WORKING Independent thinking

IN FOCUS Once children have completed this question, ask them whether they chose to count the rows or the columns. Did they take up Flo's challenge to write the two related additions?

ASSESSMENT CHECKPOINT Children should be able to independently select how they will count the array. Encourage them to find both ways by discussing their answer with a partner who has a different answer.

ANSWERS Answers for the **Reflect** part of the lesson can be found in the *Power Maths* online subscription.

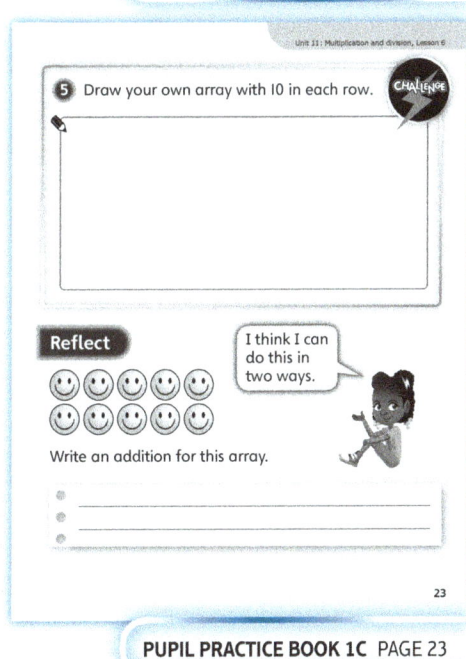

PUPIL PRACTICE BOOK 1C PAGE 23

After the lesson ⏸

- Were children more inclined to count the rows or columns or both?
- If children relied more heavily on one facet of the array, how will you develop their confidence with the other one?
- Were children able to confidently see the link between arrays and repeated addition?

Unit 11: Multiplication and division, Lesson 7

Make doubles

Learning focus

In this lesson, children will learn about the meaning of the word 'double'. Using familiar representations, they will develop their understanding of what a double is and how to find one.

Before you teach

- Where will children have met this concept before? Could you use their experiences to help introduce the lesson?
- How will you make sure children get sufficient opportunity to observe patterns and explain their generalisations?

NATIONAL CURRICULUM LINKS

Year 1 Number – multiplication and division

Solve one-step problems involving multiplication and division, by calculating the answer using concrete objects, pictorial representations and arrays with the support of the teacher.

Year 1 Non-statutory guidance

Through grouping and sharing small quantities, pupils begin to understand: multiplication and division; doubling numbers and quantities; and finding simple fractions of objects, numbers and quantities.

ASSESSING MASTERY

Children can find the double of a given number and will be able to explain what doubles are, using the appropriate vocabulary. They will be able to use representations they are familiar with to show doubles concretely and pictorially, and use repeated addition calculations to represent doubles in an abstract manner.

COMMON MISCONCEPTIONS

Children may miscount when finding the double of a number. Ask:
- How can you check your working out? What representation could you use to make sure your double is correct?

STRENGTHENING UNDERSTANDING

Introduce children to the concept of this lesson by playing a version of the game shown in **Discover**. Working in pairs, children could roll one dice each. If the dice land on the same number, the children have rolled a double. The first child to recognise this and shout 'Double!' wins a point. Children could be asked to find the total that both dice make.

GOING DEEPER

Ask children to investigate the links between numbers such as double 2 and double 20. For example, ask: *If double 2 is 4 and double 20 is 40, what is double 30?* Ask children to explain how they found the double. Challenge them to find as many doubles of 10s numbers as they can. Can they spot any patterns?

KEY LANGUAGE

In lesson: double, group

Other language to be used by the teacher: add, pair, representation, addition, working out, ten frame

STRUCTURES AND REPRESENTATIONS

Ten frame, arrays

RESOURCES

Optional: large printed ten frame, 6-sided dice, 1–20 number tracks, different coloured counters

 In the eTextbook of this lesson, you will find interactive links to a selection of teaching tools.

Quick recap

Remind children of the counting in 2s sequence before they begin their work on doubles. As a class, ask children to count in 2s from 0 to 20. You could all sing a counting song. If children are confident in counting on in 2s, ask them to count back in 2s from 20 to 0.

Discover

WAYS OF WORKING Pair work

ASK
- Question 1 a): *What totals have been made by each pair of dice in the picture?*
- Question 1 b): *How will you know if you have thrown a double? What other doubles can you make using the two dice? How else can you make a double? What resources could you use to show a double?*

IN FOCUS In question 1 a), children are asked to identify which of the two children has rolled a double. Reinforce the understanding that 'double' means two of a number. In question 1 b), children can use the number track to help them work out what double 4 is.

PRACTICAL TIPS Encourage children to play a similar game to begin the lesson by giving groups of children pairs of dice. Encourage children to model the doubles in different ways, especially the use of an array, making the link with the previous lesson.

ANSWERS

Question 1 a): Tariq rolled double 4.

Question 1 b): Double 4 is 8.

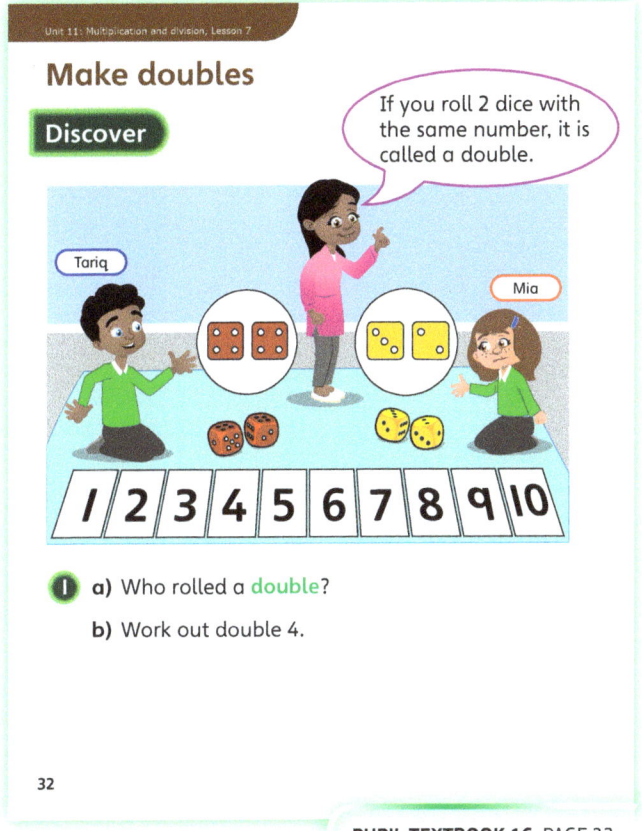

PUPIL TEXTBOOK 1C PAGE 32

Share

WAYS OF WORKING Whole class teacher led

ASK
- Question 1 b): *How did you show a double? How is your representation similar to that shown in the picture? How is it different? How does the ten frame help to show the doubles?*
- Question 1 b): *Can you find the doubles of all the numbers from 1 to 10? What do you notice about the doubles you have found?*

IN FOCUS In question 1 b), draw children's attention to the similarities and differences between the two representations of doubles. What is the same and different about them? Discuss with children why the ten frame shows two different colours of counters.

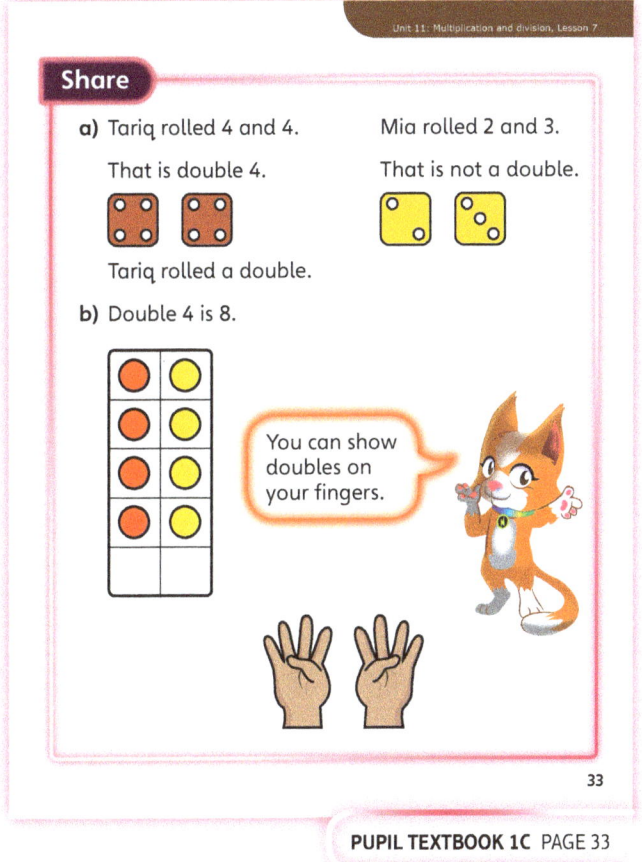

PUPIL TEXTBOOK 1C PAGE 33

71

Unit 11: Multiplication and division, Lesson 7

Think together

WAYS OF WORKING Whole class teacher led (I do, We do, You do)

ASK
- Question ①: *What models can you use to help you work these out?*
- Question ③: *Why are two ten frames shown here, instead of just one?*

IN FOCUS The questions in this part of the lesson help scaffold and reinforce the link between the concrete, pictorial and abstract representations of doubles. Encourage children to make the representations shown on the page to help them understand that doubles are all in the 2s count.

STRENGTHEN For children who are not finding doubles accurately, use ten frames to help reinforce the importance of counting carefully.

To help practise the skill of finding doubles, give pairs of children a 1–20 number track, a dice and different coloured counters each so they can play a game. Children roll a dice and find the double of the number shown. Once they have found the double, they can move their counter that far along the track. The first child to reach 20 wins. Children can then design their own doubles game.

DEEPEN While completing question ③, deepen children's understanding by asking them to find a different way of representing each double, perhaps with a number line or an array. Ask: *Do you see any patterns? What is special about all the double numbers?*

ASSESSMENT CHECKPOINT Children should be able to explain that 'double' means 'two of the same number'. They should be able to show double a number using several representations, in particular through the use of a ten frame. Encourage the link with arrays from the previous lesson.

ANSWERS

Question ① a): Double 2 is 4.
b): Double 3 is 6.

Question ②: Double 1 is 2.
Double 2 is 4.
Double 3 is 6.
Double 4 is 8.
Double 5 is 10.

Question ③: Double 6 is 12.
Double 7 is 14.
Double 8 is 16.
Double 9 is 18.
Double 10 is 20.

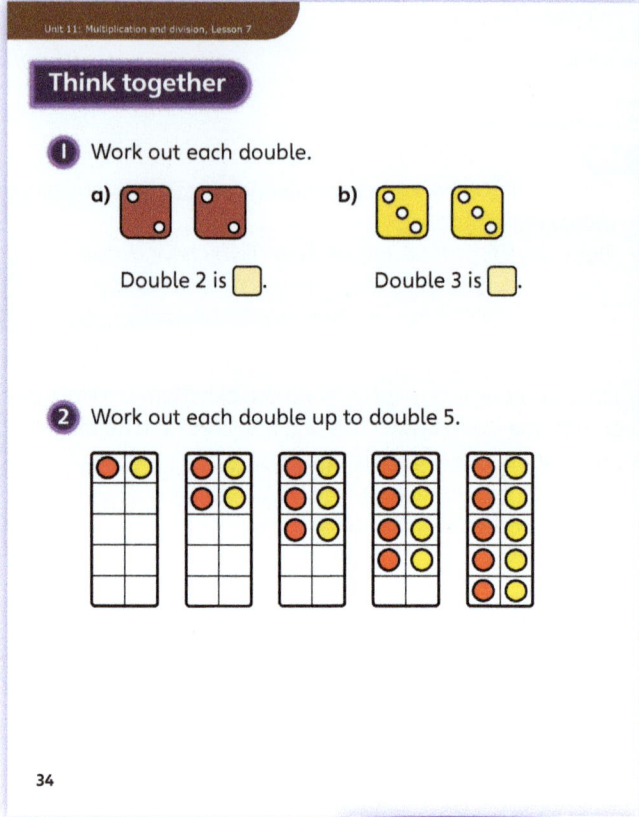

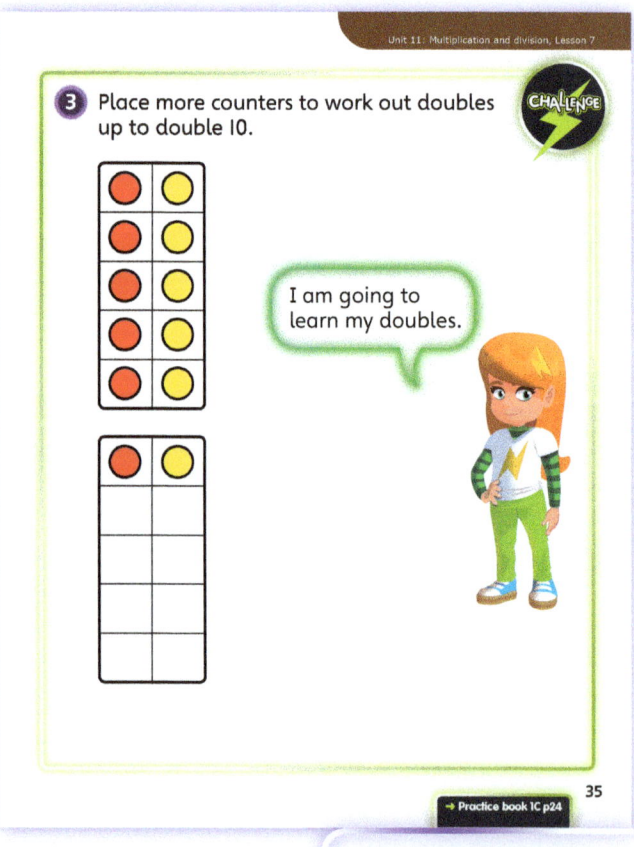

Unit 11: Multiplication and division, Lesson 7

Practice

WAYS OF WORKING Independent thinking

IN FOCUS Questions 1 to 3 help children recognise and understand pictorial representations of doubles. It would be beneficial to have concrete resources to match the questions, for children to manipulate.

Questions 3 b), c) and d) informally use the concept that the inverse of a double is a half.

STRENGTHEN For question 5, provide children with ten frames to explore the method suggested or allow them to use counters to make arrays of two rows, counting in 2s to find the doubles.

DEEPEN Question 5 offers a good opportunity to develop children's pattern recognition and ability to generalise. Ask: *What happens each time you find a new double? Do you think that will happen forever?* Challenge children to predict the next five doubles and to explain their predictions.

ASSESSMENT CHECKPOINT Children should be able to reliably find the double of a given number up to 10, knowing some of them by heart. They should be fluent at using concrete and visual representations of doubles to support their reasoning.

ANSWERS Answers for the **Practice** part of the lesson can be found in the *Power Maths* online subscription.

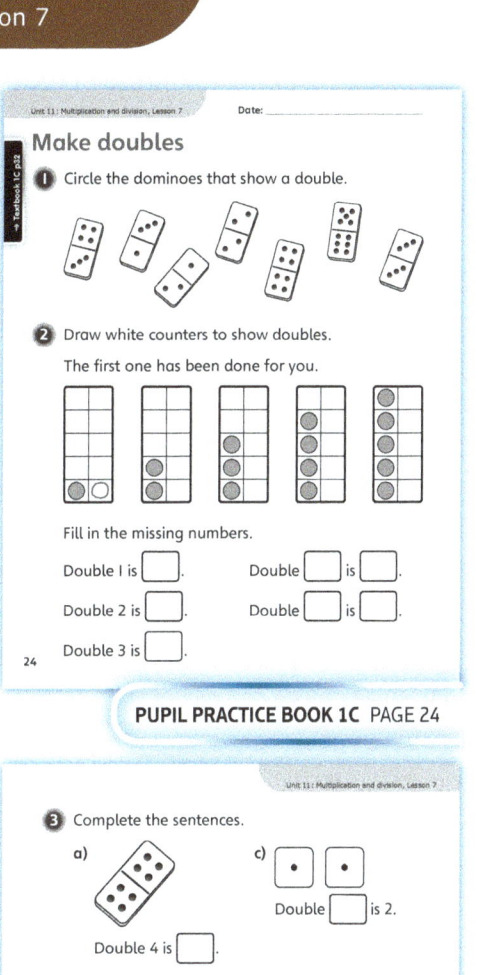

PUPIL PRACTICE BOOK 1C PAGE 24

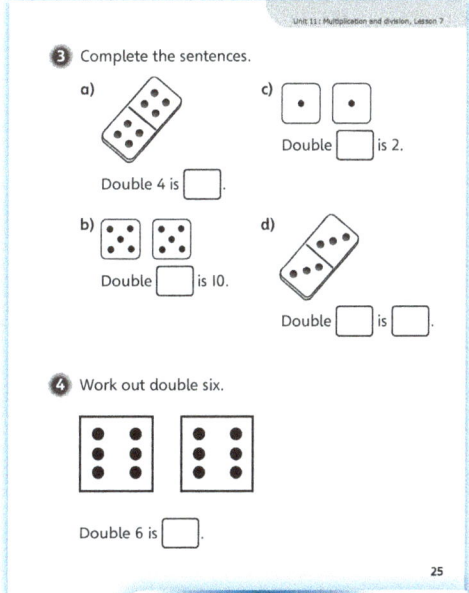

PUPIL PRACTICE BOOK 1C PAGE 25

Reflect

WAYS OF WORKING Pair work

IN FOCUS Give children time to self-assess and then feed back their ideas to the class. Ask: *How did you decide which doubles you knew? Are there any easy ways of remembering your doubles? What could you do to get better at the doubles you still find tricky? How could you practise your doubles later on today?*

ASSESSMENT CHECKPOINT Look for children able to explain what doubles they knew and which they found tricky. Check how children respond to the doubles they have identified as knowing: do they count or work them out in some way or do they just know them? Children should be able to put forward some ideas for the games they could play or resources they could use to practise their doubles, thereby demonstrating their understanding of how these activities link to finding doubles.

ANSWERS Answers for the **Reflect** part of the lesson can be found in the *Power Maths* online subscription.

After the lesson

- How capable were children at moving from simple pattern recognition to a generalisation they could apply more widely?
- What games or activities could you play that provide opportunities for further doubling practice?

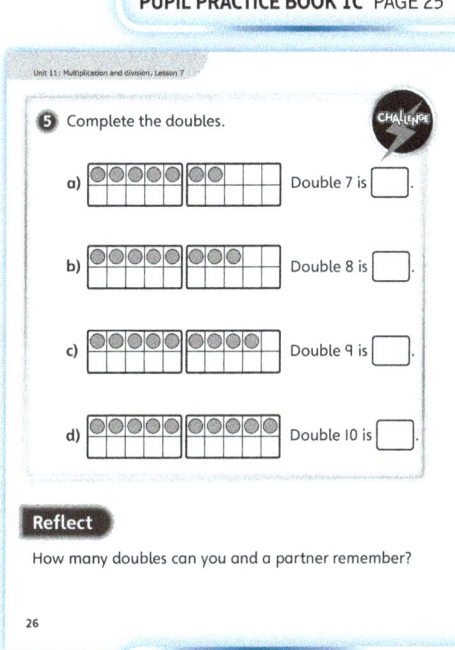

PUPIL PRACTICE BOOK 1C PAGE 26

73

Unit 11: Multiplication and division, Lesson 8

Grouping

> **Learning focus**
>
> In this lesson, children will develop their understanding of equal groups. They will recognise when groups are equal and when they are not, and how many equal groups are needed to make a whole number.

> **Before you teach**
>
> - What elements of the previous lessons on multiplication could be used to enhance children's learning in this lesson?
> - How will you incorporate these into the lesson?

NATIONAL CURRICULUM LINKS

Year 1 Number – multiplication and division

Solve one-step problems involving multiplication and division, by calculating the answer using concrete objects, pictorial representations and arrays with the support of the teacher.

ASSESSING MASTERY

Children can recognise when groups are equal and when they are not. Children can say how many equal groups make a whole number and model the groups using different representations.

COMMON MISCONCEPTIONS

Children may not recognise groups as being equal because they do not look exactly the same. Show groups of children who are wearing different clothes. Ask:
- *How many children are in each group?*
- *Does what the children are wearing change the number of children in each group? Explain why not.*

Children may transpose the numbers describing the number of groups and the amount each group represents. For example, they may label 5 groups of 2 as 2 groups of 5. Ask children to count how many objects are in each group, and also to circle each group of objects and count how many groups they have circled. Ask:
- *Can you explain how many groups of objects you have?*

STRENGTHENING UNDERSTANDING

Children could practise the skill of making equal groups through normal classroom activities and routines, for example, when making up snacks, handing out resources or grouping children into small teams.

GOING DEEPER

Ask children to investigate how many different equal groups they can make with a given number of children. Challenge them further by stating the number of equal groups needs to be in the 2s count (the term 'even' will be introduced in Year 2). For example, ask: *You are going to play some games of football. There are 12 children. How many ways could you split the children into teams? Why would three teams of four children not work for our games of football?*

KEY LANGUAGE

In lesson: equal, group, different, altogether

STRUCTURES AND REPRESENTATIONS

Arrays

RESOURCES

Mandatory: blank pieces of card, multilink cubes

Optional: counters or other countable objects (toy people, toy animals or pictures of these), printed rectangles to represent tables, printed circles to represent groups, paper clips

 In the eTextbook of this lesson, you will find interactive links to a selection of teaching tools.

> **Quick recap**
>
> Give children 20 blank pieces of card. Ask them to put the pieces of card into piles of 2. Ask: *How many piles can you make?* Now ask children to put the pieces of card into piles of 5 and then piles of 10.

Discover

WAYS OF WORKING Pair work

ASK
- Question 1 a): *How can you tell if the groups are equal?*
- Question 1 a): *Which group is the biggest? Which is the smallest?*
- Question 1 b): *How can you represent groups of 2 children? Can you make different equal groups using the number of children in the picture?*

IN FOCUS Question 1 encourages children to explore using 10 to make equal groups. Ask them to investigate what equal groups they can make with 10. Discuss what happens when 10 is used to make groups of 4. What do children notice?

PRACTICAL TIPS Use counters or cubes to replicate the scenario.

ANSWERS

Question 1 a): There are 10 children in total.
　　　　　　　1 group of 2.
　　　　　　　1 group of 3.
　　　　　　　1 group of 5.

Question 1 b): Children show 5 equal groups of 2.

Share

WAYS OF WORKING Whole class teacher led

ASK
- Question 1 a): *Were the children in equal groups? How did you prove your thinking?*
- Question 1 b): *How did you represent the equal groups of 2?*
- Question 1 b): *Did you find any other equal groups?*

IN FOCUS Use question 1 b) to ensure children are confident when observing and explaining the difference between, for example, 5 groups of 2 and 2 groups of 5: show children representations of the two sets of groups and discuss the similarities and differences.

Unit 11: Multiplication and division, Lesson 8

Think together

WAYS OF WORKING Whole class teacher led (I do, We do, You do)

ASK
- Question ❶: *What can you use to represent the children? How can you put them into groups of 4?*
- Question ❷: *How can you put the socks into groups of 2? Are there any left? How many groups did you get?*

IN FOCUS These questions help children develop the vocabulary needed to support their reasoning and mathematical discussion. Make sure all children are given the opportunity to try to say their answers as full sentences such as 'There are 3 groups of 4 children'.

Questions ❶ and ❷ are designed to help tackle the misconception of confusing the group size and the number of groups. Children should make groups of 4 children and 2 socks, using counters to represent the children and socks. Look at how children approach this. They should not share their counters into groups; instead they need to make groups of 4 or 2. Question ❸ looks at different ways you can share 20 by building towers. Give children time to explore this concept.

STRENGTHEN To support children in answering question ❶, give them concrete representations of the 12 children (toy people, or something more abstract like counters). Support them in making groups of 4. Read aloud the result of the grouping, for example: 'there are 3 groups of 4 children'.

DEEPEN Extend question ❸ by encouraging children to explore the similarities and differences between 4 groups of 5 and 5 groups of 4. Ask: *How are 4 groups of 5 the same as 5 groups of 4? How are they different? Can you show me what you have noticed using different representations?* Can they link their understanding today to their work on arrays?

ASSESSMENT CHECKPOINT Assess whether children can put counters or other concrete objects into equal groups. Can they describe the groups accurately, stating both the number of groups and the number of objects in each group?

ANSWERS

Question ❶: There will be 3 groups of 4 children.

Question ❷: There are 6 groups (pairs) of 2 socks.

Question ❸ a): 4 towers of 5 cubes.

Question ❸ b): 2 towers of 10 cubes.

Question ❸ c): 5 towers of 4 cubes.

Question ❸ d): 10 towers of 2 cubes.

Question ❸ e): There are not enough cubes to make equal towers. There are 6 towers of 3 cubes, with 2 cubes left over.

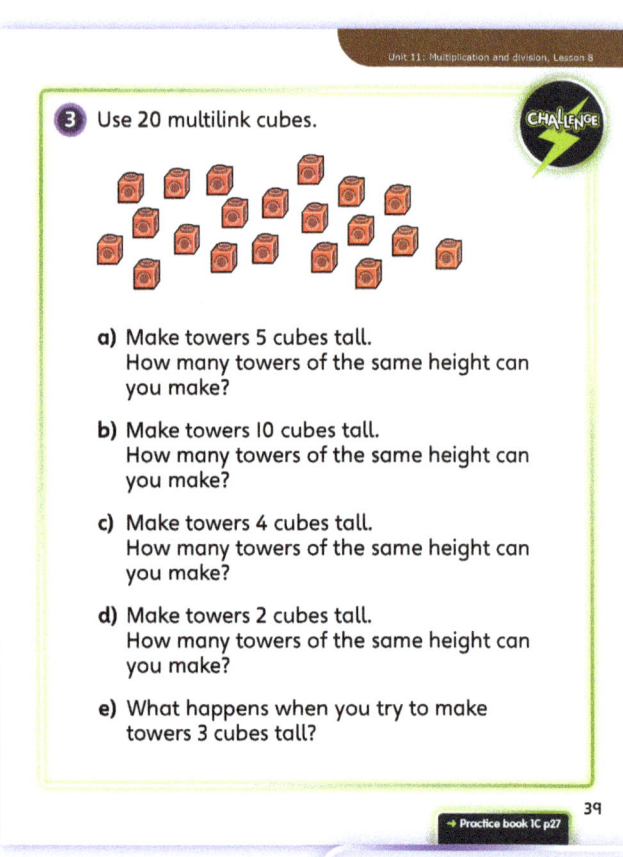

Unit 11: Multiplication and division, Lesson 8

Practice

WAYS OF WORKING Independent thinking

IN FOCUS In question ❶, children practise circling groups of 2. The objects and number of groups differ, however, the number in each group remains the same. In questions ❷ and ❸, children circle different-sized groups. Unlike in question ❶, this time the groups are not as clear and children will have to think a bit more carefully. For question ❸, children can link back to their knowledge of arrays. In question ❹, children draw their own groups based on given information and work out how many groups they need to draw.

STRENGTHEN If children find grouping objects tricky, give them concrete resources to put into equal groups. You could give those who need further support printed circles to represent the groups; children can then arrange the resources in the circles.

DEEPEN Use a question similar to the **Reflect** question to deepen children's understanding. Ask them if they can put 20 into equal groups of 2, 3, 4 and 5. Ask: *Which ones can you do? Which ones do not work?*

ASSESSMENT CHECKPOINT Use the questions in this section to assess whether children can circle groups of objects. They should be able to work out how many groups they have circled. Some children will be able to say or write down their own sentence describing what they have done.

ANSWERS Answers for the **Practice** part of the lesson can be found in the *Power Maths* online subscription.

Reflect

WAYS OF WORKING Pair work

IN FOCUS This question is best presented in a concrete way. Give pairs of children 20 cubes and allow them time to make groups of 5 cubes. Ask: *What do you notice? Can you see the 4 groups you have made? Can you say what you have made?*

ASSESSMENT CHECKPOINT Check whether children can fluently put objects into groups of a given size. Ensure that children are not sharing. Listen to children's justifications of their method: they should talk about how they grouped the cubes and describe how they ensured that each group had the correct number of cubes in it. They should be able to tell you their answer using a full sentence.

ANSWERS Answers for the **Reflect** part of the lesson can be found in the *Power Maths* online subscription.

After the lesson

- Were children able to see the link between, for example, 3 lots of 5 and 5 lots of 3?
- Did any children make the link between this lesson and their understanding of arrays?

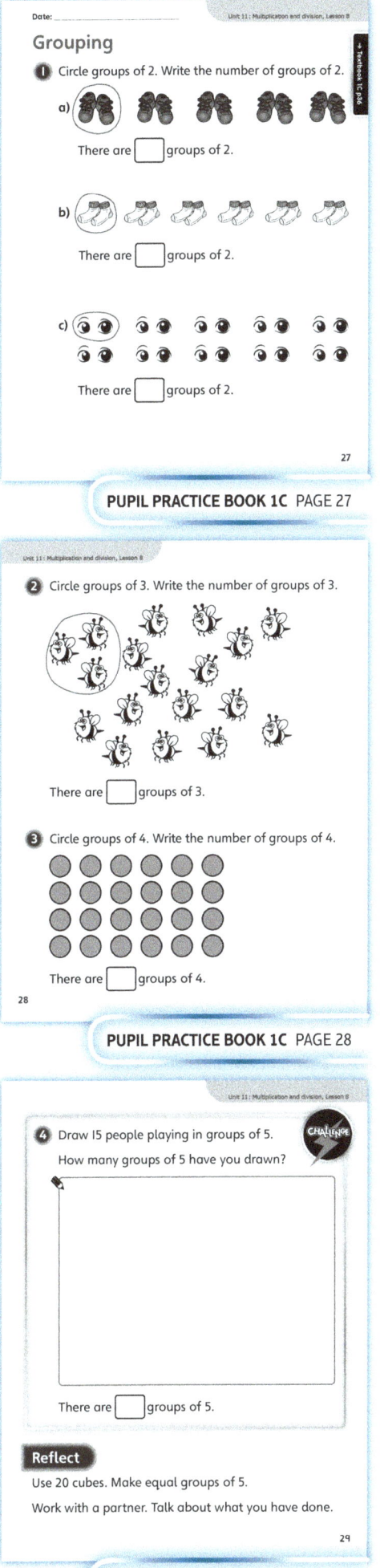

Unit 11: Multiplication and division, Lesson 9

Sharing

Learning focus

In this lesson, children will develop their understanding of equal groups. They will recognise when groups are equal and when they are not, and how many equal groups are needed to make a whole number.

Before you teach

- How will you make the difference between 'making groups of a certain number' and 'sharing between groups of a certain number' explicit in the lesson?
- Are all children secure with the idea of equal groups?

NATIONAL CURRICULUM LINKS

Year 1 Number – multiplication and division

Solve one-step problems involving multiplication and division, by calculating the answer using concrete objects, pictorial representations and arrays with the support of the teacher.

ASSESSING MASTERY

Children can recognise and explain sharing as 'one each' shared to each group over and over again until the total amount has been exhausted. Children can use this concept to share numbers into equal groups and solve simple problems.

COMMON MISCONCEPTIONS

Children may confuse making 'groups of a certain number' with 'sharing between a number of groups'. Show children two pictures, one of a number shared into equal groups of 3 and one of the same number shared into 3 groups. Ask:
- *What is the same and what is different about these two pictures?*
- *Which picture shows the number shared into 3 equal groups? How do you know?*
- *How has the number been shared in the other picture?*

STRENGTHENING UNDERSTANDING

Children could practise this skill of equal sharing through games and role play. For example, they could share out the counters between players in a board game or they could role-play a party, sharing the slices of a cake between all the party goers.

GOING DEEPER

Ask children to pick two digit cards and make a 2-digit number. When children have made their numbers, ask them to explore what equal groups they can share their number into, and what equal groups do not work. Challenge children to prove their ideas.

KEY LANGUAGE

In lesson: share, equally, same amount, group

Other language to be used by the teacher: fairly

STRUCTURES AND REPRESENTATIONS

Arrays

RESOURCES

Optional: counters or other countable objects or pictures of these, playing cards, marbles, printed circles to represent groups, digit cards

 In the eTextbook of this lesson, you will find interactive links to a selection of teaching tools.

Quick recap

Ask children to share out a group of objects fairly. For example, show 10 pencils and ask children to share them equally between 5 people. Ask: *Are there any left over? Did you use all the pencils? How many did each person get?*

Discover

WAYS OF WORKING Pair work

ASK

• Question 1 a): *How can the children share the strawberries equally? What could you use to represent the strawberries?*
• Question 1 b): *How can you share the counters? How many counters will each person get? How is this similar to the strawberries?*

IN FOCUS Question 1 a) is meant as more of a discussion for the class. Pose the question and listen to the ideas that children may have. Some children may know that they each get 5 strawberries. Encourage the method of sharing one by one: first child gets one, then second gets one and so on until all the strawberries have been shared equally. Question 1 b) asks children to share 10 counters between 2 children by giving each other 1 counter at a time. Relate this back to the strawberries in question 1 a).

PRACTICAL TIPS Use counters or cubes to replicate the scenario.

ANSWERS

Question 1 a): 10 shared between 2 is 5 each.

Question 1 b): 10 shared between 2 is always 5 each.

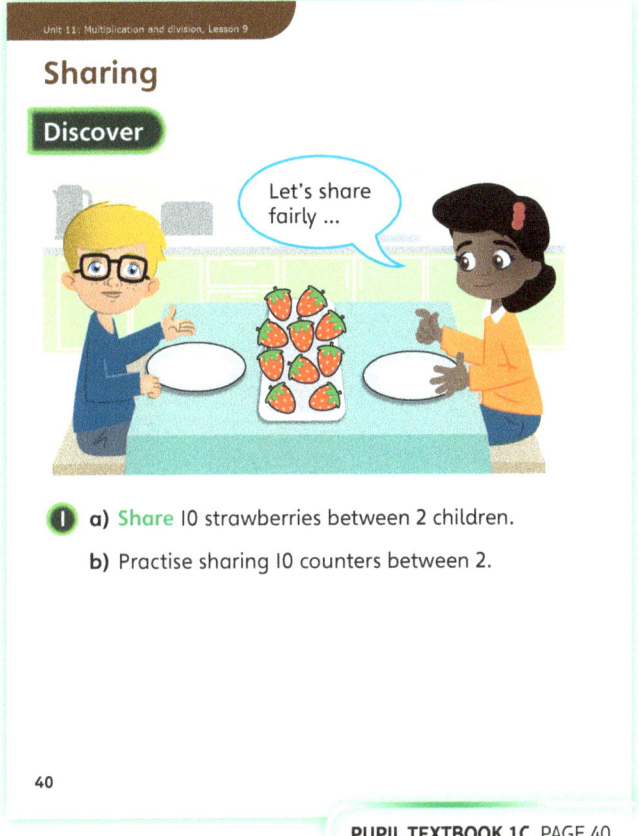

PUPIL TEXTBOOK 1C PAGE 40

Share

WAYS OF WORKING Whole class teacher led

ASK

• Question 1 a): *What are the children doing? How many strawberries do they each get?*
• Question 1 b): *Did you share the counters like the children shared the strawberries? How did you share them?*

IN FOCUS The main focus of questions 1 a) and b) is to introduce the method of sharing with children. The strawberries in question 1 a) can be used to explain how the children shared the fruit and then how this can be applied to sharing the counters in question 1 b). As a class, demonstrate how 2 children can share the 10 counters between them. Use language that is being used by the children to share the strawberries.

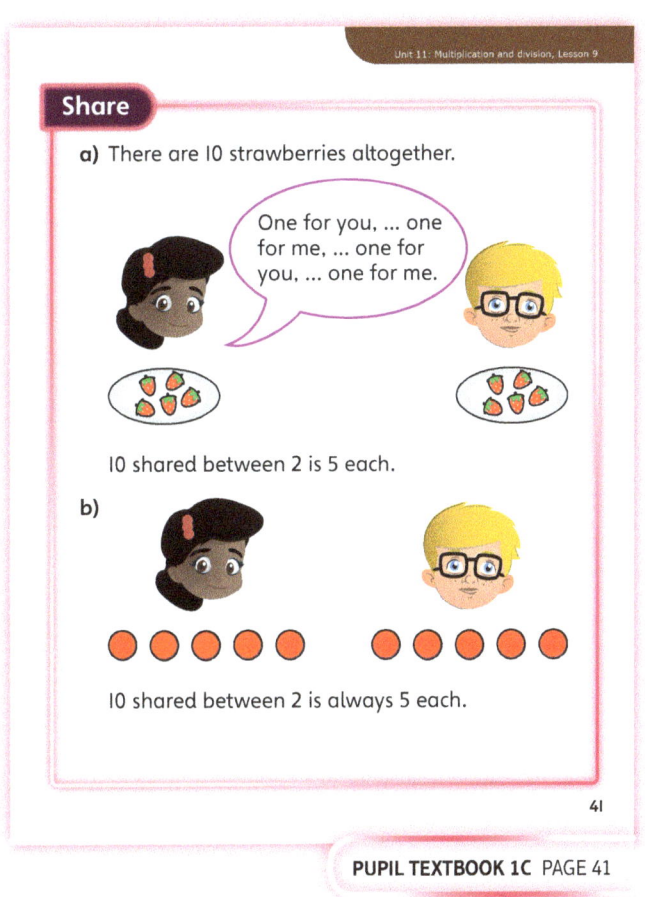

PUPIL TEXTBOOK 1C PAGE 41

Unit 11: Multiplication and division, Lesson 9

Think together

WAYS OF WORKING Whole class teacher led (I do, We do, You do)

ASK
- Question ❶: *How many counters are you sharing? How many groups are you sharing them into?*
- Question ❷: *How many counters are you sharing? How many groups are you sharing them into? How many counters are in each group?*

IN FOCUS In questions ❶ and ❷, children share 6 counters between different numbers of groups. Look to use the method where children put one counter at a time in each circle until they are all shared. Ask children to check that the groups now look equal. Question ❸ returns to a similar question that was in the **Challenge** section of the previous lesson; this time instead of grouping, children are now sharing.

STRENGTHEN Use concrete objects to support sharing and carefully demonstrate the process of sharing objects one by one. Model the method and then ask children to copy the sharing model.

DEEPEN Ask children to share 20 cubes between 2, 4, 5 and 10 groups. Then ask them to put the 20 cubes into groups of 2, 4, 5 and 10. Ask: *What do you notice? Is there a connection between sharing and grouping?*

ASSESSMENT CHECKPOINT Assess whether children can explain what is meant by sharing and how to share into equal groups. Can they share different numbers into different-sized groups and represent this in both a concrete and a pictorial manner?

ANSWERS

Question ❶: 6 shared between 2 is 3.

Question ❷: 6 shared between 3 is 2.

Question ❸ a): 12 cubes shared between 2 is 6 cubes each.

Question ❸ b): 12 cubes shared between 3 is 4 cubes each.

Question ❸ c): 12 cubes shared between 4 is 3 cubes each.

Question ❸ d): 12 cubes shared between 5 is 2 cubes each with 2 left over.

Question ❸ e): 12 can be shared equally between 2, 3 and 4 but not between 5.

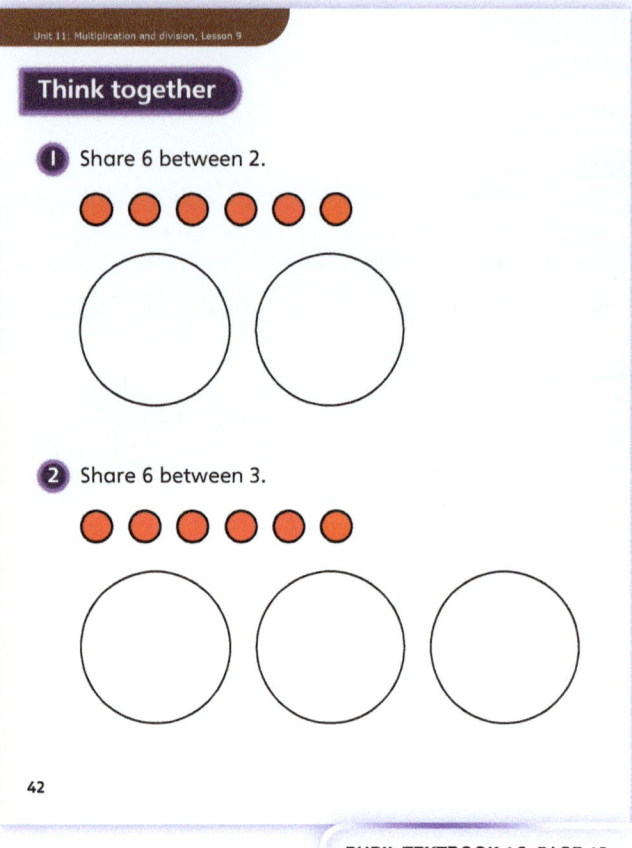

PUPIL TEXTBOOK 1C PAGE 42

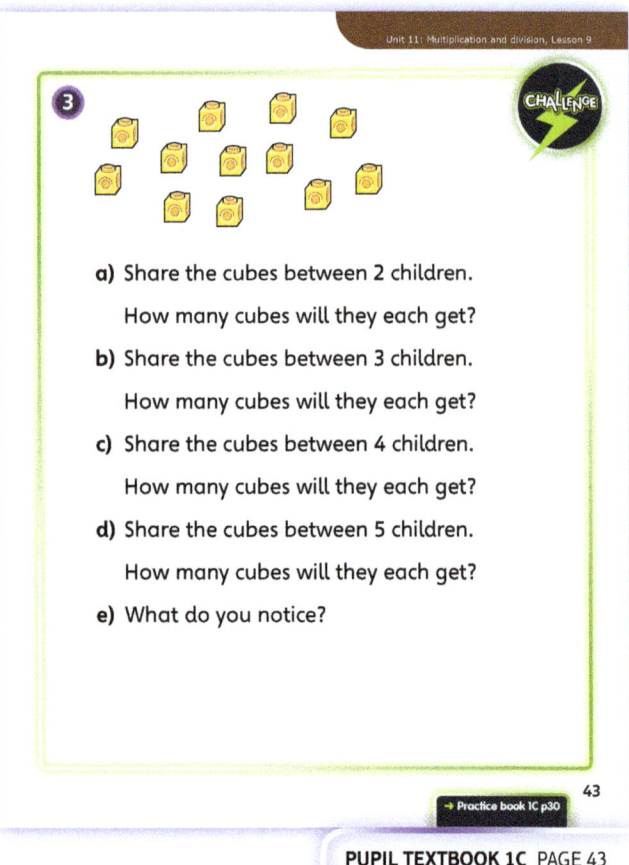

PUPIL TEXTBOOK 1C PAGE 43

Practice

WAYS OF WORKING Independent thinking

IN FOCUS Questions ❶ to ❹ are sharing questions where children are given some objects and different numbers of groups and are asked to share the objects between the groups. Question ❺ links sharing to an array. Children are asked to see how the array can help them answer the question, without having to go through the process of sharing.

STRENGTHEN Use concrete objects such as counters and cubes to support sharing and carefully demonstrate with children the process of sharing objects one by one. Model the method and then ask children to copy the sharing model.

DEEPEN Ask children to share 20 cubes between 2, 4, 5 and 10 groups. Then ask them to put the 20 cubes into groups of 2, 4, 5 and 10. Ask: *What do you notice? Is there a connection between sharing and grouping?*

ASSESSMENT CHECKPOINT Assess whether children are fluent at sharing when given a total number and a number of groups to share into. Can they represent the sharing in both a concrete and a pictorial manner?

ANSWERS Answers for the **Practice** part of the lesson can be found in the *Power Maths* online subscription.

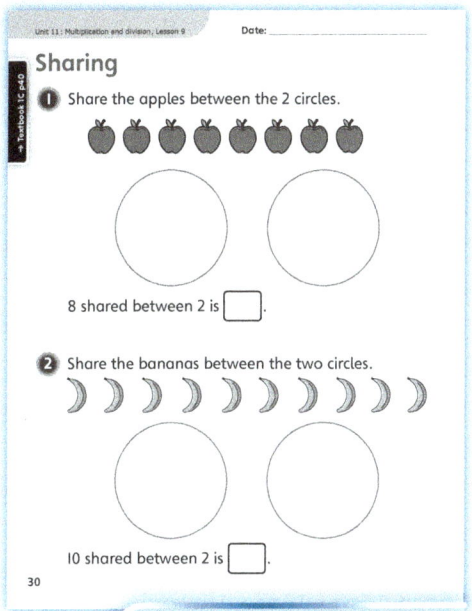

PUPIL PRACTICE BOOK 1C PAGE 30

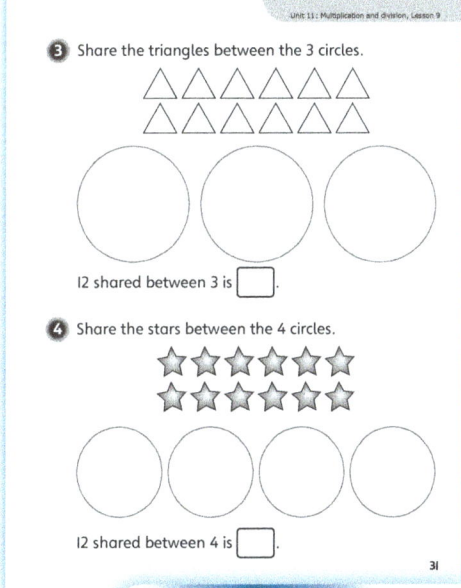

PUPIL PRACTICE BOOK 1C PAGE 31

Reflect

WAYS OF WORKING Pair work

IN FOCUS This question requires children to share 6 cars between 3 children. Allow children to use resources and draw pictures in order to carry out the sharing, before writing their explanation. They could then share their ideas with a partner or the class. Did all children use the resources in the same way? Having heard other children's explanations, would they change their own explanation?

ASSESSMENT CHECKPOINT Check that children can share the objects between the given number of groups.

ANSWERS Answers for the **Practice** part of the lesson can be found in the *Power Maths* online subscription.

PUPIL PRACTICE BOOK 1C PAGE 32

After the lesson ⏸

- Can children share some objects between a given number of groups?
- Do children know the difference between grouping and sharing?

Unit 11: Multiplication and division

End of unit check

Don't forget the unit assessment grid in your *Power Maths* online subscription.

WAYS OF WORKING Group work adult led

IN FOCUS
- Question ① assesses whether children can differentiate between the number of groups and the number of objects in each group.
- Question ② assesses whether children can skip count in 10s to find 4 groups of 10.
- Question ③ assesses children's understanding of doubles. It requires children to find both double and half of a given number.
- Question ④ assesses whether children are confident with grouping objects equally.

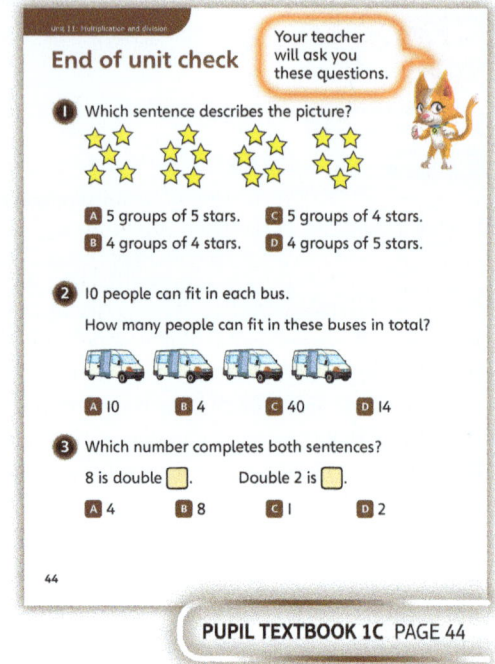

PUPIL TEXTBOOK 1C PAGE 44

Think!

WAYS OF WORKING Pair work

IN FOCUS This question assesses children's understanding of arrays. It is important to note that Joe, Sara and Poppy are all correct. To make sure children realise this, ask:
- Can you explain why each child has said what they did?
- Can you create the array? Can you use it to demonstrate each child's thinking?
- Is any one child correct or incorrect? Explain why.

ANSWERS AND COMMENTARY Children will demonstrate mastery in this question by realising that each child is correct. They will be able to explain why each child has said what they have and demonstrate each child's thinking using appropriate representations.

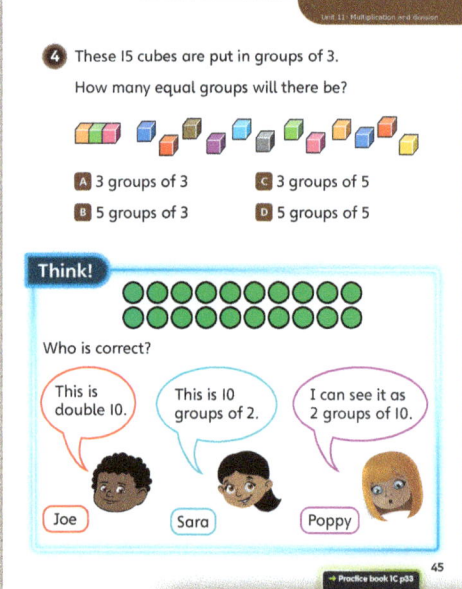

PUPIL TEXTBOOK 1C PAGE 45

Q	A	WRONG ANSWERS AND MISCONCEPTIONS	STRENGTHENING UNDERSTANDING
1	D	Choosing A, B or C could suggest that children are unsure of the distinction between the number of groups and the number of objects in each group.	**Finding and making equal groups:** Give children a collection of countable objects. Ask: *Can you sort these into equal groups of x?* For children who find this tricky, provide grids of the required group size to arrange the objects into. Ask: *What do you notice about each group? How can you tell they are equal?* **Understanding arrays:** Provide flashcards with the vocabulary of arrays. Encourage children to find arrays in their environment and to describe the elements of their arrays with appropriate mathematical vocabulary.
2	C	Choosing A may indicate that children have not recognised that they need 4 groups of 10; B suggests that children have not recognised that each bus represents 10 people.	
3	A	Choosing B, C or D would indicate that children are unsure about doubles and how to find them.	
4	B	Choosing C suggests that children may have shared the cubes into 3 groups, rather than grouping them into groups of 3 cubes each. Choosing A or D suggests that children are unsure of the concept of dividing by either sharing or grouping.	

Unit 11: Multiplication and division

My journal

WAYS OF WORKING Independent thinking

ANSWERS AND COMMENTARY

Children may draw the following groups:
- 2 groups of 10 to show that Joe is right.
- 10 groups of 2 to show that Sara is right.
- 2 groups of 10 to show that Poppy is right.

If children are struggling to find why any of the children are correct, ask:
- *What do you know about doubles? Does anything you know match what you can see in the picture?*
- *Can you see any groups of 10 or 2 in the picture?*
- *Does the number of groups match what the children have said? How do you know?*

PUPIL PRACTICE BOOK 1C PAGE 33

Power check

WAYS OF WORKING Independent thinking

ASK
- *Do you think you are more confident at solving multiplication problems now?*
- *Do you think you would be able to recognise arrays elsewhere?*
- *How confident are you that you could find doubles of numbers up to 10? 20? 50? 100?*

Power play

WAYS OF WORKING Independent thinking

IN FOCUS Use this **Power play** to assess children's ability to group objects equally. If children are struggling to group the dots, ask: *How many dots are there? How many groups are you putting them into?*

ANSWERS AND COMMENTARY If children are unable to solve the **Power play** it will be important to diagnose whether this is because they are unsure about how to group objects into equal groups or because they do not understand what is being asked of them. If it is the case that children are unsure how to group objects equally, provide them with the opportunities mentioned in the **Strengthen** sections throughout the unit to help secure their understanding.

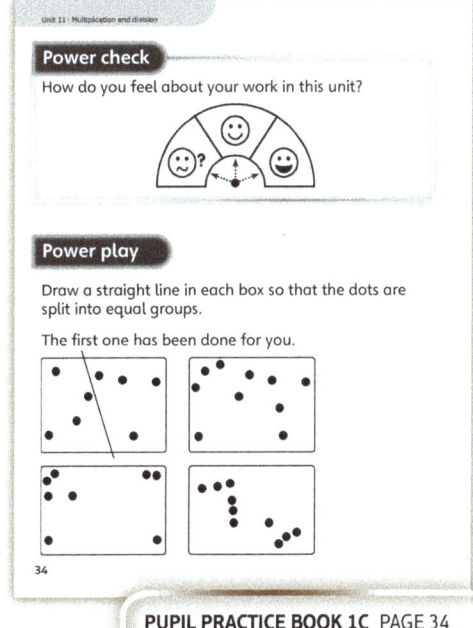

PUPIL PRACTICE BOOK 1C PAGE 34

After the unit

- How could you implement children's skills of finding doubles into the wider curriculum? Could you give them opportunities to find doubles of numbers up to and beyond 100?
- How will you use the learning and multiplicative reasoning children have developed in this unit to support their understanding of grouping and sharing?

Strengthen and **Deepen** activities for this unit can be found in the *Power Maths* online subscription.

Unit 12
Fractions

Mastery Expert tip! 'I found that lots of open-ended questions encouraged my class to think more deeply about halves and quarters. This definitely had an impact on their understanding of the unit.'

Don't forget to watch the Unit 12 video!

WHY THIS UNIT IS IMPORTANT

This unit is important as children will learn how to find halves and quarters of both shapes and groups of objects. It lays the foundations for later learning about fractions and, therefore, the foundations need to be secure.

WHERE THIS UNIT FITS

→ Unit 11: Multiplication and division
→ **Unit 12: Fractions**
→ Unit 13: Position and direction

This unit builds on simple sharing completed in earlier units during the year. The unit focuses on strategies to find halves and quarters. Following this unit, children will move on to learning about position and direction – including half and quarter turns.

Before they start this unit, it is expected that children:
- can share objects into two groups
- can share objects into four groups
- know the importance of equal sharing
- have a simple understanding of what splitting an object in half means.

ASSESSING MASTERY

Children who have mastered this unit will be able to use efficient strategies to find halves and quarters of shapes and groups of objects. Children will work accurately and confidently to find equal parts and they will know how the equal parts relate to the whole. Children will be able to work in reverse: being told what a quarter is and calculating what the whole would be. Finally, children will use the correct vocabulary and reasoning when explaining their methods.

COMMON MISCONCEPTIONS	STRENGTHENING UNDERSTANDING	GOING DEEPER
Children may not understand the concept of equal parts.	Do practical sharing activities such as cutting a cake or dividing sweets between children.	Ask children to think of where they can find halves and quarters in everyday life.
Unequal sharing is common, such as not giving the same quantity to each group.	Represent fractions with drawings and apparatus (tables and counters work well).	Ask children to explain methods of halving and quartering to deepen thinking.
Children may think quarters always look identical and not allow for different layouts.	Show children different ways to represent quarters. Compare these ways.	Ask children to start with a half or a quarter and find out what the whole should be.

Unit 12: Fractions

UNIT STARTER PAGES

Go through the unit starter pages of the Textbook with children. Talk through the key learning points (that the characters mention) and the key vocabulary.

STRUCTURES AND REPRESENTATIONS

It will help to represent halves and quarters visually on shapes.

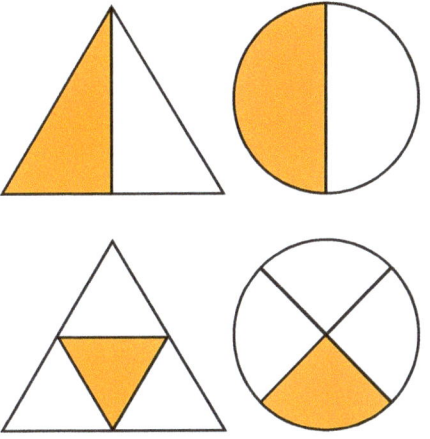

Lining up groups of objects in rows will make sharing easier.

KEY LANGUAGE

There is some key language that children will need to know as part of the learning in this unit.

- half, halves, quarter
- equal
- share, split
- part, whole

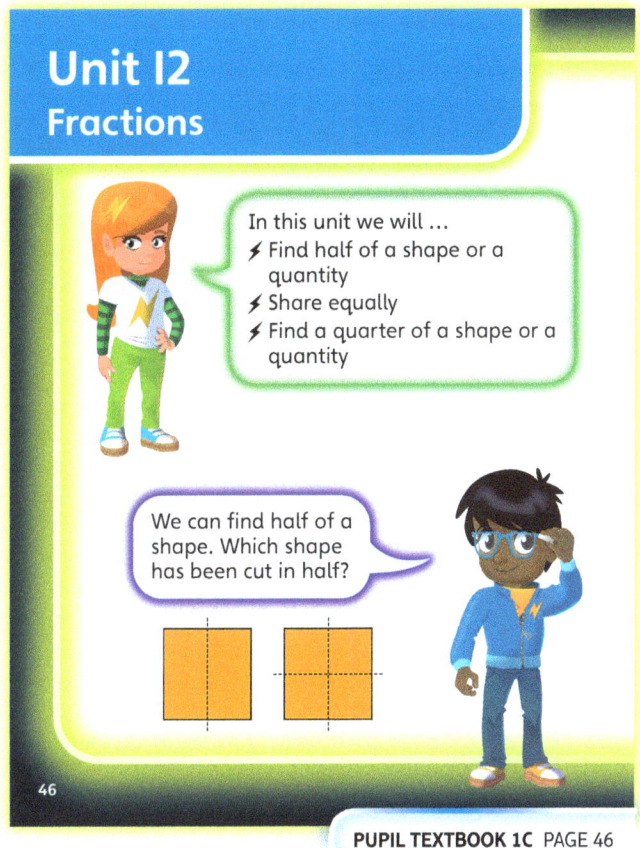

PUPIL TEXTBOOK 1C PAGE 46

PUPIL TEXTBOOK 1C PAGE 47

85

Unit 12: Fractions, Lesson 1

Recognise and find a half of a shape

Learning focus
In this lesson, children will recognise what a half is. They will apply their knowledge by halving shapes and objects.

Before you teach
- How did the previous division lesson go?
- How many children will need support with halving equally?
- Could you create a classroom display to reinforce key vocabulary and concepts?

NATIONAL CURRICULUM LINKS

Year 1 Number – fractions

Recognise, find and name a half as one of two equal parts of an object, shape or quantity.

ASSESSING MASTERY

Children can confidently find half of shapes and objects. Children can explain their method of finding half.

COMMON MISCONCEPTIONS

Children may not understand the concept of equal parts. Show a cake unfairly split in two. Ask:
- *If you have this part and I have the other part, do we have the same amount?*

STRENGTHENING UNDERSTANDING

Some children may need some support when halving objects into equal parts. Cutting or folding paper shapes with children in intervention groups would work well to strengthen understanding.

GOING DEEPER

To deepen understanding, give children half of a shape (irregular shapes could be used here). Ask them to draw the other half to create a whole and then ask them to explain their methodology.

KEY LANGUAGE

In lesson: half, halves, share, equal, split, whole, part

Other language to be used by the teacher: fair, unfair

STRUCTURES AND REPRESENTATIONS

Diagrams of shapes, half shaded, to promote visual understanding of what a half is

RESOURCES

Mandatory: shapes, paper copies of the shapes in **Think together** question ❷ (rectangles, hearts, ovals, butterflies)

Optional: mirrors, play food that can be cut into halves, string, rulers

 In the eTextbook of this lesson, you will find interactive links to a selection of teaching tools.

Quick recap

As a class, ask children to count in 2s from 0 to 50. How far can they go without stopping? Can they count in 2s from any multiple of 2 between 0 and 50?

Unit 12: Fractions, Lesson 1

Discover

WAYS OF WORKING Pair work

ASK

• Question 1 a): *What does the word 'half' mean? Does your folded paper look the same as your partner's?*

IN FOCUS Question 1 a) introduces children to the term 'half'. Having a display will lead to children discovering what it means.

Listen carefully to children's discussions in pairs for question 1 a). Do children understand how to fold the paper in half? The notion of fairness might form part of the discussion.

PRACTICAL TIPS To help children understand the term 'half', you could show play food being cut into two.

ANSWERS

Question 1 a):

Question 1 b):

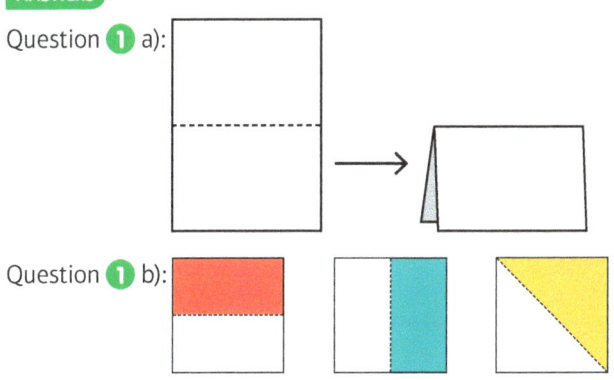

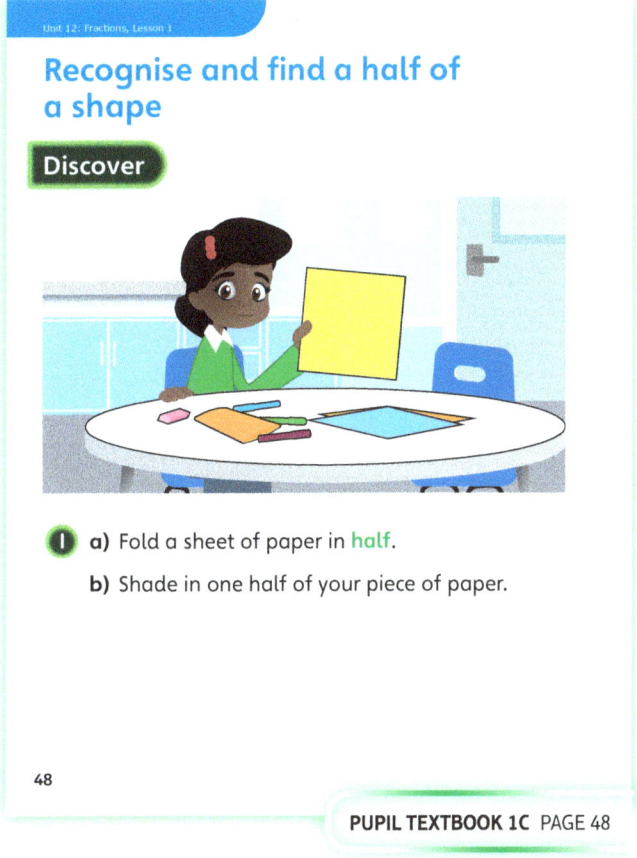

Recognise and find a half of a shape

Discover

1. a) Fold a sheet of paper in **half**.

   b) Shade in one half of your piece of paper.

PUPIL TEXTBOOK 1C PAGE 48

Share

WAYS OF WORKING Whole class teacher led

ASK

• Question 1 b): *Why is there more than one way to fold the paper into halves?*

IN FOCUS Look at question 1 b) together. Ask children if they can think of any other possible ways to split the paper into equal halves (for example, if the piece of paper is a square, you can split it diagonally from the top left-hand corner to the bottom right-hand corner).

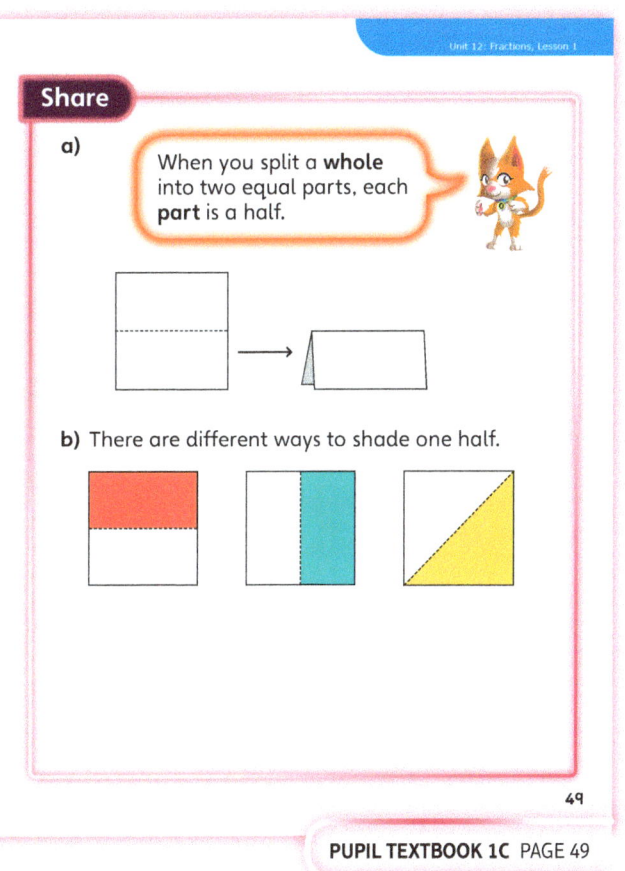

Share

a) When you split a **whole** into two equal parts, each **part** is a half.

b) There are different ways to shade one half.

PUPIL TEXTBOOK 1C PAGE 49

87

Think together

WAYS OF WORKING Whole class teacher led (I do, We do, You do)

ASK
- Question ❶: *Does the triangle have half shaded? How do you know? What would the circle look like if half of it was shaded?*
- Question ❷: *How can you split each shape into two parts? Do the parts have to be equal?*
- Question ❸: *How can you find the middle of a line? Does that help you with this question?*

IN FOCUS Throughout these questions, children recognise and find a half of each different shape. In question ❶, children recognise if shapes have been divided into two equal parts. You might want to discuss fairness if you were dividing a pizza, for example. Ask: *if the circle was a pizza and it was shared between two people, would they get the same amount?* In question ❷, children may need support to divide the shape into two. Have paper copies of the shapes available for children to fold. In question ❸, provide string or something similar to get children to think about how they can divide a line into two equal parts.

STRENGTHEN To strengthen learning in this section, run an intervention on what to do when you have a variety of pieces of paper that are shaped differently (although each must be able to be folded in half). Work with children to find a way to fold each one in half.

DEEPEN Ask children to think of real-life situations in which finding half of something is important, such as cutting a sandwich or half-time in a football match.

ASSESSMENT CHECKPOINT Questions ❶ and ❷ will allow you to assess if children can recognise half of a shape. Do children know by sight which shapes have equal halves or do they need to spend more time working it out, perhaps using paper? Question ❸ will allow you to assess if children can apply their knowledge of halves to length.

ANSWERS

Question ❶: Shape A is half shaded.

Question ❷:

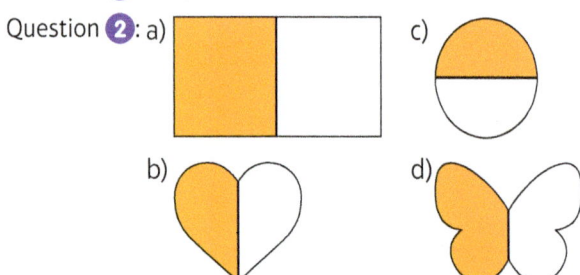

Question ❸: Various responses are possible, such as measuring the string or folding it in half.

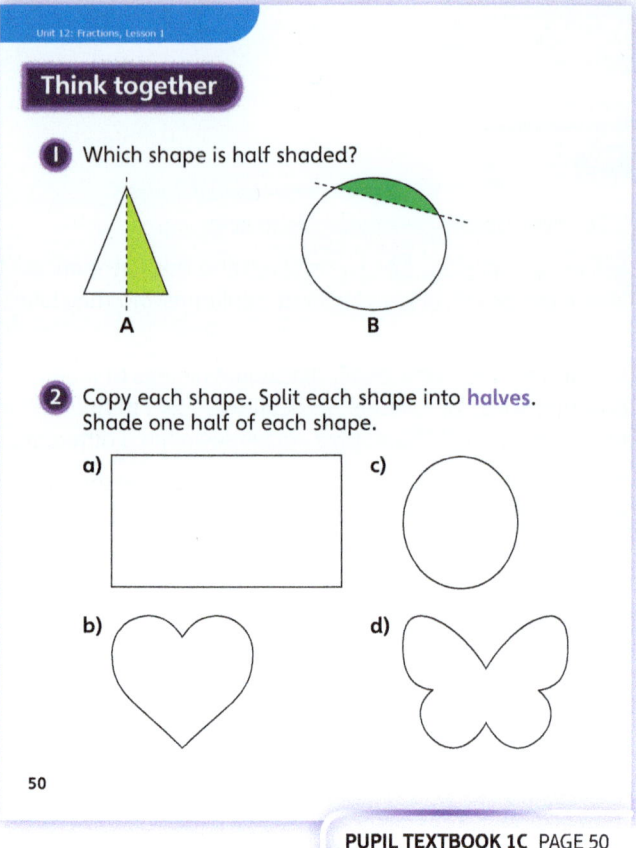

PUPIL TEXTBOOK 1C PAGE 50

PUPIL TEXTBOOK 1C PAGE 51

Unit 12: Fractions, Lesson 1

Practice

WAYS OF WORKING Independent thinking

IN FOCUS In question ①, children are given shapes that have been split into two equal parts and they are asked to shade in one half of each shape. In question ②, children need to split each shape into two equal parts first, before shading one half. Support children, where necessary, to divide shapes into two by discussing how they can do this – for example, by sight or with a ruler. Some children may want to make copies of the shapes on paper, so that they can cut out each shape and fold it in half. In question ④, children have to find half a length. They learned to measure using a ruler in Unit 9. Although they will not learn to divide by 2 until year 2, ask children to measure the length of the line. As a class, you could then try to find half of this number (perhaps by using counters and sharing into groups of 2).

STRENGTHEN To strengthen understanding, make a matching game for children to revise halves following this **Practice** section. Cut a range of shapes in half and then jumble the halves up. Children can then find the pairs that make a whole. To provide support for question ⑤, have children place a mirror along the dotted line so they can see what the missing half looks like before they try to draw it.

DEEPEN Children could find halves of lengths of objects around the classroom. Ask them to measure something using string – for example, a table. Children can cut the string to the same length as the object and then fold the string in half to find the middle (half the length).

THINK DIFFERENTLY Use question ③ to assess whether children can recognise whether half of a shape is shaded. There are different shadings of the same shape, allowing children to compare examples of shapes with two equal parts shaded and two unequal parts shaded. Discuss this as a class and perhaps ask children to draw their own shapes with equal halves shaded.

ASSESSMENT CHECKPOINT Use question ① to assess whether children know what a half is and question ③ to determine whether children can recognise a half. Question ② can be used to assess whether children can divide a shape into two equal parts.

ANSWERS Answers for the **Practice** part of the lesson can be found in the *Power Maths* online subscription.

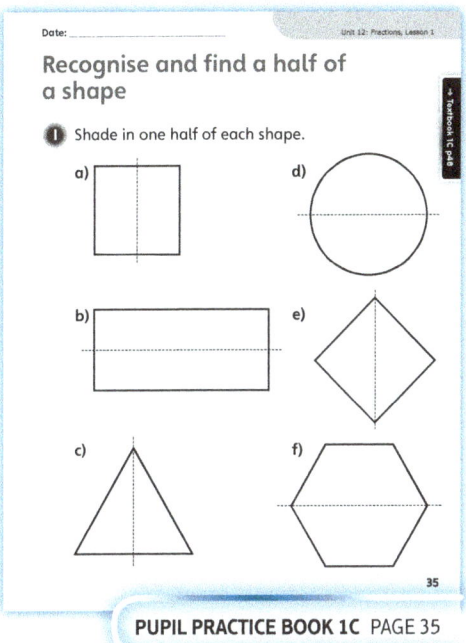

PUPIL PRACTICE BOOK 1C PAGE 35

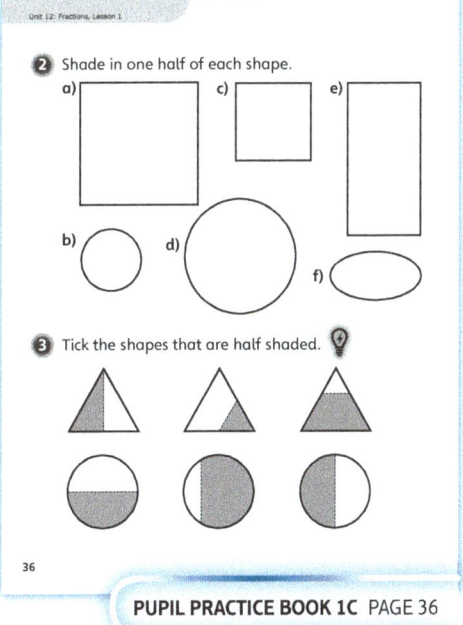

PUPIL PRACTICE BOOK 1C PAGE 36

Reflect

WAYS OF WORKING Independent thinking

IN FOCUS Encourage children to draw or trace around a familiar shape, such as a square, rectangle or circle, and then divide this shape into two equal parts. Some children may need help thinking about what shape to draw. Once they have drawn a shape, let them independently shade in half of the shape.

ASSESSMENT CHECKPOINT This question assesses whether children understand the meaning of a half of an object.

ANSWERS Answers for the **Reflect** part of the lesson can be found in the *Power Maths* online subscription.

After the lesson

- Which children need to strengthen their understanding?
- Do these children need more activities around practical halving?
- Will the key vocabulary need to be recapped at the start of the next lesson?

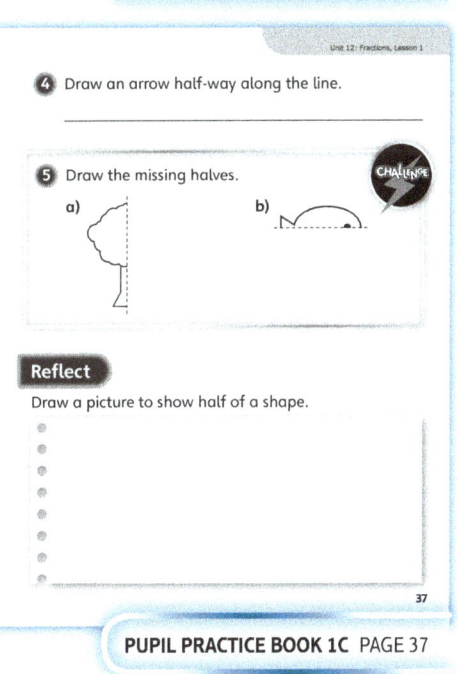

PUPIL PRACTICE BOOK 1C PAGE 37

89

Unit 12: Fractions, Lesson 2

Recognise and find a half of a quantity

Learning focus

In this lesson, children will find half of groups of objects, using their knowledge of sharing between two.

Before you teach ⏸

- How can you make links between the previous lesson and this lesson?
- Do you have enough counters ready?
- Should key words be recapped?

NATIONAL CURRICULUM LINKS

Year 1 Number – fractions

Recognise, find and name a half as one of two equal parts of an object, shape or quantity.

ASSESSING MASTERY

Children can confidently divide objects into two equal groups, making the link between the previous lesson and this lesson, halving one object or shape and halving a group of objects or shapes. Children can recognise that the total group is the 'whole' and each smaller group is 'half' and, furthermore, can count each smaller group to check they are equal.

COMMON MISCONCEPTIONS

Unequal sharing might happen in this lesson. Children often do 'one for me, one for you, one for me', and so on. However, sometimes children may make the mistake of doing 'one for me, one for you, one for you, one for me, one for you'.

STRENGTHENING UNDERSTANDING

To strengthen understanding, work with a group of children and ask them to show you a given number of fingers or counters. Ensure that you give them an even number to show you. Then ask them to put down half their fingers or hide half the counters. Repeat with different multiples of 2.

GOING DEEPER

Ask children to halve nine counters. This will really get them thinking about the left-over counter. See if some children suggest that this left-over counter could be split in half.

KEY LANGUAGE

In lesson: half, halves, each, how many?, share, equal, parts, split, whole, group, doubles

Other language to be used by the teacher: fair, unfair, unequal

STRUCTURES AND REPRESENTATIONS

Counters to visually represent groups of objects, bar model

RESOURCES

Mandatory: counters

Optional: sorting hoops, 8 apples or other items that can be shared between children, cubes

 In the eTextbook of this lesson, you will find interactive links to a selection of teaching tools.

Quick recap 🔍

Recap counting in 2s, 5s and 10s. Start from 0 and ask children to practise counting in 2s, 5s and 10s to 50. They could write the numbers they say on a whiteboard then ask a partner to check if they have written down all the numbers.

Unit 12: Fractions, Lesson 2

Discover

WAYS OF WORKING Pair work

ASK
- Question 1 a): *Should you count the total number of apples first?*
- Question 1 b): *How could you use counters to help you share? Can you see what half of 8 is without having to share one by one?*

IN FOCUS Both parts of this question are reinforcing the same principle; finding half of 8. In question 1 a), children are asked to use the method of sharing to work out how many apples each horse gets. In question 1 b), children may adopt other ways to find half of 8. Ensure there are counters available for children who need support with their counting skills and prompt them to use them if needed.

PRACTICAL TIPS You could recreate the scenario in the classroom by sharing 8 apples or other objects between 2 children.

ANSWERS

Question 1 a): There are 8 apples.
8 shared equally between 2 is 4.

Question 1 b):  The whole is 8.
One half is 4.
4 is half of 8.

Recognise and find a half of a quantity

Discover

1 a) Share the apples equally between two horses.
   b) Show how to find half of 8 apples.

PUPIL TEXTBOOK 1C PAGE 52

Share

WAYS OF WORKING Whole class teacher led

ASK
- Question 1 a): *How has Dexter shared the apples? Can you show me using counters or cubes?*
- Question 1 b): *How can you arrange your counters so you can easily see half of 8?*

IN FOCUS In this **Share** section, children consider two ways of finding half of 8. In question 1 a), use Dexter's speech to explain how the apples could be shared one by one, by giving one horse one apple and then the other horse one apple, and so on until the apples have all been shared equally between the horses. Ask children to do this with you using counters or cubes to show the method of sharing. Use the image in question 1 b) to explain that sometimes it is possible to see half of a quantity by looking, as opposed to having to share one by one. Ask if, looking at the image, children can see why half of 8 is equal to 4.

Share

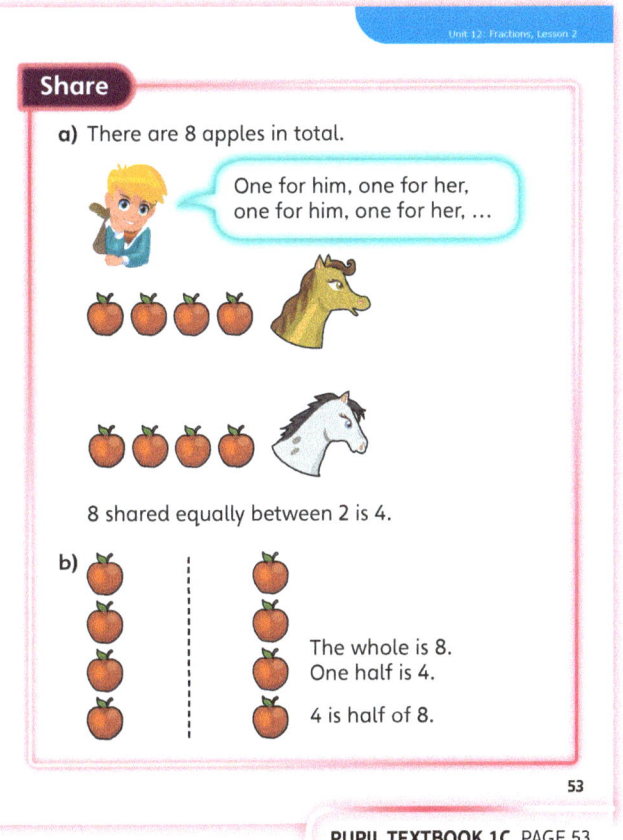

PUPIL TEXTBOOK 1C PAGE 53

Think together

WAYS OF WORKING Whole class teacher led (I do, We do, You do)

ASK
- Question ❶: *How could you use counters to help you?*
- Question ❷: *What are you looking out for in each picture?*
- Question ❸: *Can you explain your doubling and halving strategy to a partner?*

IN FOCUS Look at question ❸ together. Ask children how using counters on a ten frame can help them with doubling and halving. Talk through the fact that the ten frame has two rows of five – making it easy to 'cut' a number of counters in half. Reinforce that half of 10 is 5 and that two lots of 5 is 10.

STRENGTHEN To strengthen understanding, play some sharing games in which groups of counters are shared equally between 2 children.

DEEPEN Tell children that a certain number is half of a quantity of counters. Ask them to find out how many counters the whole would have been. Repeat with different quantities of counters.

ASSESSMENT CHECKPOINT Question ❸ will let you assess whether children can double and halve different quantities, represent this in a visual structure.

ANSWERS

Question ❶: Half of 6 is 3.

Question ❷: A and D

Question ❸: Various answers are possible:
- Double 2 is 4.   Half of 4 is 2.
- Double 3 is 6.   Half of 6 is 3.
- Double 4 is 8.   Half of 8 is 4.
- Double 5 is 10.   Half of 10 is 5.

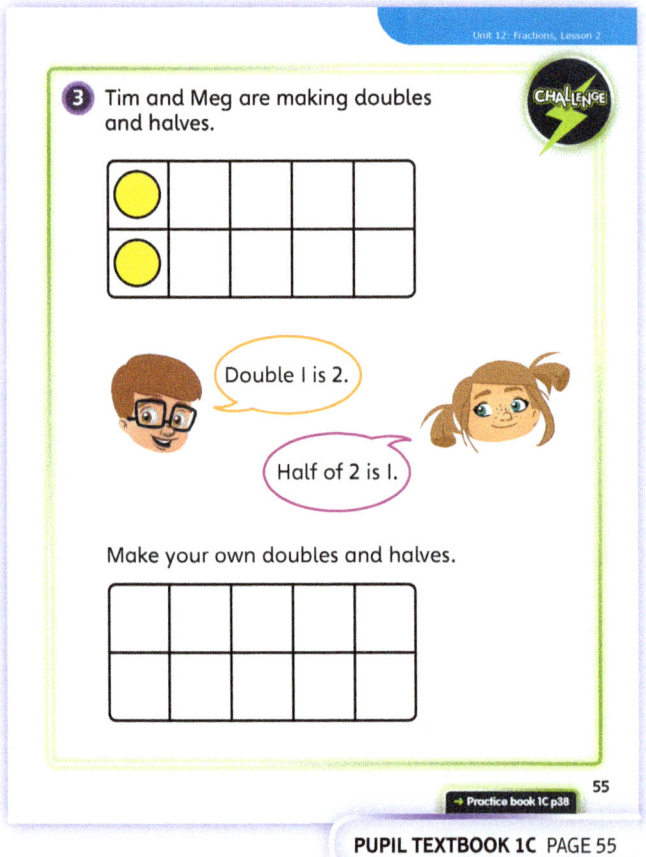

PUPIL TEXTBOOK 1C PAGE 55

92

Unit 12: Fractions, Lesson 2

Practice

WAYS OF WORKING Independent thinking

IN FOCUS In question ❶, children are presented with a number of objects and are required to shade in half of them. They should do this by using the dotted line to help them see that the objects have been split into 2. This links to the work that they did in the previous lesson. In question ❷, children are presented with a number of objects without the dividing line. They first have to work out where to put the line. In question ❷ a), children may struggle at first as the line they should draw is horizontal rather than vertical. In question ❸, children practise their knowledge of halves by using the pairs representations of counters on the ten frames to help them.

STRENGTHEN Support children by using concrete objects to help them see a half without necessarily counting one by one. Make arrangements of counters in an array and ask if they can make two equal groups. Ask them to check, by counting, that the groups are equal. This will help them with questions ❶ and ❷.

DEEPEN Ask children to find half of some even numbers greater than 20 – for example, 24. They might use a sharing method or they may want to make a rectangular array and find a way of dividing it into two equal parts. Ask them to explore both methods.

ASSESSMENT CHECKPOINT Use questions ❶ and ❷ to assess if children can recognise half of a group of objects. Use question ❸ to help children consolidate their halves and doubles.

ANSWERS Answers for the **Practice** part of the lesson can be found in the *Power Maths* online subscription.

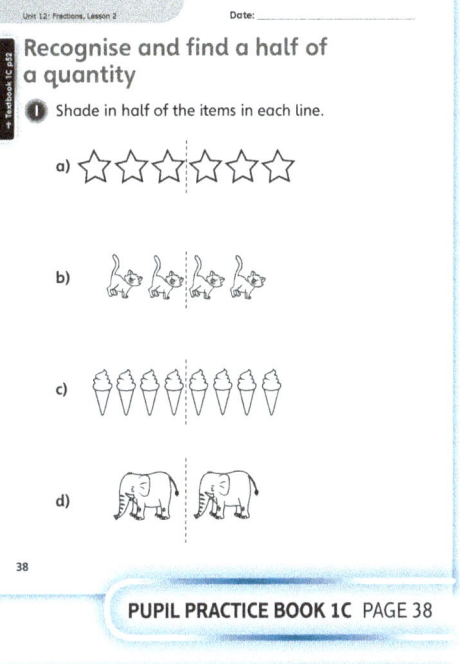

PUPIL PRACTICE BOOK 1C PAGE 38

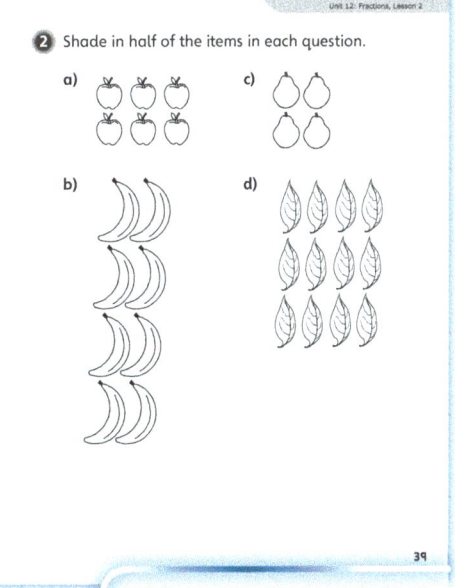

PUPIL PRACTICE BOOK 1C PAGE 39

Reflect

WAYS OF WORKING Whole class

IN FOCUS This question will allow children to draw on their knowledge on the different methods of doubling and halving. To finish this lesson, ask the class to stand and sort themselves into two halves. Even if you have an odd number of children, it will be interesting to see what happens and how children approach this.

ASSESSMENT CHECKPOINT This question will allow you to check children's understanding of the methods they use to double and halve a group of objects. Note if they are using appropriate mathematical vocabulary.

ANSWERS Answers for the **Reflect** part of the lesson can be found in the *Power Maths* online subscription.

After the lesson ⏸

- How many children mastered the lesson?
- Did children make links between Lesson 1 and Lesson 2?
- What methods did children use to find half of groups of objects?

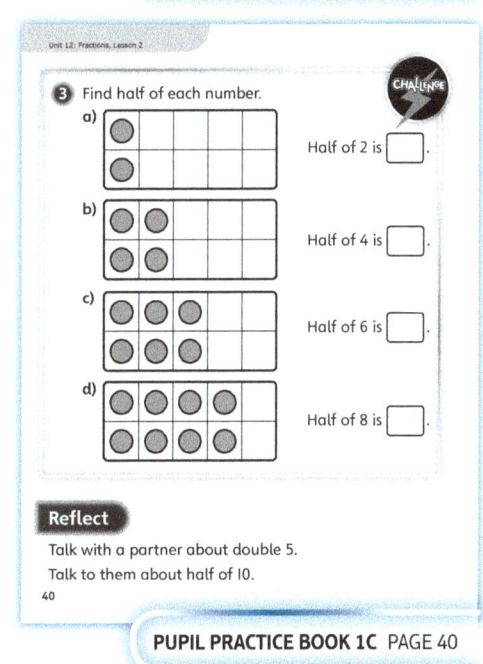

PUPIL PRACTICE BOOK 1C PAGE 40

93

Unit 12: Fractions, Lesson 3

Recognise and find a quarter of a shape

Learning focus

In this lesson, children will recognise what a quarter is. They will apply their knowledge by finding quarters of shapes.

Before you teach ⏸

- How will you introduce the new word 'quarter'?
- How many children mastered halving in the last lesson?
- How will you link halving to quartering?

NATIONAL CURRICULUM LINKS

Year 1 Number – fractions

Recognise, find and name a quarter as one of four equal parts of an object, shape or quantity.

ASSESSING MASTERY

Children can confidently find a quarter of each shape and explain their method. Secure children can do the inverse too: be given a quarter of a shape and work out the original shape.

COMMON MISCONCEPTIONS

Children may think that all quarters must look the same, because they are equal parts. However, this is not always true. Ask:
- *Can you draw a shape and split it into quarters, where not all the parts look the same? Look at the last shape in the* **Share** *section. Do all the parts look the same?*

STRENGTHENING UNDERSTANDING

Give children different paper shapes and ask them to fold them into quarters. This practical work will strengthen understanding.

GOING DEEPER

Provide children with a range of irregular shapes on paper and ask them to find a quarter. They will begin by folding, but realise that the parts are not equal. Discuss this and come up with strategies to find quarters.

KEY LANGUAGE

In lesson: **quarters**, split, equal, parts, whole

STRUCTURES AND REPRESENTATIONS

Shaded shapes are used throughout this lesson

RESOURCES

Mandatory: paper squares, rectangles and circles
Optional: multilink cubes

 In the eTextbook of this lesson, you will find interactive links to a selection of teaching tools.

Quick recap

Show children images of shapes with some divided into equal parts and some divided into unequal parts. Ask children to sort the shapes into those that show equal parts and those that do not. You could take examples from Year 1 Unit 11 lessons.

Unit 12: Fractions, Lesson 3

Discover

WAYS OF WORKING Pair work

ASK
- Question 1 a): *Can you see that the board has been split into four equal parts?*
- Question 1 b): *What do you have to remember when splitting a shape into quarters? How many parts will there be? Do the parts have to be equal?*

IN FOCUS In question 1 b), children are introduced to a new word, 'quarter'. Some children will use inference from the clues provided to discover the meaning whilst others may need some prompting.

PRACTICAL TIPS Hand out paper squares and ask children to fold them into quarters.

ANSWERS

Question 1 a): There are 4 equal parts.
Each part is one quarter.

Question 1 b):

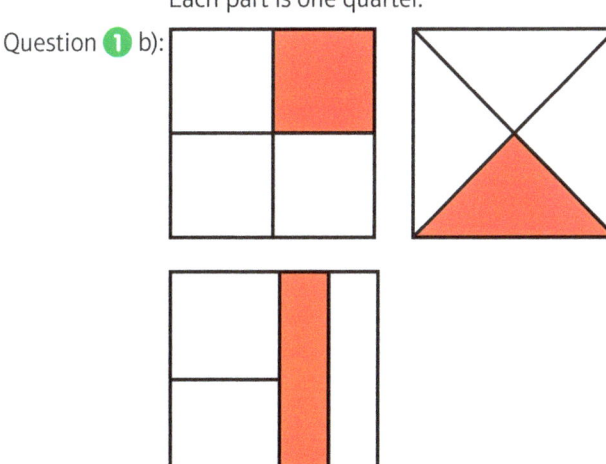

There are 4 equal parts.
Each part is 1 quarter.

Share

WAYS OF WORKING Whole class teacher led

ASK
- Question 1 a): *What does the word 'equal' mean? Are the parts equal? What is each part worth?*
- Question 1 b): *Do each of these show equal parts? What fraction is shaded in each shape? In the last one, are equal parts shaded?*

IN FOCUS For question 1 a), discuss that the board is divided into equal parts and explain that each part is a quarter of the board. In question 1 b), children see other representations of a quarter. Discuss that the final shape is still split into quarters even though all the parts do not look the same. Explain that the shape has been divided in half first and then each half divided into a half, but in a different way. Do not worry at this stage if children do not understand the reason why; the key message is that equal parts can look different.

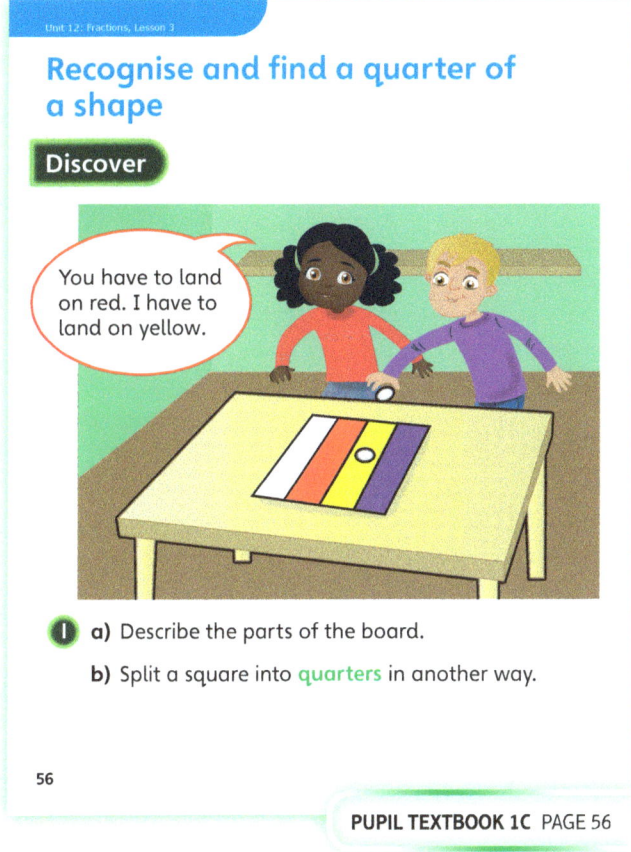

PUPIL TEXTBOOK 1C PAGE 56

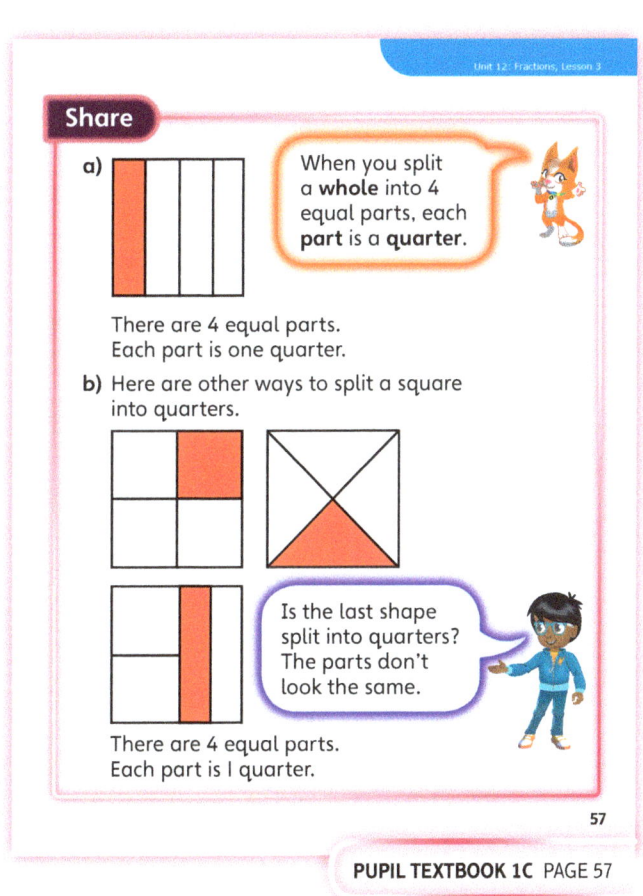

PUPIL TEXTBOOK 1C PAGE 57

Unit 12: Fractions, Lesson 3

Think together

WAYS OF WORKING Whole class teacher led (I do, We do, You do)

ASK
- Question ❶: *How many parts should there be for quarters? Which boards have been split into quarters?*
- Question ❷: *How many parts has the circle been split into? Are the parts equal? Is the circle split into quarters?*

IN FOCUS In question ❶, children first look at which boards have been split into equal parts. Take the opportunity to discuss why the boards either show quarters or they do not. For example, for shape A, children should note that the board has been split into three equal parts instead of four and so does not show quarters. Question ❷ shows a circle split into four parts and children need to discuss that the parts are not quarters as they are not equal. Question ❸ encourages the discussion of whether it is possible to find a quarter by halving and halving again.

STRENGTHEN Support children with finding quarters of shapes by providing paper squares, rectangles and circles. Ask them to fold the shapes into half and half again. Focus on shapes where children can divide them into quarters in different ways.

DEEPEN Ask children to divide a square into quarters, but where the quarters are not all the same shape. For example, split the square in two. Then, divide each half in two a different way, such as:

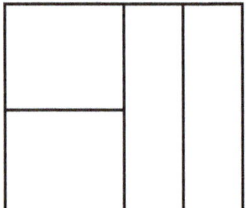

ASSESSMENT CHECKPOINT Questions ❶ and ❷ assess whether children have an understanding of quarters; that they know a shape has been divided into four equal parts.

ANSWERS

Question ❶: A and B are split into equal parts.
B is split into quarters.

Question ❷: No. There are 4 parts, but they are not equal, so they are not quarters.

Question ❸: There will be 4 equal parts.

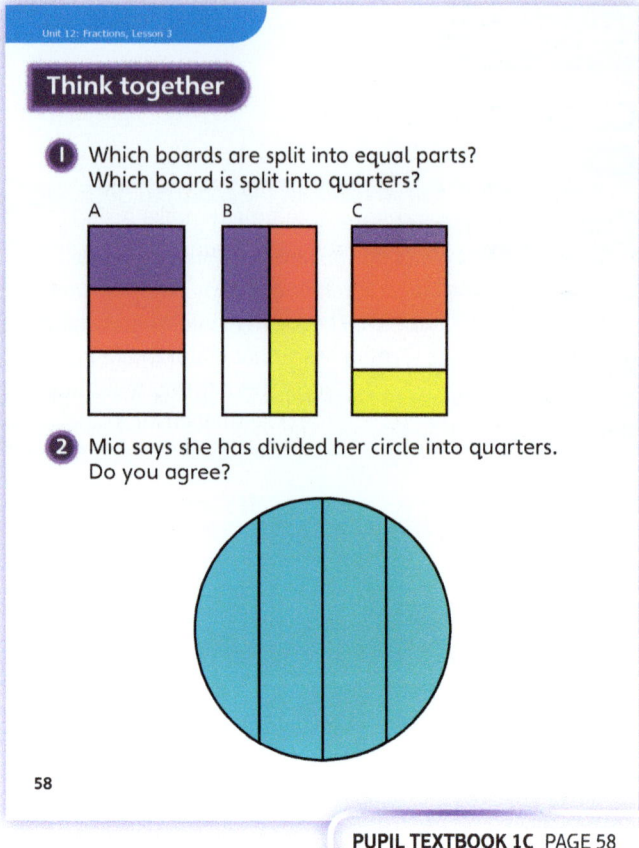

PUPIL TEXTBOOK 1C PAGE 58

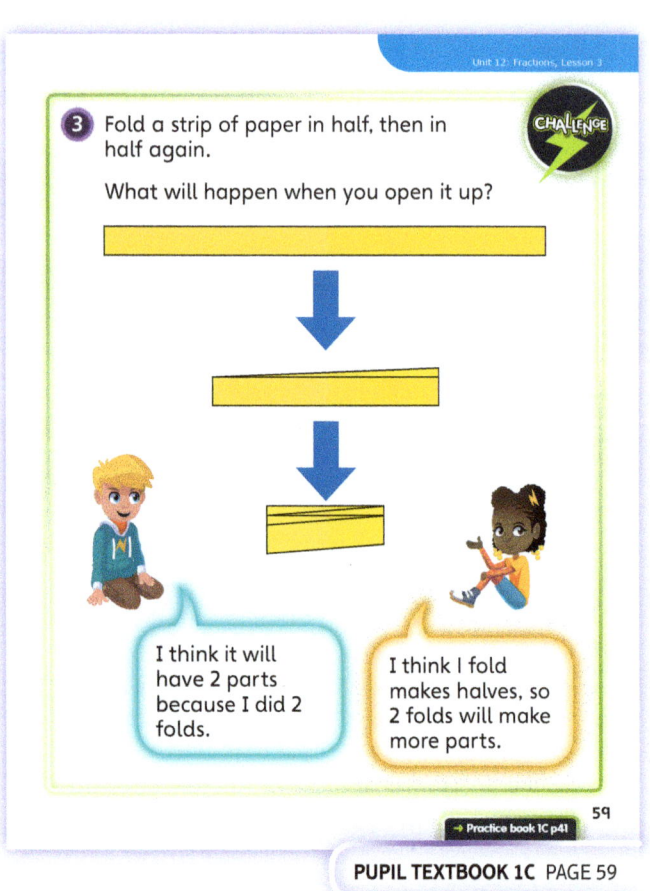

PUPIL TEXTBOOK 1C PAGE 59

Unit 12: Fractions, Lesson 3

Practice

WAYS OF WORKING Pair work

IN FOCUS In question ❶, children are presented with a number of shapes and are required to shade in a quarter of each of them. They should do this by using the dotted lines to help them see that the shapes have been split into four equal parts. In question ❷, children are presented with a number of shapes without the dividing lines. They first have to work out where to draw the lines and then shade in a quarter of each shape. Encourage children to quarter the shapes in a range of different ways.

For question ❸, the first shape only has three parts and they are unequal. Discuss this with children and see if they realise they can split the larger section to make four equal parts altogether.

STRENGTHEN Print some shapes similar to the ones in questions ❶ to ❸ onto a sheet of paper (enlarge them and include the dividing lines). Ask children to cut the shapes out, count the parts and then see if they are equal in size. This will help children revise their strategies.

DEEPEN To deepen understanding, ask children to do some more work with quartering irregular shapes, such as those in question ❺ (this will call upon estimation skills too). Another idea is to get some pictures that can have lines drawn on them and ask children to divide them into quarters. For example, the pictures could be of various food items such as an 8-square chocolate bar, a sandwich, a ring donut, a cake or a pizza. In this way, children begin to appreciate the concept of sharing into equal parts.

THINK DIFFERENTLY Question ❹ requires children to apply what they know about finding quarters to a real-life context. They need to think about how the wall could be split into four equal parts.

ASSESSMENT CHECKPOINT Question ❺ will allow you to assess whether children can split irregular shapes into quarters in different ways.

ANSWERS Answers for the **Practice** part of the lesson can be found in the *Power Maths* online subscription.

Reflect

WAYS OF WORKING Independent thinking

IN FOCUS This activity asks children to demonstrate their knowledge of dividing a shape into quarters. You may want children to use paper copies that they can fold. If children find one way, then ask them if they can show quarters in a different way.

ASSESSMENT CHECKPOINT This question will allow you to assess which children can show quarters of a shape without scaffolding and can reason why.

ANSWERS Answers for the **Reflect** part of the lesson can befound in the *Power Maths* online subscription.

After the lesson

- Do you need to schedule some same-day intervention for certain children?
- Do children understand that quarters do not always look identical?
- Could you do some artwork making four-part patterns, for example using stamps?

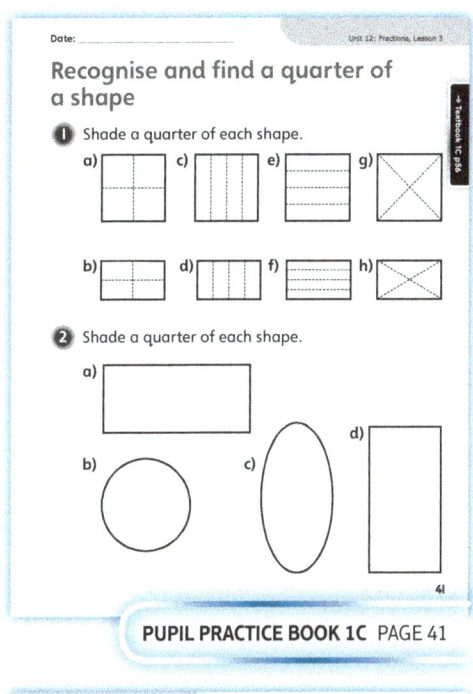

PUPIL PRACTICE BOOK 1C PAGE 41

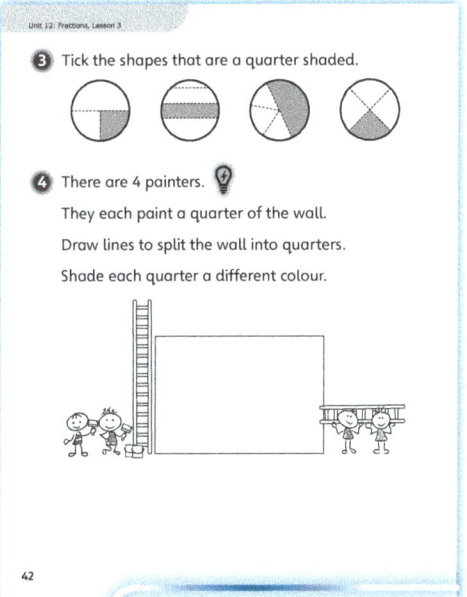

PUPIL PRACTICE BOOK 1C PAGE 42

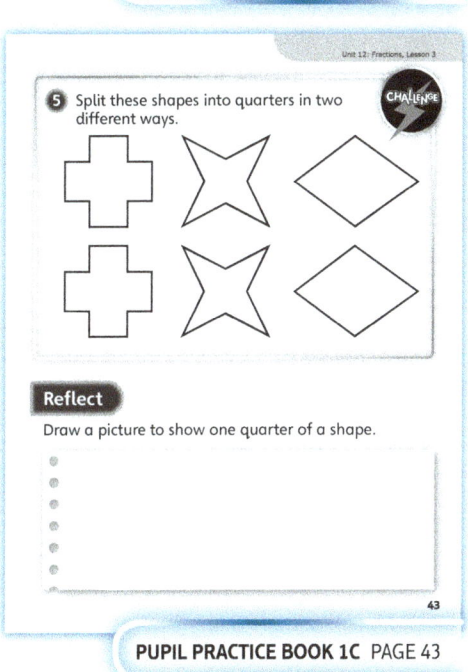

PUPIL PRACTICE BOOK 1C PAGE 43

Unit 12: Fractions, Lesson 4

Recognise and find a quarter of a quantity

Learning focus

In this lesson, children will find a quarter of a small quantity using equal sharing.

Before you teach

- Are children confident with finding quarters of shapes?
- How will you link the previous lesson to this one?
- Could you show children some real-life examples of finding one quarter of a small quantity (sharing fruit, grouping children)?

NATIONAL CURRICULUM LINKS

Year 1 Number – fractions

Recognise, find and name a quarter as one of four equal parts of an object, shape or quantity.

ASSESSING MASTERY

Children can securely find a quarter of a small quantity by sharing between four and using mental strategies to do this.

COMMON MISCONCEPTIONS

Care should be taken if using the 'quarter is a half of a half' explanation as this may generate a misconception about repeated halving for all fractions later on. Ask:
- *Can you use the repeated halving method to split something into five groups?*

STRENGTHENING UNDERSTANDING

To strengthen understanding, practise sharing counters between four children. Focus on giving each child a counter in order and checking that each child has the same number of counters at the end to ensure they have shared equally.

GOING DEEPER

Ask children to find a quarter of various quantities (using real-life examples). For example, children could find one quarter of 8 litres or 16 biscuits.

You could also ask children to draw what one and a quarter might look like.

KEY LANGUAGE

In lesson: quarter, whole, quarter full, split, share, equal

Other language to be used by the teacher: fair, unfair, unequal

STRUCTURES AND REPRESENTATIONS

Children will use counters or draw dots to support their sharing

RESOURCES

Mandatory: counters

 In the eTextbook of this lesson, you will find interactive links to a selection of teaching tools.

Quick recap

Ask pairs or small groups of children to take 12 counters and to make a 3 × 4 array. Ask them if they can split the array in half straight away, without sharing the counters one by one. Can they take half of the counters without counting?

Unit 12: Fractions, Lesson 4

Discover

WAYS OF WORKING Pair work

ASK

- Question 1 a): *How do you find a quarter of a group of objects?*
- Question 1 a): *Do you remember finding half of a group of objects – could this help?*
- Question 1 b): *What strategy could you use to solve this and find the whole?*

IN FOCUS For question 1 a), look carefully at the sharing strategies children use. As in the previous lesson, children may try to cut each orange into 4 pieces. Some children will count each one – watch if they make any markings such as dots.

PRACTICAL TIPS You could recreate the scenario in the classroom using 8 objects in a tray, shared between 4 children.

ANSWERS

Question 1 a): Each child gets 2 oranges.

Question 1 b): There are 3 oranges in each quarter. The whole is 12 oranges.

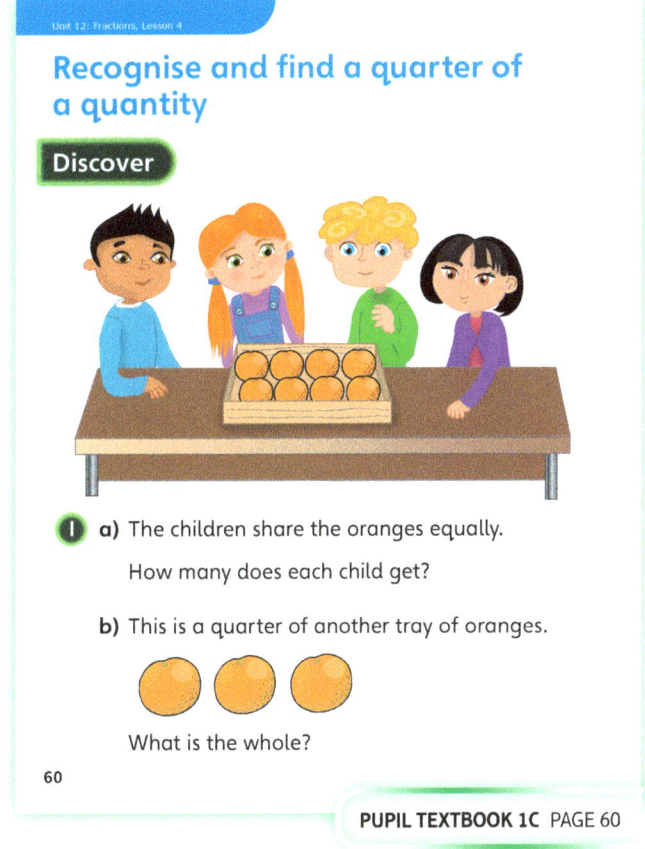

PUPIL TEXTBOOK 1C PAGE 60

Share

WAYS OF WORKING Whole class teacher led

ASK

- Question 1 a): *Can you explain why the children divide the number of oranges into quarters in order to share them equally?*
- Question 1 a): *Can you think of any other way of splitting this group into quarters?*
- Question 1 a): *What if the 8 oranges were arranged in a line? Where would you split these into quarters?*
- Question 1 b): *Why does the grid have four parts?*

IN FOCUS In question 1 a), children should observe that there are 4 children, so they should split the oranges into four equal groups. This is the same as splitting them into quarters. Question 1 b) has a grid of four squares to show children a visual strategy to help them find the answer. If children find this challenging, show them a similar grid, but with 1 orange in the top left square. Tell them that 1 orange is a quarter of the whole, and help them deduce that the whole is 4 oranges. This should help children understand the method of replicating a quarter so that it shows 4 times to get the whole.

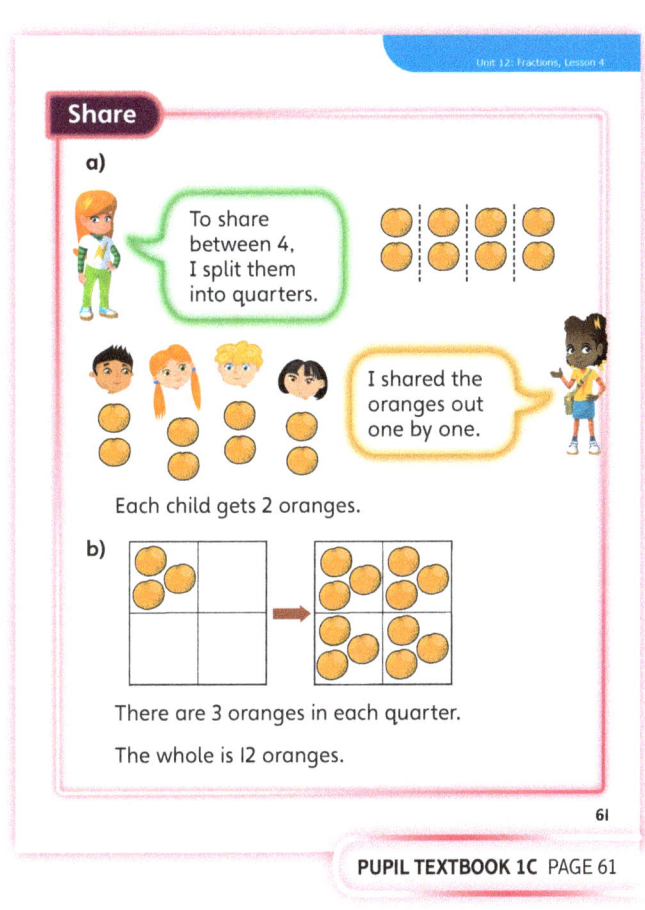

PUPIL TEXTBOOK 1C PAGE 61

Think together

WAYS OF WORKING Whole class teacher led (I do, We do, You do)

ASK
- Question ①: *How could you use a drawing to help you work this out?*
- Question ③ b): *Can you use objects or draw a picture to help you?*

IN FOCUS In question ②, children are given a quarter and are asked to find the whole. Check that they recognise why the grid has been split into four parts and how many pears would go in each part.

STRENGTHEN Some children may need support with their sharing strategies. Model sharing counters or making dots on paper.

DEEPEN Do some work linking halves to quarters. Children could find one half and then one quarter of 20 and then see how the answers relate to one another.

Ask children if they know how many quarters make a half. Can they draw a visual representation of this?

ASSESSMENT CHECKPOINT Question ③ b) will allow you to see which children can apply their knowledge of finding quarters abstractly.

ANSWERS

Question ①: 1 is a quarter of 4.

Question ②: The whole is 16, 4 groups of 4.

Question ③ a): 5 is a quarter of 20.

Question ③ b): 6 is a quarter of 24.

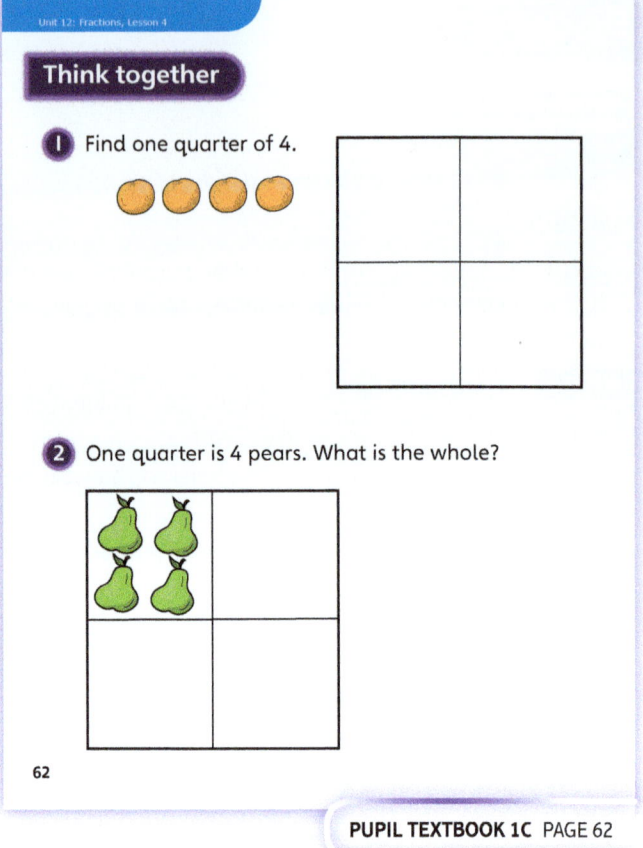

PUPIL TEXTBOOK 1C PAGE 62

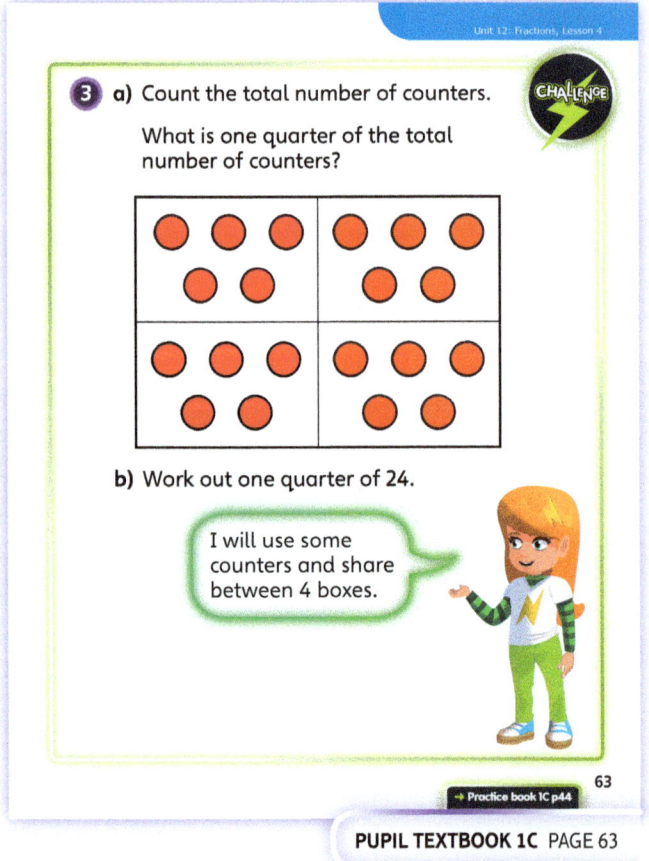

PUPIL TEXTBOOK 1C PAGE 63

100

Unit 12: Fractions, Lesson 4

Practice

WAYS OF WORKING Independent thinking

IN FOCUS In question ①, children practise splitting quantities into quarters by dividing images of objects into 4. In question ②, children share the birds one by one to find a quarter. They use the same four-box representation as in question ① to help them. By now, children are getting used to seeing a quarter, so, in question ③, they need to draw the other quarters to show a whole. In question ④, children are asked to tick the representations that show a quarter. Throughout these questions, you could encourage children to say or write sentences such as 'A quarter of … is …', which they have worked out from their diagrams.

STRENGTHEN Support children by using concrete objects to help them see or find a quarter. Give small groups of children a sheet of paper split into quarters and some cubes. Ask them to make arrangements of counters in an array and then ask if they can make four equal groups. Get them to check, by counting, that the groups are equal.

DEEPEN Provide children with more questions similar to question ⑤, where the objects being quartered are not divided in a standard 2×2 array. For example, ask children to make an 8×3 array of 24 counters and ask them to find a quarter.

ASSESSMENT CHECKPOINT Use questions ① to ④ to assess whether children understand how to find a quarter of a set of objects using both sharing methods and recognising a quarter.

ANSWERS Answers for the **Practice** part of the lesson can be found in the *Power Maths* online subscription.

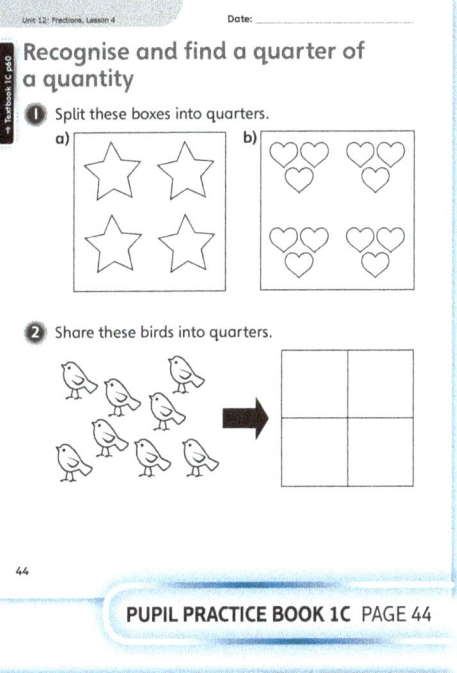

PUPIL PRACTICE BOOK 1C PAGE 44

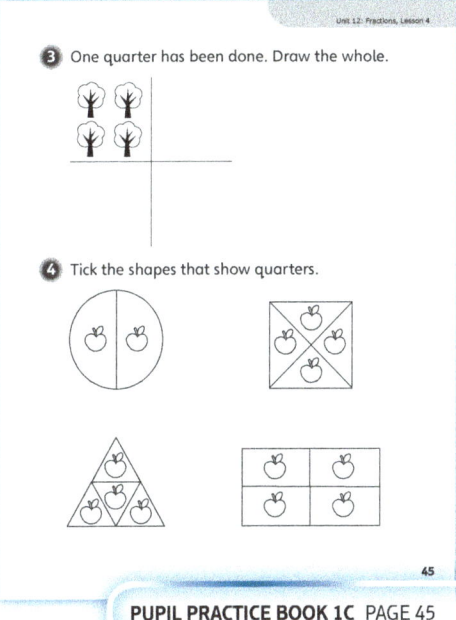

PUPIL PRACTICE BOOK 1C PAGE 45

Reflect

WAYS OF WORKING Whole class

IN FOCUS Ask the class to discuss their answers to this question in pairs first. Then ask children to come to the board and draw their pictures of finding a half and a quarter of 12. Afterwards, see if any children have a different way to show halving and quartering 12 and see if they are able to recognise the link between halves and quarters.

ASSESSMENT CHECKPOINT This question will tell you if children have grasped the method and vocabulary associated with finding quarters of small quantities.

ANSWERS Answers for the **Reflect** part of the lesson can be found in the *Power Maths* online subscription.

After the lesson

- Could you set a home learning activity in which children have to find quarters of small quantities?
- Do you need to re-model any sharing strategies?

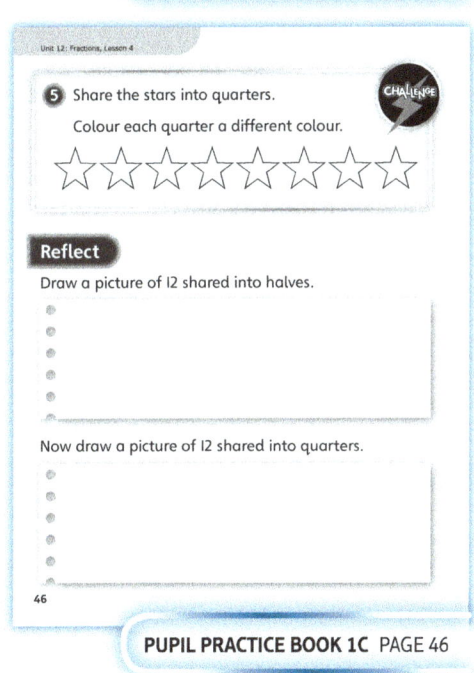

PUPIL PRACTICE BOOK 1C PAGE 46

101

End of unit check

Unit 12: Fractions

> Don't forget the unit assessment grid in your *Power Maths* online subscription.

WAYS OF WORKING Group work adult led

IN FOCUS Look at question ❶ with some children you want to assess in more detail. Can children tell you why they have chosen the answer they have? Can they say why the other answers are **not** cut into halves? Use question ❷ to assess whether children are confident in finding half a quantity. Use question ❸ to assess whether children are confident to recognise when a shape has been split into four equal parts. Use question ❹ to assess whether children can find a half and a quarter of a quantity.

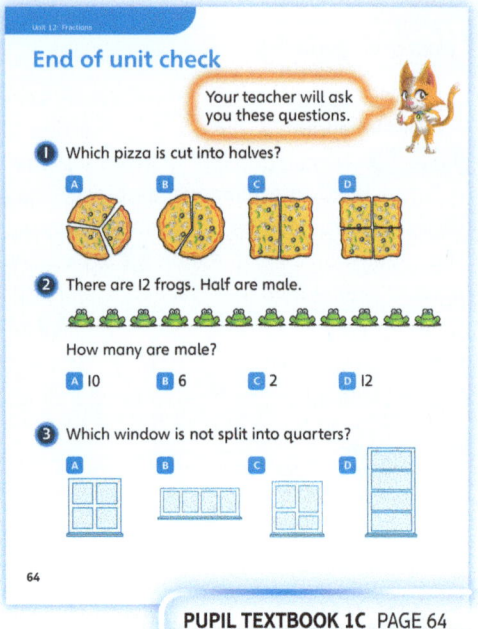

PUPIL TEXTBOOK 1C PAGE 64

Think!

WAYS OF WORKING Pair work or small groups

IN FOCUS This question will encourage children to think deeply. Various responses are possible, but should indicate that the whole is 10 so can be halved (5). However, 10 cannot be split into 4 equal parts, so Eva cannot have a quarter of 10. Encourage children to think through or discuss this section before writing their answer in **My journal**.

ANSWERS AND COMMENTARY Children will demonstrate mastery in this section by finding half of a group of objects and recognising that they must share equally. They will also demonstrate mastery by solving word problems about halves and quarters, and will know what 'part' and 'whole' mean.

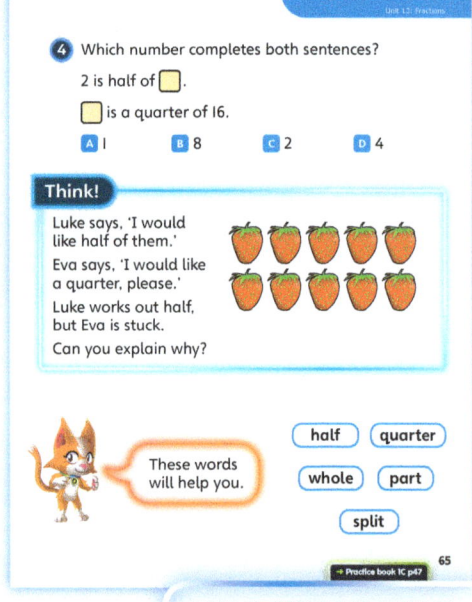

PUPIL TEXTBOOK 1C PAGE 65

Q	A	WRONG ANSWERS AND MISCONCEPTIONS	STRENGTHENING UNDERSTANDING
1	C	Choosing B suggests that the child does not understand that the parts must be equal.	You may need to run an intervention where children undertake extra sharing activities with counters. Vocabulary practice might be needed – this could be a good home learning opportunity. Some children may need to practise reading questions carefully.
2	B	D indicates that children have repeated the whole rather than finding half.	
3	C	Any other answer suggests that children may not have read the question carefully and therefore look for the windows that are split into quarters, rather than the odd one out.	
4	D	Answering B may show that children have not remembered the difference between halves and quarters.	

102

Unit 12: Fractions

My journal

WAYS OF WORKING Independent thinking

ANSWERS AND COMMENTARY

It is easier for Luke because you can share the strawberries equally into two halves.

It is harder for Eva because you cannot share the strawberries equally into quarters. There will be two left over.

Children may know the correct answer, but may find it challenging to explain the reason why. Support them with vocabulary and structuring an answer that includes examples.

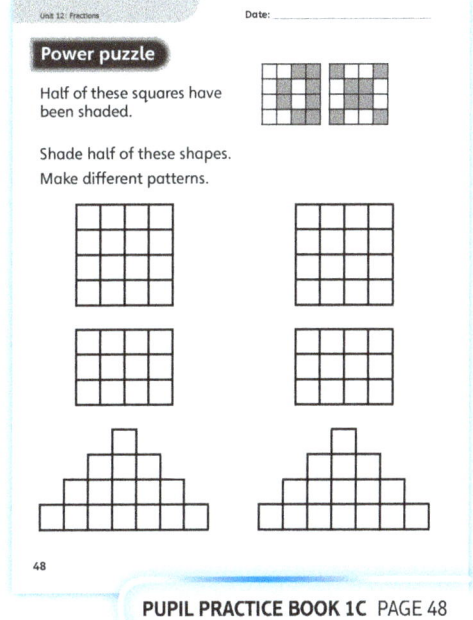

PUPIL PRACTICE BOOK 1C PAGE 47

Power check

WAYS OF WORKING Independent thinking

ASK
- What can you do now that you could not do at the start of the unit?
- What do you know now that you did not know at the start of the unit?
- Can you write down any new words you have learnt and what they mean?

Power puzzle

WAYS OF WORKING Pair work or small groups

IN FOCUS How can children start the puzzle? It is useful to remind children to carefully count the total number of boxes in each grid and then find half to shade. Observe whether children use their fingers to calculate half or whether they make marks on the grids.

Some children could also shade a quarter using a different colour.

ANSWERS AND COMMENTARY Look at whether the correct numbers of boxes have been shaded. Children should not be shading the same pattern twice. Children should recognise that each box represents a part of a whole and that shading any boxes within the shape that amount to half of the total will be correct, whether the shaded boxes are next to each other or not.

PUPIL PRACTICE BOOK 1C PAGE 48

After the unit
- Which children need further support with halves and quarters?
- Are all children ready for the next unit?
- Which teaching strategies helped children master this unit?

Strengthen and **Deepen** activities for this unit can be found in the *Power Maths* online subscription.

Unit 13
Position and direction

Mastery Expert tip! 'I found that taking children outside to give and follow instructions really helped to secure their understanding of the vocabulary used in this unit.'

Don't forget to watch the Unit 13 video!

WHY THIS UNIT IS IMPORTANT

This unit gives a practical application to children's learning from the previous unit on fractions. They will learn to describe rotations as quarter, half, three-quarter and whole turns, and will combine turns with lateral movement to give and follow route instructions. Children will also learn to describe the position of an object in relation to other objects, using the words 'above', 'below', 'left', 'right' and 'between'.

Children will develop these skills further in Year 2, when they will be introduced to the terms 'clockwise' and 'anticlockwise' and will follow more complex routes. The ability to identify fractions of a turn will also provide children with the groundwork for telling the time in Unit 16. Following a sequence of instructions will develop children's procedural fluency, in preparation for future units involving number and, in particular, calculations.

WHERE THIS UNIT FITS

→ Unit 12: Fractions
→ **Unit 13: Position and direction**
→ Unit 14: Numbers to 100

In this unit, children will apply their knowledge of fractions to contextual and practical problems. Being able to identify and describe position and movement will help children to develop their spatial awareness and reasoning. Unit 14 will look at numbers to 100.

Before they start this unit, it is expected that children:
- can give and follow a simple series of instructions with two or three steps
- understand the concept of a whole, halves and quarters, especially in relation to a circle
- understand 'turn' as rotation around a point.

ASSESSING MASTERY

Children who have mastered this unit will be able to describe the direction and fraction of a turn using the words 'left' and 'right', 'quarter', 'half', 'three-quarters' and 'whole'. They will be able to identify and describe a route to a desired goal, and to give and follow a series of instructions. They will also be able to describe the position of an object in relation to one or more other objects using the words 'left', 'right', 'above', 'below' and 'between'.

COMMON MISCONCEPTIONS	STRENGTHENING UNDERSTANDING	GOING DEEPER
Children may confuse left and right when describing turns, particularly if the object is oriented differently from them.	Give children objects they can rotate. Ask them to turn their object left and right, with the object in different orientations. You could label the sides of the object 'left' and 'right' to support them.	Ask children to draw a treasure map on a grid or create one in the playground, with objects to avoid. Challenge them to write a series of instructions to reach the treasure for a partner to follow.
Children may not understand quarter and half turns and the relationship between them.	Give children paper circles folded into quarters. Children could shade in a quarter, half, three-quarters and whole turns in both directions, and use the results as a reference throughout the unit.	

Unit 13: Position and direction

UNIT STARTER PAGES

Draw children's attention to the vocabulary on the page. Discuss what they think the words mean in the context of position and direction. Can children use the vocabulary to describe the arrows? Can they use the vocabulary to describe the positions of the animals?

STRUCTURES AND REPRESENTATIONS

Although there are no set mathematical structures and representations for this unit, curved arrows can be used to illustrate the direction and fraction of a turn.

KEY LANGUAGE

There is some key language that children will need to know as part of the learning in this unit:

→ turn, position, direction
→ half turn, quarter turn, three-quarter turn, whole turn
→ left, right, in between
→ forwards, backwards
→ above, below
→ top, middle, bottom
→ up, down
→ first, second, third, fourth

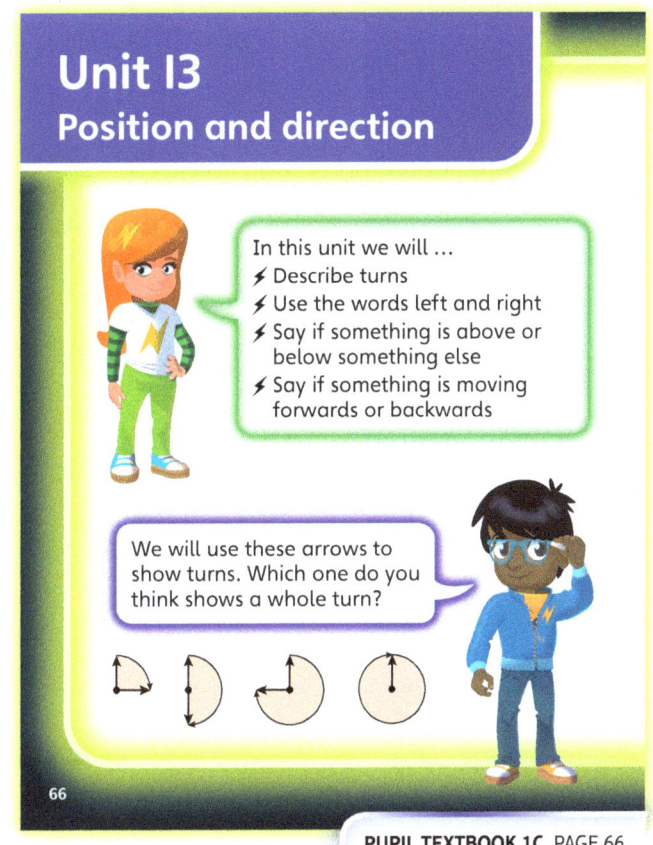

PUPIL TEXTBOOK 1C PAGE 66

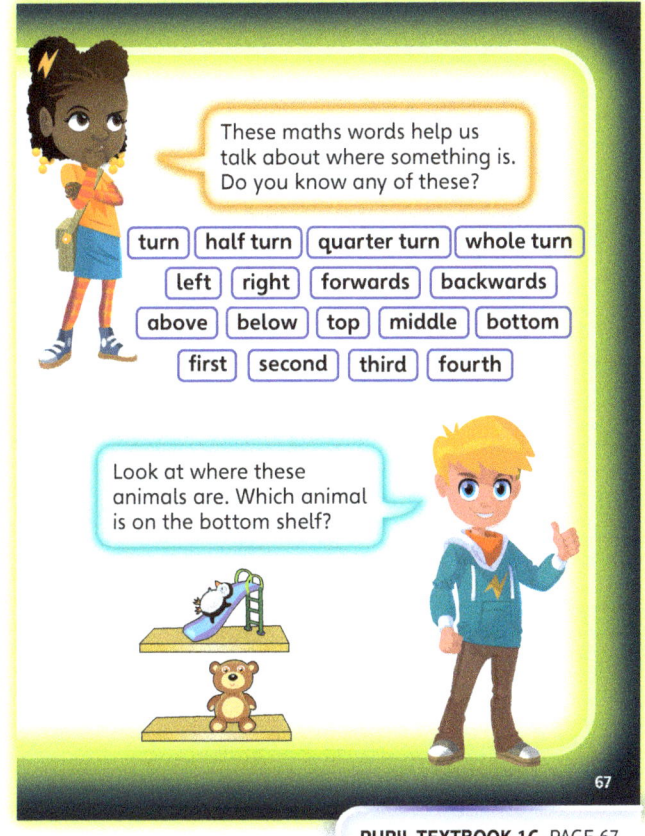

PUPIL TEXTBOOK 1C PAGE 67

Unit 13: Position and direction, Lesson 1

Describe turns

Learning focus

In this lesson, children will learn to describe turns as quarter, half, three-quarter or whole turns.

Before you teach

- Are children secure with halves and quarters?
- What practical opportunities will you provide within the lesson?
- Would children benefit from having the key vocabulary displayed?

NATIONAL CURRICULUM LINKS

Year 1 Geometry – position and direction

Describe position, direction and movement, including whole, half, quarter and three-quarter turns.

ASSESSING MASTERY

Children can describe the turn of an object using the words 'quarter', 'half', 'three-quarter' and 'whole'. Children can recognise that a half turn will result in facing the same way, regardless of which direction they turn in, and that a quarter turn in one direction will give the same result as a three-quarter turn in the opposite direction.

COMMON MISCONCEPTIONS

Children may not have secure understanding of quarters and halves. Ask children to stand up and do a quarter turn or a turn towards the window. Ask:
- *How much have you turned? How many quarters are there in a half? How many quarters are there in a whole?*

STRENGTHENING UNDERSTANDING

Ask children to carry out the rotations in the problems practically, either using objects or turning themselves, so that they develop a conceptual understanding of quarter and half turns. Provide children with paper circles and ask them to fold them into half and half again. They can shade in a quarter, half and three-quarters and keep the result as a reference for use throughout the lesson.

GOING DEEPER

Challenge children to complete a series of turns. Ask them to rotate and stop to face different directions and then record the turns they completed. Extend by going beyond a whole turn or changing which direction they turn within a series of turns.

KEY LANGUAGE

In lesson: half turn, facing, turn, **quarter turn**, three-quarter turn, position, whole turn

Other language to be used by the teacher: direction

RESOURCES

Optional: paper circles, colouring pencils, objects to rotate such as toys, programmable toys or a toy car

 In the eTextbook of this lesson, you will find interactive links to a selection of teaching tools.

Quick recap

Ask children to draw a circle on mini whiteboards and to shade in half of the circle. Look for children dividing the shape into two equal parts first and then shading one half. Discuss correct and incorrect responses. If children find this easy, repeat for quarters.

Unit 13: Position and direction, Lesson 1

Discover

WAYS OF WORKING Pair work

ASK

• Question 1 a): *Which way is Tom facing? What happens if he makes a half turn? What will he be facing? What does it mean to do half a turn? Does it matter which way he turns?*
• Question 1 b): *Choose an object to face. What happens when you make a half turn?*

IN FOCUS Question 1 a) asks children to think about what happens if Tom makes a half turn. Ask them to consider whether it matters which way Tom turns. Question 1 b) is about practising half turns. You might give children arrows, ask them to make a half turn and then to try to generalise what they notice (that they are always facing the opposite way to the direction they started in). Ask them to rehearse their explanation with a partner.

PRACTICAL TIPS Recreate the scenario in the classroom or playground. Put out four objects and ask children to make different half turns from a central point.

ANSWERS

Question 1 a): Tom will point to the tree.

Question 1 b): A half turn will make you look directly behind you each time.
For example, if Tom is looking at the pond, a half turn will mean he is looking at the playground. If he is looking at the tree, then a half turn will mean he is looking at the rock.

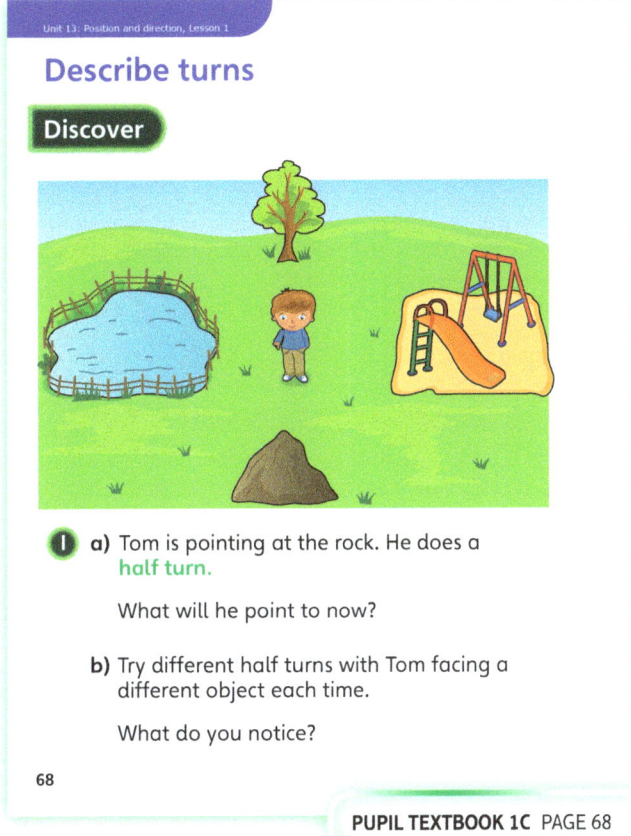

PUPIL TEXTBOOK 1C PAGE 68

Share

WAYS OF WORKING Whole class teacher led

ASK

• Question 1 a): *Which way is Tom facing before he turns? Where is he facing after he turns?*
• Question 1 b): *Which way could you point Tom? Which way is he facing after half a turn? What happens when you make any half turn?*

IN FOCUS Question 1 a) illustrates that the end result of a half turn is the same regardless of direction. Discuss whether the direction needs to be stated for half turns. Question 1 b) illustrates different half turns and that Tom will be facing the opposite direction with each half turn that he takes. Use Sparks to help children explain what happens. You might want to ask the class to stand up and show this.

PUPIL TEXTBOOK 1C PAGE 69

107

Think together

WAYS OF WORKING Whole class teacher led (I do, We do, You do)

ASK
- Question ❶: *What does a quarter turn mean? Can you draw a quarter turn? Does the direction you turn make a difference to the way you are facing at the end?*
- Question ❷: *Which way is Tom facing? What happens if Tom makes a quarter turn? Is there more than one answer?*

IN FOCUS In question ❶, ask children to face the front and to then make a quarter turn. Discuss what they notice and relate it to quarters covered in the previous unit on fractions. You may also want to discuss that the direction they turn matters. In question ❷, discuss how there are two possible answers. At this stage, you do not have to discuss the language 'left' and 'right'. Question ❸ draws out what happens if children make multiple quarter turns, such as noticing that 2 quarter turns make a half turn. This can be related to their work on fractions in the previous unit.

STRENGTHEN Practise what happens if children make different turns. Provide children with a paper circle folded into quarters. Encourage them to use toys to make different turns and see where the toy is facing after each turn.

DEEPEN Have four children sit around a table so that each child is sitting in one of the four quadrants. Place a programmable toy in the centre of the table, facing one child. This child then says the name of one of the other children and describes the turn that the toy has to make in order to face them. The first child then inputs the instructions to make the toy face the named child. It is then the turn of the child that the toy is now facing. (If no programmable toy is available, children can manually turn an item such as a toy car.)

ASSESSMENT CHECKPOINT Use questions ❶ and ❷ to determine whether children are confident with quarter turns and they understand that the direction in which they turn matters. Use question ❸ to assess whether children understand that 2 quarter turns make a half turn, and it does not matter which direction you do them as you will end up facing the opposite direction to the one you started in. You can also assess whether children understand that 3 quarter turns in one direction are equivalent to 1 quarter turn in the opposite direction, and that 4 quarter turns brings to you back to the position you started in.

ANSWERS

Question ❶: Check children can make quarter turns in different directions. They should understand that if they change the way they turn, they will end up facing in a different direction.

Question ❷: Tom will face either the tree or the rock, but it depends which direction he makes his quarter turn in.

Question ❸: Observe children as they make whole, half and quarter turns. Do they notice that making 2 quarter turns is the same as making a half turn? Or that making 4 quarter turns is the same as making 1 whole turn?

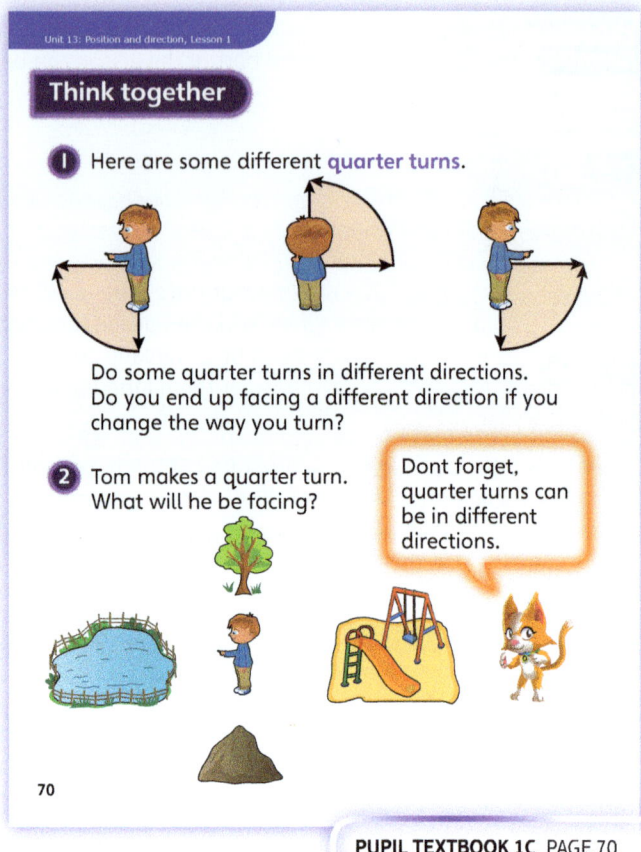

PUPIL TEXTBOOK 1C PAGE 70

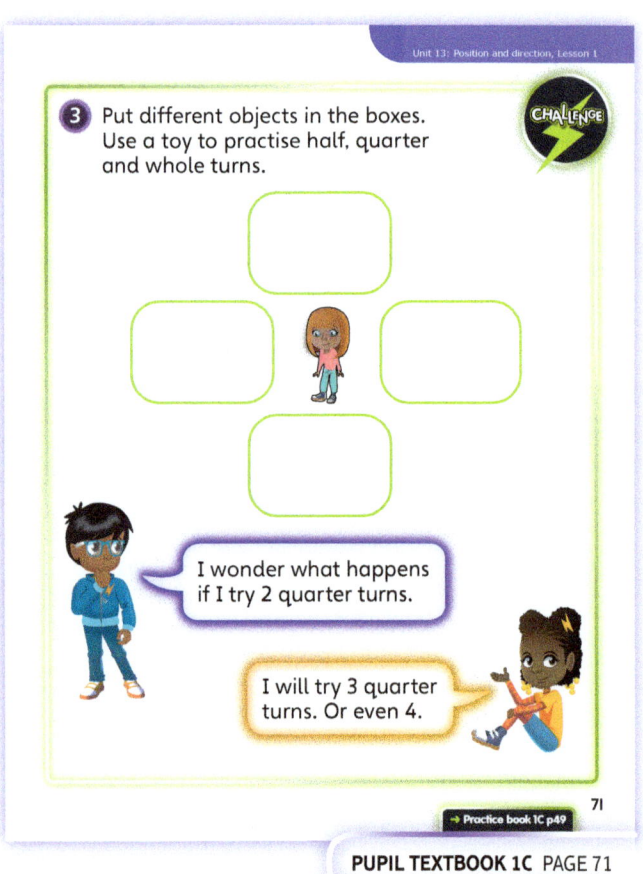

PUPIL TEXTBOOK 1C PAGE 71

Unit 13: Position and direction, Lesson 1

Practice

WAYS OF WORKING Independent thinking

IN FOCUS In questions ❶ and ❷, children are given pictorial representations and are asked to circle the result of different turns. Look out particularly for children making the quarter turns correctly. Question ❸ looks at turning objects. You might want to go through an example of this as a whole class, before asking children to have a go. You could use toy cars or printouts of arrows for children to practise with.

STRENGTHEN Provide children with objects so they can physically carry out the turns described in the problems. If they are struggling to understand the vocabulary, show children the images below. Help them connect the portion of the shape that is shaded (which they learnt in Unit 12) with the angle of turn. For example, if you shade half the shape and travel around tracing the outside of the shaded portion, you get a half turn.

Half turn   Quarter turn right   Three-quarter turn left   Whole turn

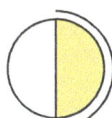 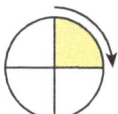 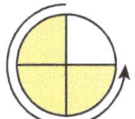 

DEEPEN Extend question ❸ by asking children to draw the tractor as if it has made different types of turns. They could start by trying to draw any turn listed in question ❸ that did not fit the description. Children could start to explore how a quarter turn in one direction is the same as a three-quarter turn in the opposite direction.

ASSESSMENT CHECKPOINT Use questions ❶ and ❷ to assess whether children understand what happens as the result of half and quarter turns. Use question ❸ to assess whether children can understand when turns made in opposite directions can be equivalent to one another (for example, a three-quarter turn left and a quarter turn right; a half turn left and half turn right).

ANSWERS Answers for the **Practice** part of the lesson can be found in the *Power Maths* online subscription.

Reflect

WAYS OF WORKING Pair work

IN FOCUS This question requires children to physically demonstrate multiple turns. In order to do this, they need to be able to visualise it for themselves, which requires a secure understanding. To help children make full quarter turns, rather than partial quarter turns, place objects at each North, East, South and West point around them. Each partner should carry out these instructions but they could start facing different ways so that they don't copy each other's movements.

ASSESSMENT CHECKPOINT Assess whether children have a secure understanding of quarter, half, three-quarter and whole turns and whether they can use the correct vocabulary.

ANSWERS Answers for the **Reflect** part of the lesson can be found in the *Power Maths* online subscription.

After the lesson

- Are children secure with quarter, half, three-quarter and whole turns?
- Can children use the correct vocabulary to describe a turn?

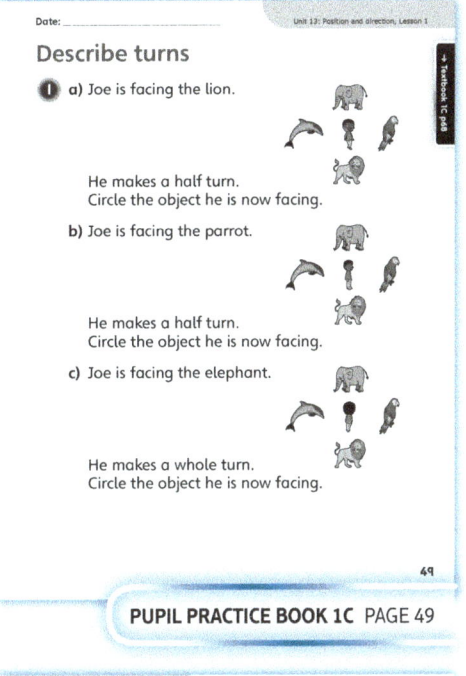

PUPIL PRACTICE BOOK 1C PAGE 49

PUPIL PRACTICE BOOK 1C PAGE 50

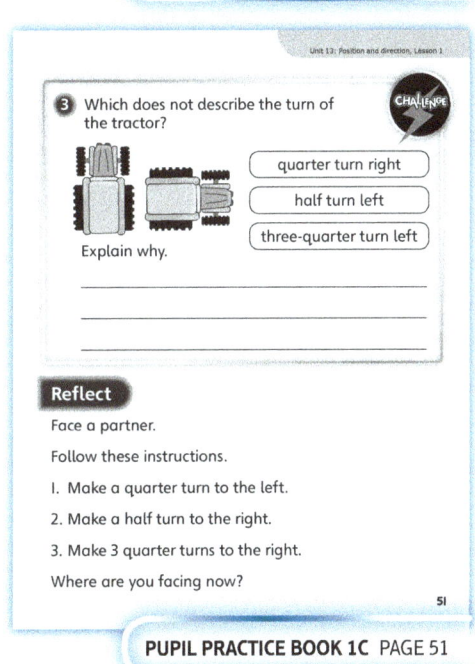

PUPIL PRACTICE BOOK 1C PAGE 51

109

Unit 13: Position and direction, Lesson 2

Describe position – left and right

Learning focus
In this lesson, children will build on their knowledge of position and direction by using the words 'left' and 'right'. They will know which direction is left and which direction is right.

Before you teach
- Do children understand half and whole turns?
- Do children understand there are two quarter turns they children can make?

NATIONAL CURRICULUM LINKS

Year 1 Non-statutory guidance

Pupils use the language of position, direction and motion, including: left and right, top, middle and bottom, on top of, in front of, above, between, around, near, close and far, up and down, forwards and backwards, inside and outside.

ASSESSING MASTERY

Children know the meaning of the words 'left' and 'right'. They should be able to communicate what they can see on the left and what they can see on the right.

COMMON MISCONCEPTIONS

Children often confuse left and right and do not have a strategy to remember it. Ask:
- *How can you remember your left from your right? Can you use hands or objects to help you?*

STRENGTHENING UNDERSTANDING

It is important that children have a strategy to help them remember their left and right. They could use their hands and hold up their thumbs and first fingers to see which make an 'L' shape.

Children should understand that it does not matter where they stand: their left and right remain the same.

GOING DEEPER

Ask children to integrate the directional language from the previous lesson with left and right. For example, ask children to make a quarter turn left and a quarter turn right. Which way do they turn? What happens if children make a half turn left and a half turn right? Ask them whether there is a difference.

KEY LANGUAGE

In lesson: left, right

Other language to be used by the teacher: turn

RESOURCES

Optional: arrow cards saying left and right

 In the eTextbook of this lesson, you will find interactive links to a selection of teaching tools.

Quick recap

Draw four basic shapes on the board (one at the top, bottom, left and right), with a stick character in the centre, facing the bottom shape. Ask: *What shape will the character be facing if they make a half turn? What shapes could the character be facing if they make a quarter turn?*

Unit 13: Position and direction, Lesson 2

Discover

WAYS OF WORKING Pair work

ASK
- Question 1 a): *Which direction is left? Can you point to the left?*
- Question 1 b): *Which direction is right? Can you point to the right?*

IN FOCUS This question focuses on children learning their left and right. Observe the class and see which children are already confident with their left and right and which children need further support and are relying on guessing. Use strategies to support children to help them remember their left and right. One good tip is to ask children to put their hands out in front of them so that the backs of their hands are facing them and fingers are pointing upwards. If they then make their two thumbs point to each other, the hand that makes an 'L' shape is the left hand.

PRACTICAL TIPS Ask children to point to their left and then to their right. Ask them who is sitting to their left and who is sitting to their right. This will help reinforce the language.

ANSWERS

Question 1 a): The cat and the tree are on the left side of the road.

Question 1 b): The house and the person are on the right side of the road.

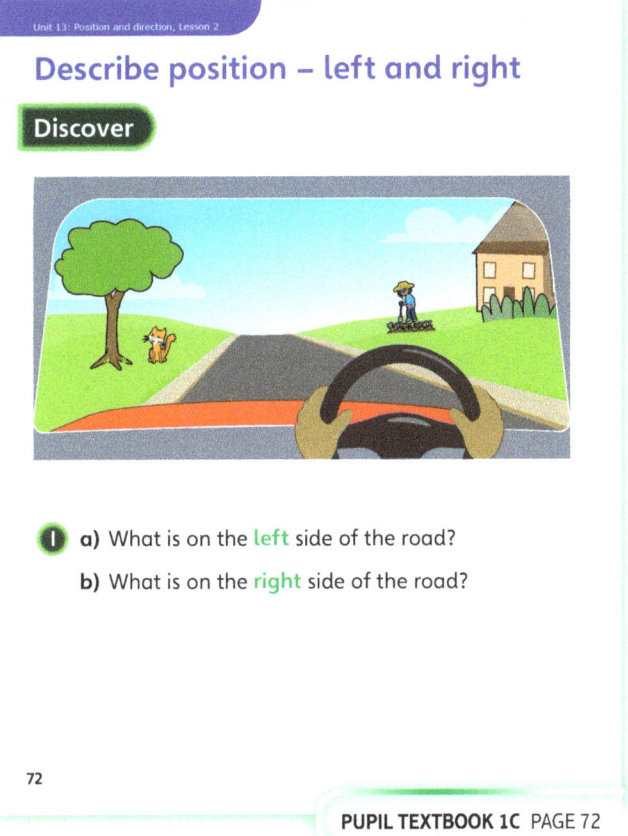

PUPIL TEXTBOOK 1C PAGE 72

Share

WAYS OF WORKING Whole class teacher led

ASK
- Question 1 a): *What is on the left-hand side of the road?*
- Question 1 b): *What can you see on the right-hand side of the road?*

IN FOCUS Use the arrows and Sparks's comments to help children understand the difference between left and right. Ask children to copy the words onto their whiteboards a few times to practise writing the words 'left' and 'right'. You might want to check understanding by asking children to point or to look left and right. Ask them what they can see.

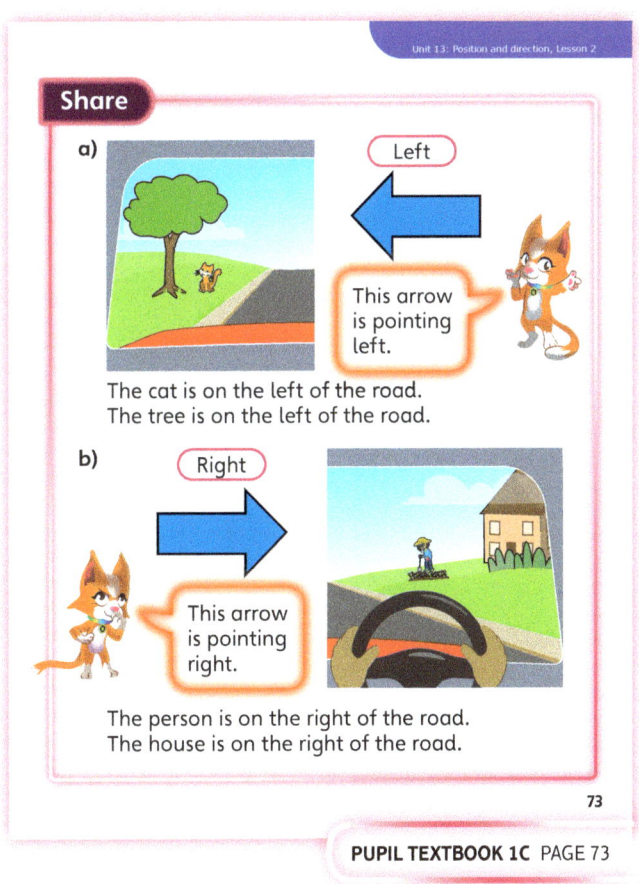

PUPIL TEXTBOOK 1C PAGE 73

111

Think together

WAYS OF WORKING Whole class teacher led (I do, We do, You do)

ASK

- Question ❷: *Which is your left hand? Which is your left foot? Which is your left eye? What do you notice?*

IN FOCUS Question ❶ is about children remembering their left and right. Ask children to look or turn left and right. For question ❷, children wave or shake their left and right feet and hands. Discuss what they notice. They see that the left hand and foot are on the left side of the body. Question ❸ tests the language and asks children to identify which animals are on the left and on the right.

STRENGTHEN One way to help children to remember left and right is to show their thumbs and first fingers and put the others down. The thumb and first finger that make an 'L' shape is left. The other side is the right hand of the child. Children need to have a method to know their right from their left.

DEEPEN Ask children to stand or sit, facing a partner. Ask them both to look left. Ask: *Why do you not look in the same direction?* Use this to help explain that left and right depend on the direction you are facing.

ASSESSMENT CHECKPOINT Use questions ❶ and ❷ to check children understand their left and right. Use question ❸ to check children can solve problems involving left and right.

ANSWERS

Question ❶: Children's answers will vary. Check they can accurately describe things to their left and their right. For example: The wall is to my left. My partner Amy is to my right.

Question ❷: Children should lift their left hand and then their right hand and point to their left foot and then their right foot.

Question ❸ a): The lion is on the left.

Question ❸ b): The zebra is on the right.

Question ❸ c): The elephant is in the middle.

Question ❸ d): The elephant is on the left of the zebra.

Question ❸ e): The elephant is on the right of the lion.

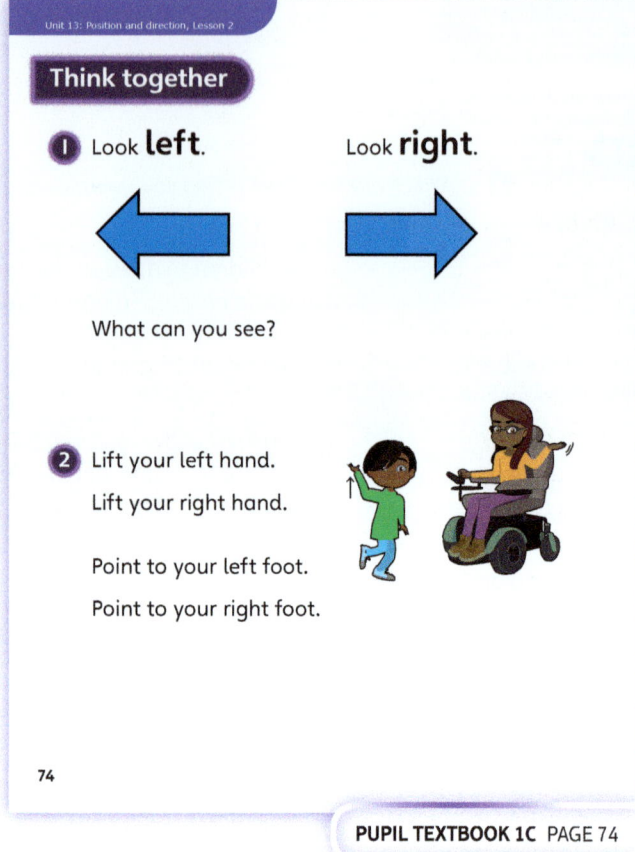

PUPIL TEXTBOOK 1C PAGE 74

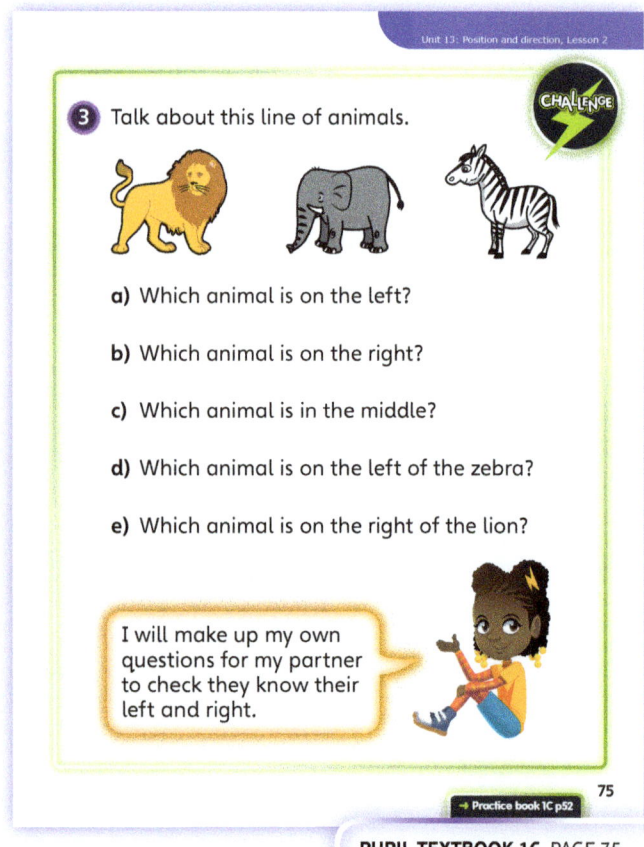

PUPIL TEXTBOOK 1C PAGE 75

Unit 13: Position and direction, Lesson 2

Practice

WAYS OF WORKING Independent thinking

IN FOCUS Questions ① and ② check that children can recognise left and right by asking them to shade in the correct arrows. Children also have the opportunity to practise writing the words left and right. Questions ③ and ④ ask children to circle the objects that are on the left and the right. In question ⑤, children need to work out which child is on the left and which child is on the right. This question could be interpreted in two ways and it is fine at this stage if children just circle the child on the left and the child on the right, without considering this from the point of view of the middle child.

STRENGTHEN One way to help children remember their left and right is to show their thumbs and first fingers and put the others down. The thumb and first finger that make an 'L' shape is left. The other side is the right hand of the child. Children need to have a method to know their right from their left.

DEEPEN For question ⑤, ask children to work out who is on Joe's left and Joe's right from Joe's point of view. Ask: How does this compare to how you look at it?

ASSESSMENT CHECKPOINT Use questions ① to ④ to check that children understand their left from their right. Use question ⑤ to assess whether children can determine which people or objects are on the left or the right of another object.

ANSWERS Answers for the **Practice** part of the lesson can be found in the *Power Maths* online subscription.

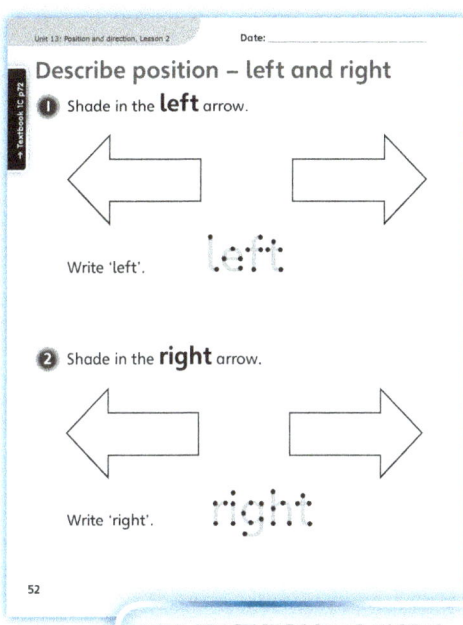

PUPIL PRACTICE BOOK 1C PAGE 52

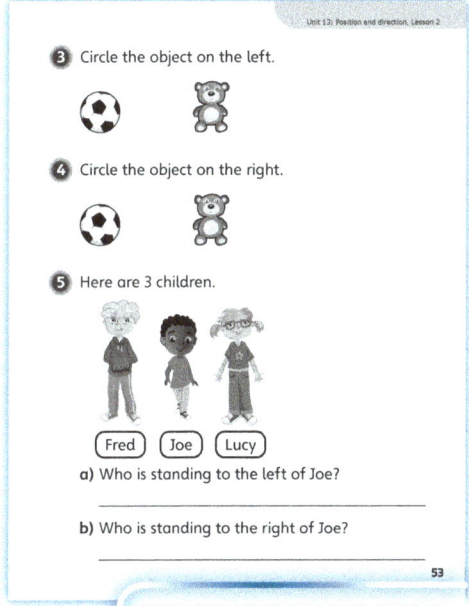

PUPIL PRACTICE BOOK 1C PAGE 53

Reflect

WAYS OF WORKING Whole class

IN FOCUS This question is designed to help secure children's understanding of the directions left and right. You can draw on their learning from the previous lesson by using language such as 'make a quarter turn right and left'. Make sure that children are partnered with someone of different ability so that one of each pair should be able to confirm the hand that is being held up.

ASSESSMENT CHECKPOINT Check that children know their left from their right by what they do and say.

ANSWERS Answers for the **Reflect** part of the lesson can be found in the *Power Maths* online subscription.

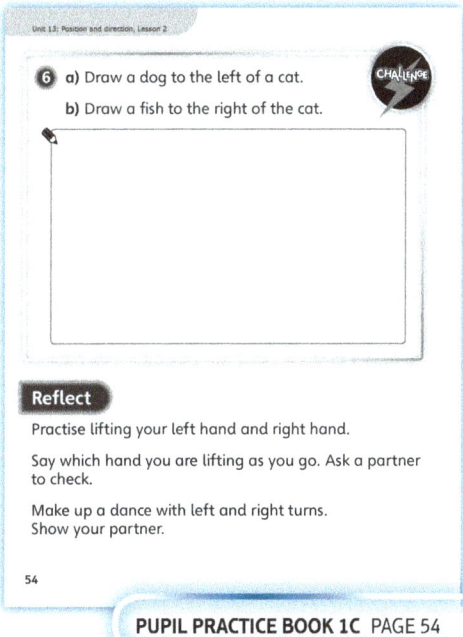

PUPIL PRACTICE BOOK 1C PAGE 54

After the lesson

- Do children know their left from their right?
- Can children look left and look right?

Unit 13: Position and direction, Lesson 3

Describe position – forwards and backwards

Learning focus

In this lesson, children will describe direction and lateral movement. They will learn how to follow and give instructions in order to reach a given goal.

Before you teach

- Are children secure describing the fraction of a turn?
- How will you support children who struggle to identify left and right?

NATIONAL CURRICULUM LINKS

Year 1 Non-statutory guidance

Pupils use the language of position, direction and motion, including: left and right, top, middle and bottom, on top of, in front of, above, between, around, near, close and far, up and down, forwards and backwards, inside and outside.

ASSESSING MASTERY

Children can use the correct vocabulary associated with rotation and lateral movement in order to describe a route. Children can follow instructions to get to a given goal and identify left and right turns, regardless of the orientation of the object.

COMMON MISCONCEPTIONS

Children may confuse left and right. They may not be able to reliably identify their own left and right or they may struggle to identify an object's left and right when its orientation is different from theirs. Ask:
- *Who can hold up their left* [or right] *hand? Who can point to my left* [or right] *hand?*

STRENGTHENING UNDERSTANDING

Provide practical opportunities for children to carry out the problems. Having objects that children can rotate can help them to identify the direction and fraction of a turn.

GOING DEEPER

Children could apply what they have learnt to create their own problems for each other to solve. For example, they could make up a treasure map with a set of instructions for a partner to follow from the starting point to the treasure.

KEY LANGUAGE

In lesson: left, right, **forwards**, quarter turn, half turn, three-quarter turn, whole turn

Other language to be used by the teacher: direction, facing, backwards

RESOURCES

Optional: objects for children to physically rotate and move, skipping ropes, PE cones, vocabulary prompts showing key words with pictures to illustrate them, blank 5×5 grids

 In the eTextbook of this lesson, you will find interactive links to a selection of teaching tools.

Quick recap

Check children's prior understanding of key vocabulary such as left, right, half turn, whole turn, quarter turn, three-quarter turn. Give instructions such as 'wave your left hand, turn 90 degrees right' using the vocabulary.

Unit 13: Position and direction, Lesson 3

Discover

WAYS OF WORKING Pair work

ASK

- Question 1 a): *In which direction does Amy walk?*
- Question 1 a): *How did you work out which direction Amy had to turn?*

IN FOCUS Question 1 b) requires children to begin to link together instructions in order to get to a desired goal. They also need to determine the direction of the turns using the words 'left' and 'right'. When Amy gets to the final cone, she is no longer facing the same way as the viewer. Some children may, therefore, struggle to determine which direction she has to turn.

PRACTICAL TIPS Recreate the scenario in the school hall or the playground using skipping ropes and cones. As a child walks along the skipping ropes, ask the other children to describe what turns the child will have to make to get to the end of the maze.

ANSWERS

Question 1 a): Amy will make a quarter turn right.

Question 1 b): Walk forwards to the next cone. Make a quarter turn left. Walk forwards to the end of the maze.

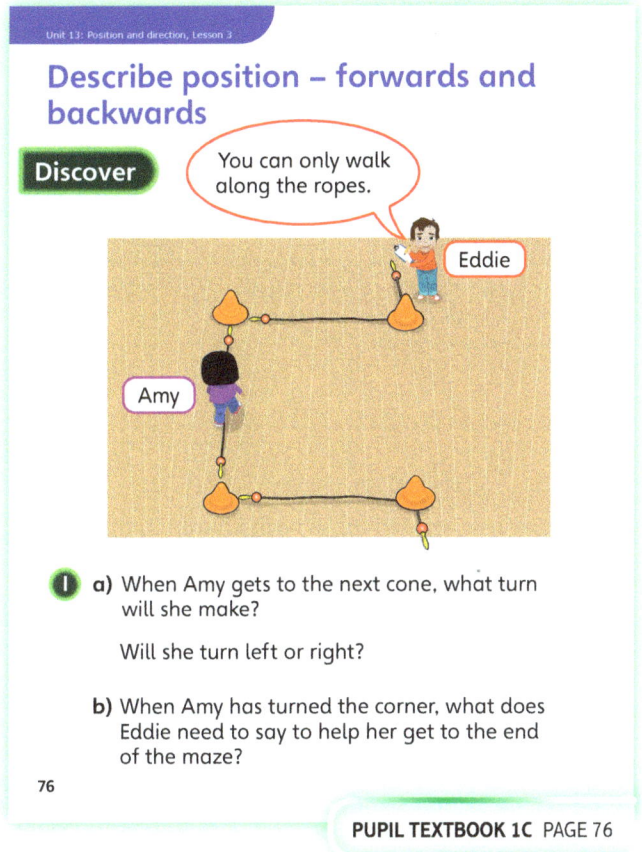

Share

WAYS OF WORKING Whole class teacher led

ASK

- Question 1 b): *From the last cone, what movements would Amy need to make to complete a square?*
- Question 1 b): *What movements has Amy already made?*
- Question 1 b): *How can Amy go back the way she came?*

IN FOCUS Question 1 a) is an opportunity to get children to carry out left and right turns in order to secure their understanding. They could also practise determining their partner's direction of turn when facing them, as they then need to consider it from their partner's perspective as opposed to their own.

In question 1 b), once children have correctly understood the directions that will take Amy to the end of the maze, use the questions above to explore directions that will take her to other places. Can children give directions to Amy to help her complete a square, or to go back the way she came? There is great opportunity for children to work in pairs to re-create this scenario. The person giving the directions is able to choose an ending point, and has to give the correct directions to help their partner arrive at that place.

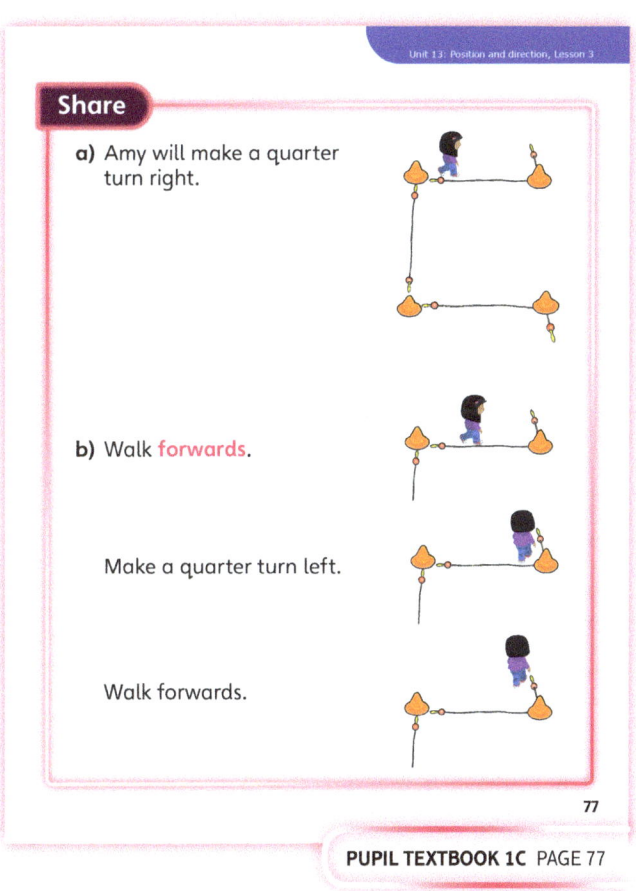

Unit 13: Position and direction, Lesson 3

Think together

WAYS OF WORKING Whole class teacher led (I do, We do, You do)

ASK
- Question ❶: *Does Amy need to turn left or right when she gets to the cone?*
- Question ❷: *Is there more than one possible answer for any of the instructions?*
- Question ❸: *How can you keep track of which way you are facing?*

IN FOCUS Question ❸ requires children to follow a series of instructions. They need to keep track of which direction they are facing after the turn in order to complete the route. Some children may interpret moving forwards as moving up the grid regardless of their current orientation.

STRENGTHEN For question ❷, you could set up a course that replicates Amy's. Children can then carry out Amy's movements before completing the instructions. They could label their left and right hands as prompts.

DEEPEN Challenge children to create their own route using the map in question ❸. They should visit two or more of the items. They could then ask a partner to test out their instructions.

ASSESSMENT CHECKPOINT Use questions ❶ and ❷ to assess whether children can apply the correct vocabulary. Can they determine the correct direction of the turns after Amy's orientation changes?

Question ❸ will determine whether children can follow a series of instructions in order to reach a desired goal. It will also highlight which children recognise left and right.

ANSWERS

Question ❶: 1. Walk forwards 4 steps;
2. At the cone, make a quarter turn left;
3. Walk forward 4 steps to the finish.

Question ❷: 1. Walk forwards 5 steps;
2. At the cone, make a quarter turn left;
3. Walk forward 3 steps;
4. At the cone, make a quarter turn left;
5. Walk forwards 8 steps to the finish.

Question ❸: You will end up at the monkey bars.
If you follow Ash's suggestion and for step 3 you walk backwards 2 steps, you will end up at the swing.

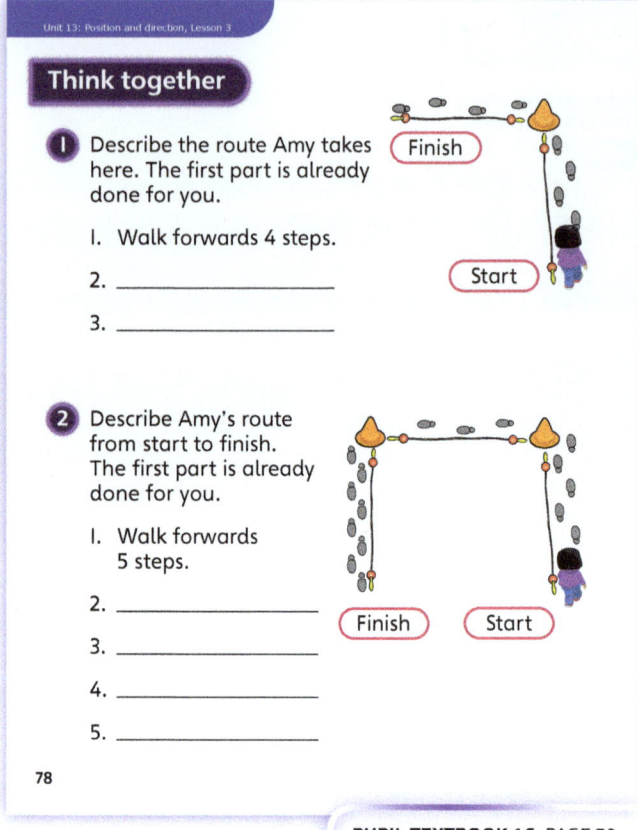

PUPIL TEXTBOOK 1C PAGE 78

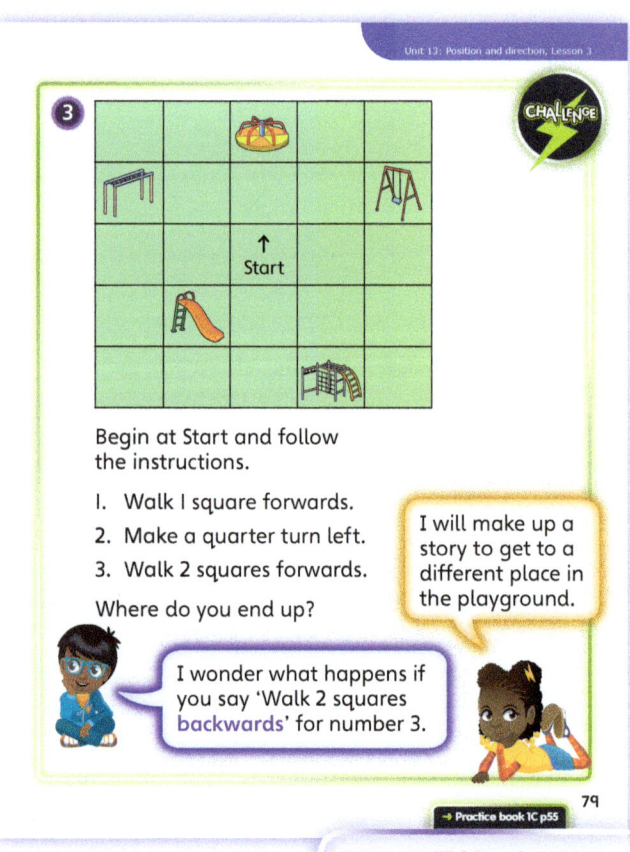

PUPIL TEXTBOOK 1C PAGE 79

Unit 13: Position and direction, Lesson 3

Practice

WAYS OF WORKING Pair work

IN FOCUS Questions ❶ and ❷ could be done in pairs. Encourage children to talk to one another using words to describe how to go through the maze and get from one object to another. You might want children to use a toy figure to help them. Ensure that children include turns to face the correct direction before walking forwards. This is sometimes a step that children miss out. This work can be linked to work they may do in computing.

STRENGTHEN Have vocabulary prompts ready showing the key words, with pictures to illustrate them, that children can refer to. Some children may still need practical opportunities to secure their understanding of rotation and linear movement. They could go outside to the playground and give each other simple instructions to follow.

DEEPEN Give each child a blank 5×5 grid. Challenge them to create a treasure map with a start point, an X for the end point and obstacles to avoid (for example, water, pirates, quicksand). Then ask children to plan a route to get to the treasure safely.

ASSESSMENT CHECKPOINT Use questions ❶ and ❷ to check that children can use accurate language to move around the maze and the map. Children could use language from this lesson and previous lessons.

ANSWERS Answers for the **Practice** part of the lesson can be found in the *Power Maths* online subscription.

Reflect

WAYS OF WORKING Pair work

IN FOCUS Children use the language from the lesson to describe how they can move from one part of the classroom or school to another. One child describes what their partner needs to do and the other follows the action. Check that children correctly do the action. Swap roles and ask the other child to describe how to get back.

ASSESSMENT CHECKPOINT Listen and check for children using the correct positional and directional language.

ANSWERS Answers for the **Reflect** part of the lesson can be found in the *Power Maths* online subscription.

After the lesson

- Can you create a PE activity in which children can apply their learning from this lesson?
- How many children can confidently and correctly apply the vocabulary used in this lesson?
- Were children able to adjust their perception of left and right as the orientation of the object they were observing changed?

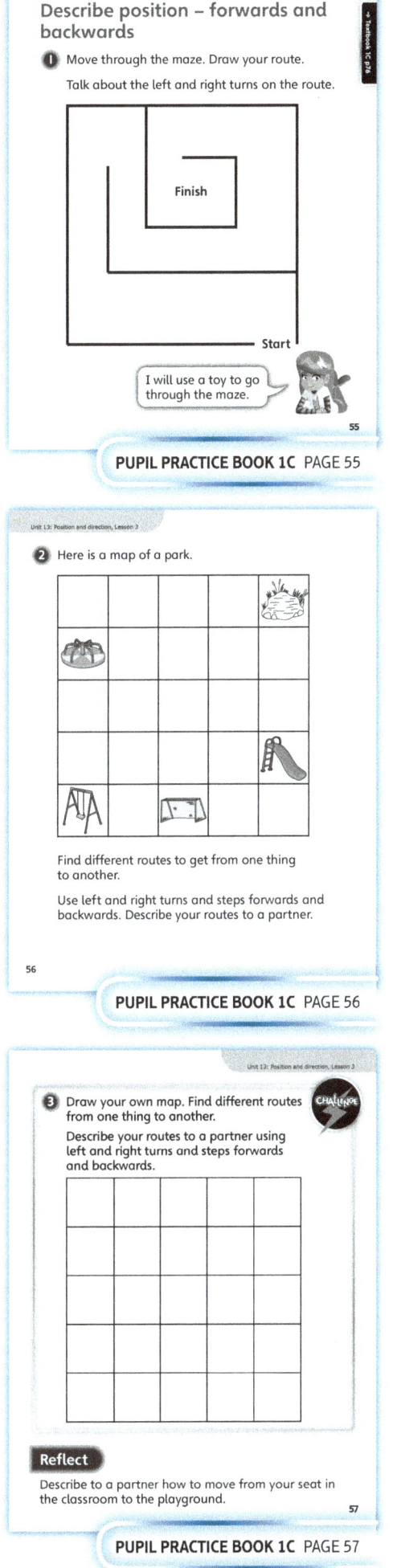

117

Unit 13: Position and direction, Lesson 4

Describe position – above and below

Learning focus
In this lesson, children will describe the position of an object based on its relation to other objects. They will apply their knowledge of left and right from the previous lessons.

Before you teach
- Are children secure in identifying left and right?
- Do children need practice in identifying position before the lesson?

NATIONAL CURRICULUM LINKS

Year 1 Non-statutory guidance

Pupils use the language of position, direction and motion, including: left and right, top, middle and bottom, on top of, in front of, above, between, around, near, close and far, up and down, forwards and backwards, inside and outside.

ASSESSING MASTERY

Children can describe the position of one object in relation to other objects using the words 'left', 'right', 'above' and 'below'. They can adapt their description of position when relating it to a different object.

COMMON MISCONCEPTIONS

Children may confuse the terms above and below. Ask:
- *Can you place your pencil below or above your book? What is above or below your desk?*

Children may also still confuse the terms left and right. Ask:
- *Who is sitting to the left of you? Who is sitting to the right of you? Is the window to the left or right of you?*

STRENGTHENING UNDERSTANDING

Give children construction equipment and ask them to follow instructions such as: *Put a blue cube above a red cube. Place a red cube to the right of a blue cube.*

GOING DEEPER

Give pairs of children multilink cubes. Ask one child in each pair to make a simple design, without showing it to their partner. Using the key vocabulary, they should then give instructions to their partner; their partner should attempt to make the design without showing it to the first child. When they have finished, children can compare their designs to see if they match.

KEY LANGUAGE

In lesson: above, position, top, middle, bottom, below, up, down, in between

Other language to be used by the teacher: left, right

RESOURCES

Optional: construction equipment, multilink cubes, vocabulary prompts showing key words with pictures to illustrate them, blank 2×3 grids, cubes or counters, different objects commonly found in the classroom such as pens and pencils

 In the eTextbook of this lesson, you will find interactive links to a selection of teaching tools.

Quick recap

As a class, ask children to look up, look down, look left and look right. Do this facing different directions in the classroom, to show that the way they are facing does not change their action.

Unit 13: Position and direction, Lesson 4

Discover

WAYS OF WORKING Pair work

ASK

- Questions ① a) and b): *Did you come up with a different answer from your partner? Who is correct?*
- Question ① b): *How many ways can you think of to describe the position of the teddy bear?*

IN FOCUS Question ① b) requires children to describe the position of the teddy bear. There are a number of ways of doing this: establish that the description changes depending on what we choose to relate the position of the teddy bear to.

PRACTICAL TIPS Recreate the scenario in the classroom with items you have available, such as pens and pencils. Ask children to describe the position of some of the items.

ANSWERS

Question ① a): The books are above the dinosaur.

Question ① b): The teddy is on the bottom shelf.
The teddy is below the stacking rings.
The teddy is to the right of the robot.

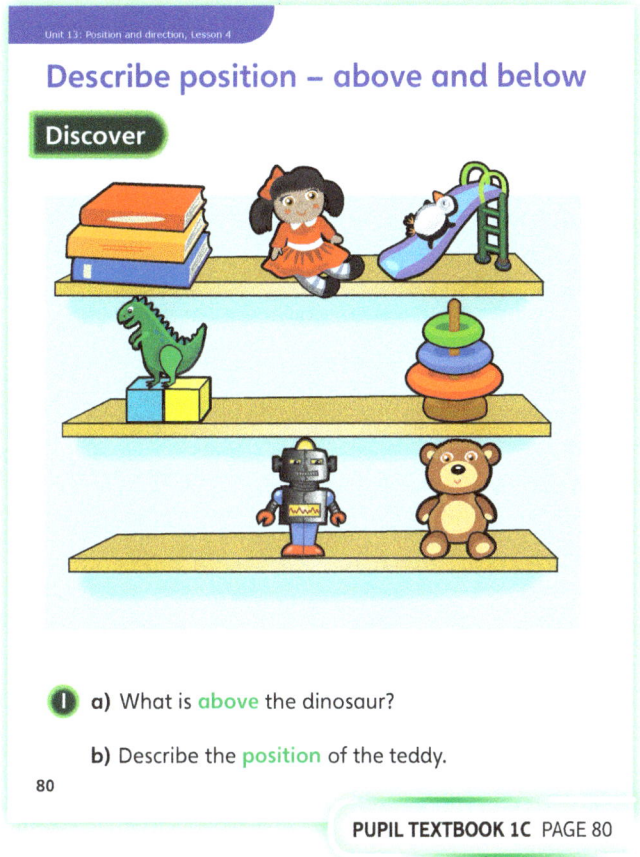

PUPIL TEXTBOOK 1C PAGE 80

Share

WAYS OF WORKING Whole class teacher led

ASK

- Question ① b): *Is the teddy bear below the slide?*
- Question ① b): *How would you describe the position of the dinosaur?*
- Question ① b): *Can you describe the position of the rings in relation to two different objects?*

IN FOCUS Question ① b) highlights how the position of the teddy bear can be described in different ways. Discuss how the position of other objects can also be described in different ways. Emphasise that, as the point of reference changes, the description of the position also changes.

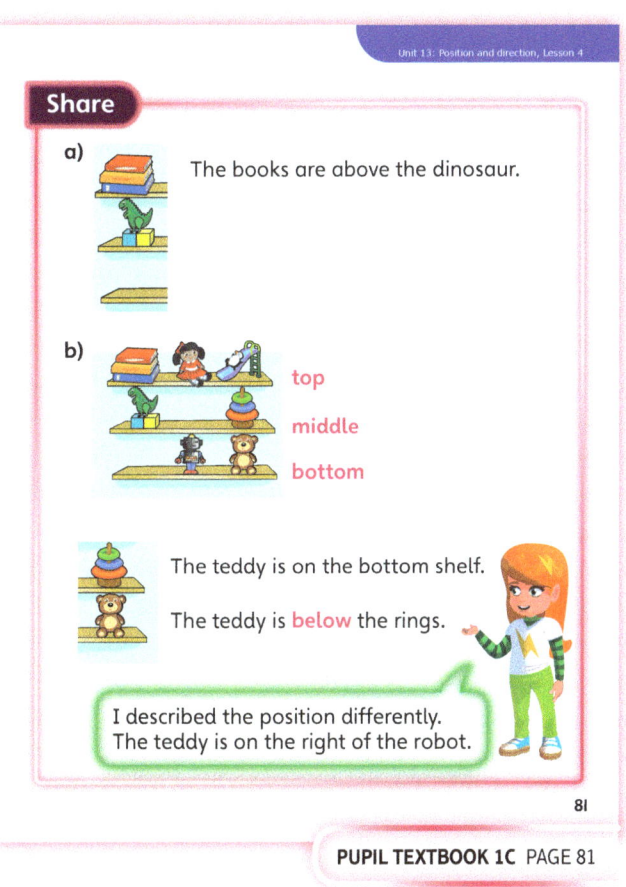

PUPIL TEXTBOOK 1C PAGE 81

119

Unit 13: Position and direction, Lesson 4

Think together

WAYS OF WORKING Whole class teacher led (I do, We do, You do)

ASK
- Question ①: *What objects can you see? Where are those objects?*
- Question ②: *Can you put a counter below the cube? If you put another counter in the bottom right-hand corner of the grid, would it still be to the right of the red counter?*
- Question ③: *Pick an object. What is above and what is below the object?*

IN FOCUS Question ① reinforces the language used in the **Share** section. Encourage children to use full sentences, saying them out loud to each other. In question ②, children are instructed to place a counter in the bottom row. There are two ways they could do this: in the bottom centre square or the bottom right square. The partner should try to tailor their description of the counter to describe where it has been placed, for example: 'below the cube and to the right of the red counter', or 'below and to the right of the cube'. In question ③, children pick an object and describe where it is, with a partner trying to work out the object they have chosen. Model a couple of rounds of the game as a whole class, to make sure children get the correct idea.

STRENGTHEN Take some different coloured cubes or counters and arrange them in different ways in a 3×3 frame, each time asking children to describe the position of a particular colour. Give children construction equipment or multilink cubes so they can practically explore positions for themselves. Have vocabulary prompts available showing the key words, with pictures to illustrate them, for children to refer to.

DEEPEN Give pairs of children multilink cubes. Ask one child in each pair to make a simple design, without showing it to their partner. Using the key vocabulary, they should then give instructions to their partner to make the item. Their partner should attempt to make the design without showing it to the first child. At the end, both children should see how similar their constructions are.

ASSESSMENT CHECKPOINT Use the questions to determine whether children can accurately use language such as top, middle, bottom, above and below. Children may also use language such as left and right from earlier lessons.

ANSWERS

Question ① a): The teddy is on the bottom shelf.

Question ① b): Children should point to the stacking rings above the teddy.

Question ① c): The books, doll and penguin on the slide are on the top shelf.

Question ②: Children should discuss the counter's position in the grid with a partner.

Question ③ a): Check children can follow instructions to work out an unknown item. For example, it is below the helicopter (bicycle).

Question ③ b): Check children can follow instructions to work out an unknown item. For example, it is to the right of the boat (train).

Think together

① Look at the shelves
   a) Which shelf is the teddy on?
   b) Point to the object above the teddy.
   c) What objects are on the top shelf?

② Here is a grid.

Put another counter in the bottom row.
Ask a partner to describe its position.

PUPIL TEXTBOOK 1C PAGE 82

③ Here are some objects.

Pick an object. Don't show your partner.

a) Tell your partner what is above or below your object.

Can they work out which is your object?

b) Your partner should pick a different object. They must tell you what is left or right of it.

Can you work out which is their object?

PUPIL TEXTBOOK 1C PAGE 83

Unit 13: Position and direction, Lesson 4

Practice

WAYS OF WORKING Pair work

IN FOCUS Question ③ requires children to think of three ways to describe the position of the socks. Therefore, they need to consider at least three different points of reference and think carefully about the vocabulary of position.

STRENGTHEN Give each child a blank 2×3 grid and six cubes or counters of different colours. Tell children where to place each cube or counter, giving one instruction at a time. For example: *Place the red cube in the middle of the top row. Now place the blue cube to the left of the red cube.* Once all squares are filled, support children in describing the position of different-coloured cubes or counters.

DEEPEN Extend question ③ by asking children to try to use all the words that they know to describe the position of the socks. Are there any words they cannot use?

Give pairs of children two blank 2×3 grids and six cubes or counters of different colours. Ask one child to place the counters or cubes on their grid, keeping it hidden from their partner. The second child should then ask questions that require only yes or no answers in order to place their counters or cubes in the same positions as the first child's. They can then compare their grids to see if they match.

THINK DIFFERENTLY Question ④ requires children to follow instructions to solve the puzzle. They need to apply what they know about positional language to work out the word from the clues.

ASSESSMENT CHECKPOINT Use question ④ to assess children's understanding of the vocabulary of position. Can they follow descriptions of position to identify specific letters?

ANSWERS Answers for the **Practice** part of the lesson can be found in the *Power Maths* online subscription.

Reflect

WAYS OF WORKING Pair work

IN FOCUS This question requires children to think carefully about how best to describe the position of a shape so that a partner can identify it. It highlights the importance of clear descriptions with correct language so another person can follow it. Note that in strict mathematical terms there are two cuboids, as a cube is also a cuboid. Children should be encouraged to use the term 'cube' and recognise the difference between the two shapes. This will also help avoid any confusion in their descriptions.

ASSESSMENT CHECKPOINT This activity will determine whether children can correctly use and understand all the vocabulary related to position.

ANSWERS Answers for the **Reflect** part of the lesson can be found in the *Power Maths* online subscription.

After the lesson

- Are children secure with the terms left and right, and above and below?
- Were children clear and concise in their descriptions of position?

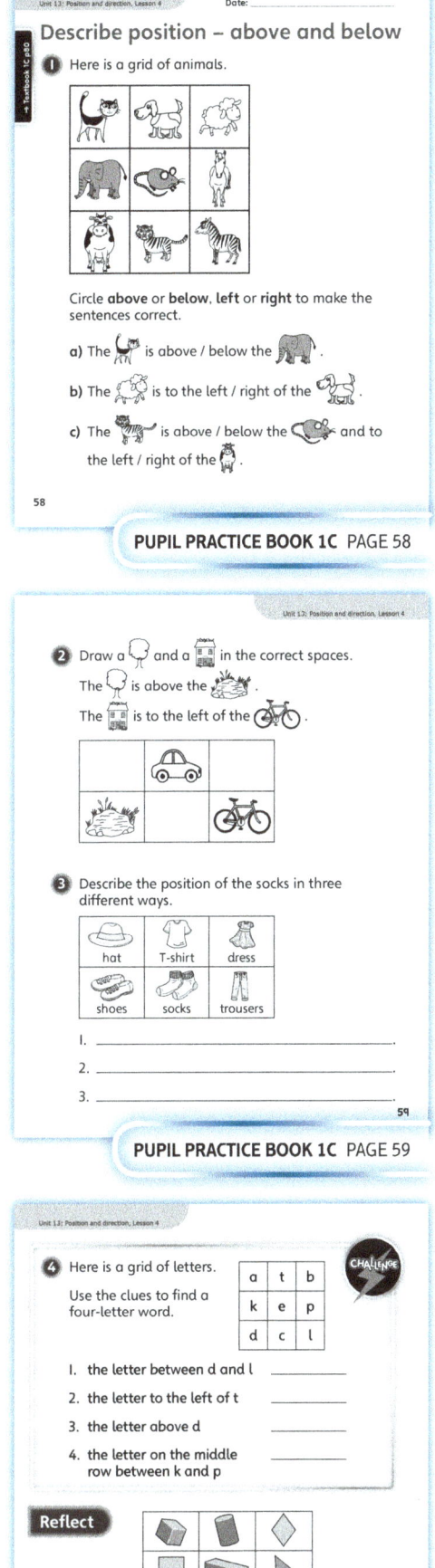

PUPIL PRACTICE BOOK 1C PAGE 58

PUPIL PRACTICE BOOK 1C PAGE 59

PUPIL PRACTICE BOOK 1C PAGE 60

Unit 13: Position and direction, Lesson 5

Ordinal numbers

Learning focus

In this lesson, children will learn to describe the order and position of objects using ordinal numbers. They will develop their understanding of the difference between a numeral representing the number of objects and the ordinal position of something.

Before you teach

- Have any children had experience with ordinal numbers outside of school, such as in competitions?
- How could you use this experience to engage children's interest in the concept from the beginning?

NATIONAL CURRICULUM LINKS

Year 1 Non-statutory guidance

Pupils practise counting (1, 2, 3 …), ordering (for example, first, second, third …), and to indicate a quantity (for example, 3 apples, 2 centimetres), including solving simple concrete problems, until they are fluent.

ASSESSING MASTERY

Children can describe the position of an object in a set using ordinal numbers. Children can confidently describe how a number representing an amount is different to a number representing an ordinal position.

COMMON MISCONCEPTIONS

Children may confuse cardinal numbers and ordinal numbers, such as 5 and 5th. For example, when asked to shade in the fifth object, children may shade in five objects. Ask:
- *Can you point to the first object in the list? Can you point to the second object in the list? When you pointed to the fifth position, how many objects were you pointing at?*

Children may start counting from the wrong end of a collection of objects. Ask:
- *Can you point to the first word of a sentence? Where is the second word?*

STRENGTHENING UNDERSTANDING

Strengthen understanding by having children take part in races outside. Ask: *How do you know who won the race? How can you describe the person who follows the winner? What about the next person after them?*

GOING DEEPER

Design simple riddles for children to solve. For example, ask: *Can you order where everyone finished in the race? Tim finished behind Milly. John finished in front of Tim. Milly was first.*

KEY LANGUAGE

In lesson: 1st, **first**, 2nd, **second**, 3rd, **third**, 4th, **fourth**, 5th, fifth, 6th, sixth, 7th, seventh, 8th, eighth, 9th, ninth, 10th, tenth

Other language to be used by the teacher: order

RESOURCES

Optional: four types of PE equipment such as a hoola hoop, a bean bag, a bench, a small hurdle for children to jump over, a selection of orderable and countable objects, number cards displaying ordinal numbers, number tracks listing ordinal numbers in order, cards

 In the eTextbook of this lesson, you will find interactive links to a selection of teaching tools.

Quick recap

Play a counting game with children. Give them a starting number between 0 and 10, ask them to count out loud until they get to 20 and then count back to the number they started with. Can they count as a class without making a mistake?

Unit 13: Position and direction, Lesson 5

Discover

WAYS OF WORKING Pair work

ASK

• Question 1 a): *What subject do you think the children are doing? How have the activities been ordered?*
• Question 1 a): *How can you tell which activity is the second activity?*
• Question 1 b): *How can you tell which activity is the fourth activity?*

IN FOCUS Questions 1 a) and b) give children their first experience of the ordinal language that they will use throughout the lesson. Discuss other places where children may have heard this language before, such as in sports competitions.

PRACTICAL TIPS You could recreate the scenario in the school hall or in the playground, using four different types of PE equipment. Ensure the activity stations are labelled in order: 1, 2, 3 and 4.

ANSWERS

Question 1 a): The second activity is throwing a bean bag into a bucket.

Question 1 b): The fourth activity is jumping over some buckets.

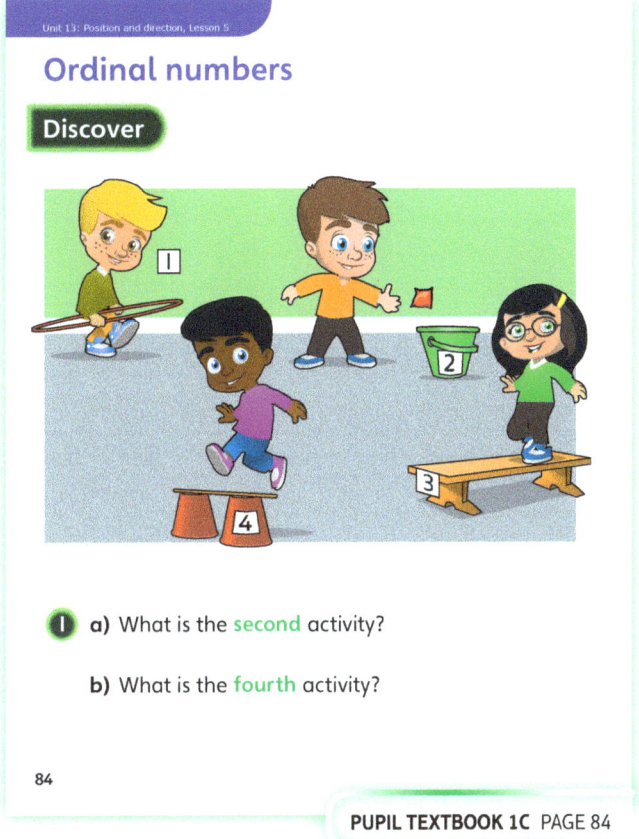

PUPIL TEXTBOOK 1C PAGE 84

Share

WAYS OF WORKING Whole class teacher led

ASK

• Questions 1 a) and b): *How has the order of the activities been made clearer in this picture? What else can you say about the order that the children are in? In what place is the child balancing on a bench?*
• Question 1 b): *How many children are doing the fourth activity?*

IN FOCUS Questions 1 a) and b) provide an opportunity to address the misconception that the ordinal number has to match the amount that is in that position. Approach this by discussing how many children are doing each activity. For example, fourth position does not have to mean that four children are doing the activity.

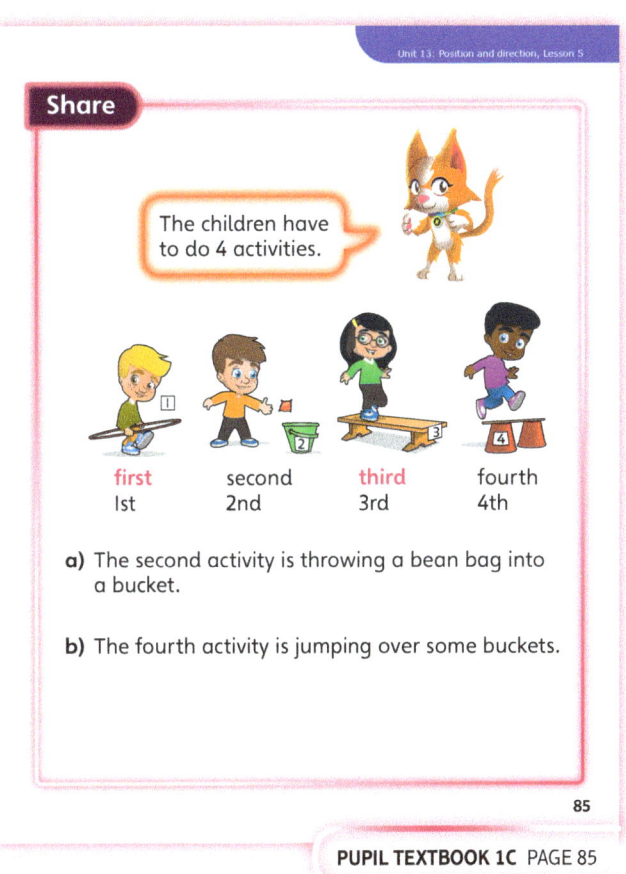

PUPIL TEXTBOOK 1C PAGE 85

Unit 13: Position and direction, Lesson 5

Think together

WAYS OF WORKING Whole class teacher led (I do, We do, You do)

ASK
- Questions ❶ and ❸: *Where do you begin counting from? Can you begin counting in the middle of the group?*
- Question ❶: *How can you tell which child is in the third place?*
- Question ❸: *How far do you need to count if you are looking for the sixth position?*

IN FOCUS Questions ❶ and ❷ develop children's recognition of and ability to count in ordinal numbers. In question ❸, ensure that children are comfortable and confident with where to begin counting on an ordered line.

STRENGTHEN Provide a number track with ordinal numbers positioned on it for children who are still developing their ability to recognise and use ordinal numbers.

DEEPEN When approaching question ❸, discuss what is meant by 'after the sixth person' as opposed to the sixth person. Elicit how Dexter is incorrect, as Tim is the sixth person. Discuss Ash's idea. Ask: *Would circling sixth place help you?* Children might suggest that anyone that comes after the circled person can then be counted as coming after the sixth person.

ASSESSMENT CHECKPOINT Question ❸ should help you to assess if children can use the ordinal numbers to describe the positions of objects within an ordered group.

ANSWERS

Question ❶: Children should point left to right in order and say, 'first, second, third, fourth, fifth'.

Question ❷: There is meat on the fourth pizza.

Question ❸: 1st – Asha; 2nd – Meg; 3rd – Milo; 4th – Joe; 5th – Ola; 6th – Tim; 7th – Jack; 8th – Lou. The 7th person comes after the 6th person. The 7th person is Jack.

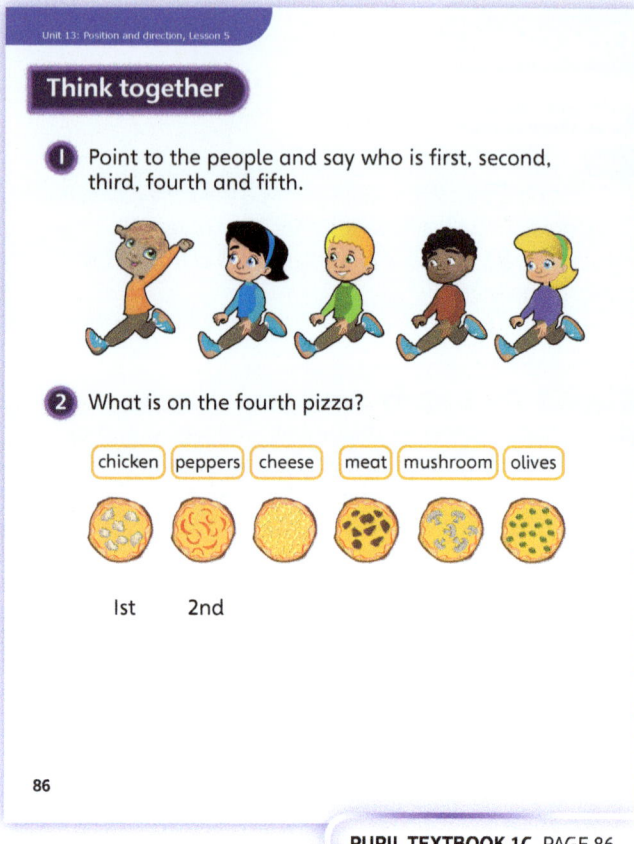

PUPIL TEXTBOOK 1C PAGE 86

PUPIL TEXTBOOK 1C PAGE 87

Unit 13: Position and direction, Lesson 5

Practice

WAYS OF WORKING Independent thinking

IN FOCUS Question ① asks children to recognise different ordinal numbers by circling the relevant object in a line. Question ② requires children to continue a sequence. Encourage them to use the language 1st, 2nd, 3rd, etc. In question ③, children draw their own sequence based on given information about the position of different objects.

STRENGTHEN Provide a number track with ordinal numbers positioned on it for children who are still developing their ability to recognise and use ordinal numbers.

DEEPEN If children use and apply their understanding of ordinal numbers correctly when answering questions ③ and ④, encourage them to create similar puzzles to challenge a partner.

THINK DIFFERENTLY Question ④ requires children to recognise that objects can be counted from different starting points. Discuss as a class where there is a difference.

ASSESSMENT CHECKPOINT Questions ① to ③ should help you assess whether children are confidently identifying which end of the group to start counting from and recognising the position of different objects, using language such as first (1st), second (2nd), etc.

ANSWERS Answers for the **Practice** part of the lesson can be found in the *Power Maths* online subscription.

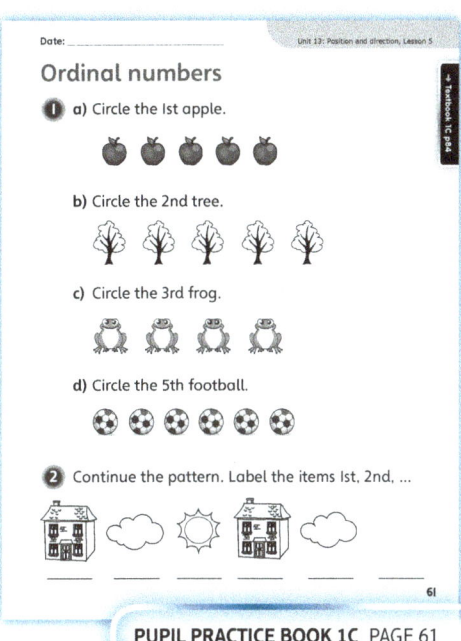

PUPIL PRACTICE BOOK 1C PAGE 61

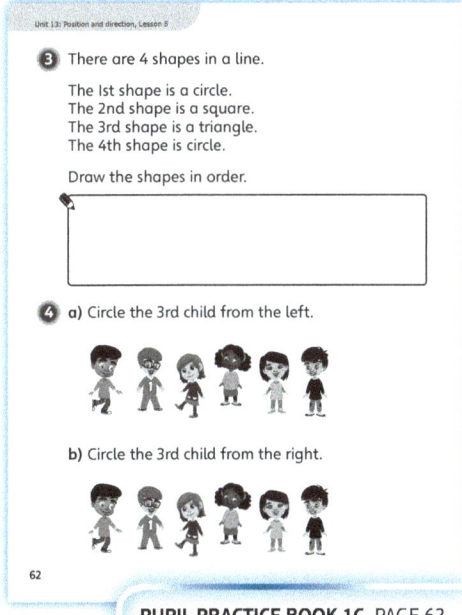

PUPIL PRACTICE BOOK 1C PAGE 62

Reflect

WAYS OF WORKING Independent thinking

IN FOCUS Children are asked to tell the story of what happens when they wake up in the morning, using the language of 'first, second, third and fourth'. The prompts help them with their story and encourage them to use the language of ordinal numbers that they have learnt in the lesson. Ask them to tell their story to a partner.

ASSESSMENT CHECKPOINT Listen for children using the correct language of first, second, third, etc. For example, ensure they do not say that first they leave the house and second they get out of bed. This shows they do not understand the order of ordinal numbers.

ANSWERS Answers for the **Reflect** part of the lesson can be found in the *Power Maths* online subscription.

After the lesson

- Are children fully able to recognise the difference between cardinal numbers and ordinal numbers?
- Did you encounter any misconceptions in this lesson that you did not expect to encounter? How will you approach these misconceptions going forward?

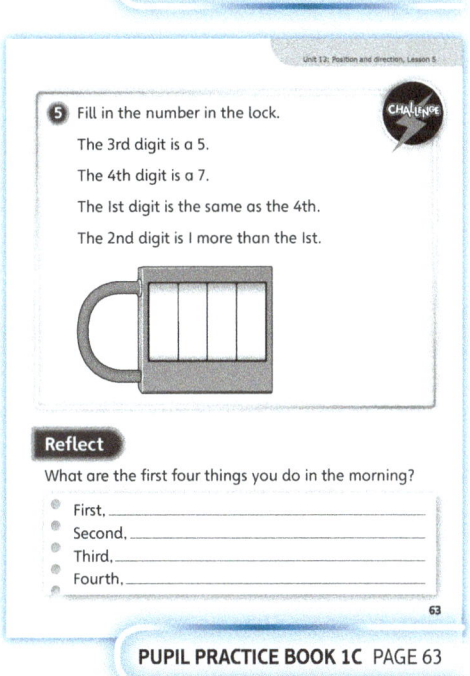

PUPIL PRACTICE BOOK 1C PAGE 63

Unit 13: Position and direction

End of unit check

Don't forget the unit assessment grid in your *Power Maths* online subscription.

WAYS OF WORKING Group work adult led

IN FOCUS
- Questions ① and ② will identify whether children understand the vocabulary associated with describing position in relation to other objects.
- Question ③ will identify whether children understand and can apply the vocabulary used to describe the direction and fraction of a turn.

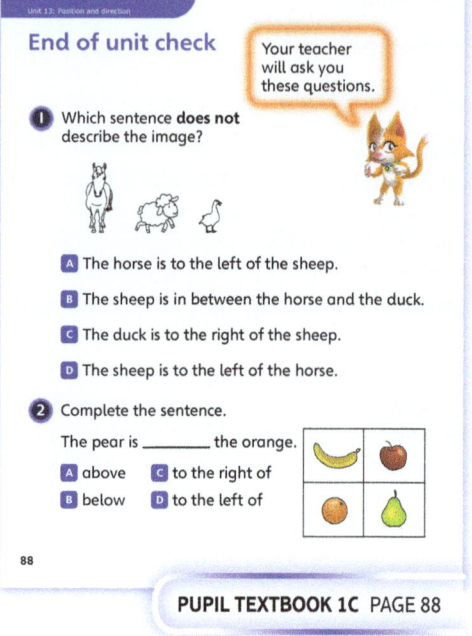

PUPIL TEXTBOOK 1C PAGE 88

Think!

WAYS OF WORKING Pair work or small groups

IN FOCUS
- This question can be used to assess whether children can identify and describe routes to a specific goal.
- Children should use the words at the bottom of the page in their answers. Ask: *What other words related to position and direction do you need to use?*
- Encourage children to think through or discuss different possible routes before writing their answer in **My journal**. You could challenge them to find the longest or shortest route, or to find all the possible routes. You could ask: *Can the mouse visit every square only once?*

ANSWERS AND COMMENTARY Children who have mastered the concepts in this unit will be able to accurately describe linear and rotational movement, and the position of objects, using the words 'forwards', 'backwards', 'left', 'right', 'quarter', 'half', 'three-quarter', 'whole', 'above', 'below' and 'between'. They will be able to follow a series of descriptions or instructions related to position and direction, and give accurate and concise instructions and descriptions.

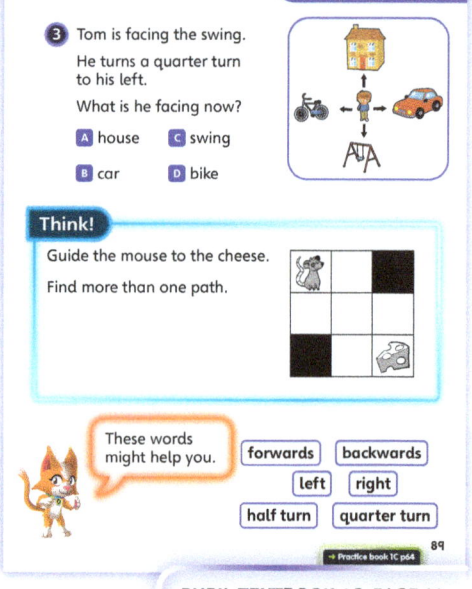

PUPIL TEXTBOOK 1C PAGE 89

Q	A	WRONG ANSWERS AND MISCONCEPTIONS	STRENGTHENING UNDERSTANDING
1	D	A or C suggests that children are not secure with identifying left and right. B suggests that children have not understood the question.	Give children practical opportunities to practise the key language. For example, ask them to follow instructions such as 'Make a quarter turn to your right' or answer questions such as 'Face the window. What turn have you made?' Similarly, ask children to create a simple design using different coloured multilink cubes by following instructions such as 'Put the red cube to the right of the blue cube' or give them a completed design and ask them to describe the position of different cubes.
2	C	A or B suggests that children do not understand positional language.	
3	B	D suggests that children have confused left and right, perhaps because of David's orientation.	

Unit 13: Position and direction

My journal

WAYS OF WORKING Independent thinking

ANSWERS AND COMMENTARY There are four possible paths:
- Go forwards 1 square; make a quarter turn right; go forwards 2 squares; make a quarter turn left; go forwards 1 square.
- Make a quarter turn right; go forwards 1 square; make a quarter turn left; go forwards 2 squares; make a quarter turn right; go forwards 1 square.
- Go forwards 1 square; make a quarter turn right; go forwards 1 square; make a quarter turn left; go forwards 1 square; make a quarter turn right; go forwards 1 square.
- Make a quarter turn right; go forwards 1 square; make a quarter turn left; go forwards 1 square; make a quarter turn right; go forwards 1 square; make a quarter turn left; go forwards 1 square.

If children are still experiencing difficulties, carry out the problem on a grid drawn on the playground with a child as the mouse. As the child moves towards the cheese, encourage them to describe each movement they make. Have prompts available (such as circles shaded to represent different fractions of a turn and pictorial representations of left and right) if necessary.

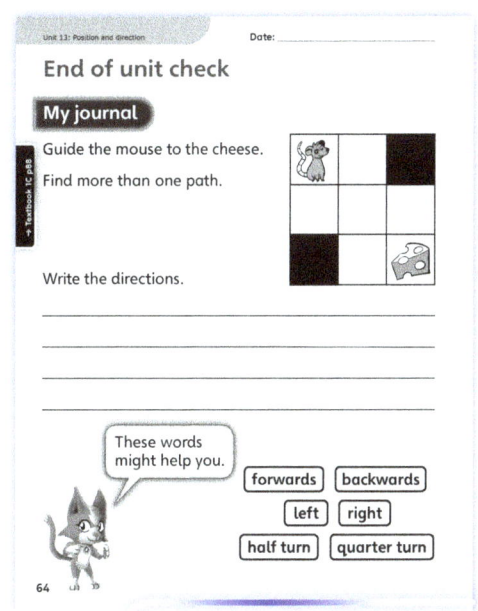

PUPIL PRACTICE BOOK 1C PAGE 64

Power check

WAYS OF WORKING Independent thinking

ASK
- How happy do you feel about following or giving instructions for a route?
- Can you use the vocabulary you learnt in this unit to describe the position of something to a partner?

Power puzzle

WAYS OF WORKING Pair work or small groups

IN FOCUS Use the **Power puzzle** to see if children have a secure understanding of the vocabulary relating to position. Can they apply this knowledge to solve a problem? Make sure children understand that they need to determine the names of the unlabelled children from the clues given. Carefully read each clue to children, emphasising the key vocabulary describing position. Ask children to work in pairs, and tell them that they must both agree before they label each child: this will encourage children to justify their decisions.

ANSWERS AND COMMENTARY If children are successful in completing the puzzle, this suggests that they have mastery in their understanding of positional language. If children are unsuccessful, then they may still be confusing some of the vocabulary.

	Maya	Quinn
Anya	Katie	Bob
Hassan	Shaan	

PUPIL PRACTICE BOOK 1C PAGE 65

After the unit

- Are children secure in their understanding of the vocabulary relating to position?
- Can you find opportunities to use positional language in everyday situations, such as asking children to find resources or organising and tidying the classroom?

Strengthen and **Deepen** activities for this unit can be found in the *Power Maths* online subscription.

127

Unit 14
Numbers to 100

Mastery Expert tip! 'When teaching this unit, I made sure that children were secure on counting, using a number line, finding one more and one less and comparing numbers within 50 before getting too involved in this unit. This turned out to be time well spent.'

Don't forget to watch the Unit 14 video!

WHY THIS UNIT IS IMPORTANT

In this unit, children will develop their understanding of numbers to 100. They will begin by counting beyond 50 and noticing the counting patterns. They will use a 100 square and other pictorial representations. They will count efficiently by counting the 10s first and then the 1s. They will position numbers on a number line and use these to help them compare two 2-digit numbers. Children will also find one more and one less than numbers to 100. An understanding of 2-digit numbers will support children's understanding of, and ability to work with, numbers and the number system.

WHERE THIS UNIT FITS

→ Unit 13: Position and direction
→ **Unit 14: Numbers to 100**
→ Unit 15: Money

This unit builds on children's previous number work, in particular Unit 8: Numbers to 50 and Units 3 and 4: Addition and subtraction within 10, in which they explored number bonds to 10. This unit focuses on the structure of 2-digit numbers that children will count beyond 50. Unit 15 will focus on money.

Before they start this unit, it is expected that children:
- can recognise and use ten frames
- can recognise and use different representations of 10 and 1
- can count on and back in 10s and 1s from 0.

ASSESSING MASTERY

Children who have mastered this unit will be able to confidently count on and back to and from 100. They will be able to confidently discuss the patterns found when counting in 10s and 1s and show them using multiple representations. Children will know that a 2-digit number is made up of 10s and 1s. They will find one more than and one less than a number. They will be able to place numbers on a number line and use their position on a number line or number track to compare numbers.

COMMON MISCONCEPTIONS	STRENGTHENING UNDERSTANDING	GOING DEEPER
Children may miscount when counting across the next 10: for example, saying 78, 79, 70-10 (seventy-ten). They may confuse sevenTEEN and sevenTY.	If children are making these mistakes, provide a 100 square or number track and count aloud with them. When you reach the next 10, say and write the words clearly. You may also want to make the numbers using counters and ten frames or cubes.	Give children riddles that require an understanding of place value. For example: *My number has 6 tens and 7 ones. What could my number be?*
Children may forget to align the start of their concrete or pictorial representations of numbers when comparing them, thereby skewing the comparison.	Offer children something that provides a clear starting point for their comparisons, such as the edge of a page or a blank number line.	

Unit 14: Numbers to 100

UNIT STARTER PAGES

Use these pages to introduce the unit focus to children. You can use the characters to explore different ways of working too.

STRUCTURES AND REPRESENTATIONS

Ten frame: This model will help children visualise 10. It can be used to help them count in 10s and to recall number bonds to 10.

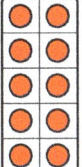

Number line: A number line helps children represent the order of numbers. It can be used to help children count on and back from a given starting point and help them identify patterns within the count.

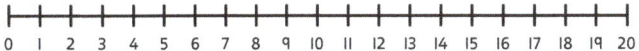

Number track: Like a number line, a number track helps children represent the order of numbers. A number track can support children in counting and in comparing and ordering numbers.

51	52	53	54	55	56	57	58	59	60

100 square: This model will help children visualise 100 and see patterns such as those created when counting one more or one less, or ten more or ten less.

1	2	3	4	5	6	7	8	9	10
11	12	13	14	15	16	17	18	19	20
21	22	23	24	25	26	27	28	29	30
31	32	33	34	35	36	37	38	39	40
41	42	43	44	45	46	47	48	49	50
51	52	53	54	55	56	57	58	59	60
61	62	63	64	65	66	67	68	69	70
71	72	73	74	75	76	77	78	79	80
81	82	83	84	85	86	87	88	89	90
91	92	93	94	95	96	97	98	99	100

KEY LANGUAGE

There is some key language that children will need to know as part of the learning in this unit:

→ 100 square, number square
→ pattern, same, different
→ one less, less than, fewer, smaller, less, (<)
→ one more, greater than, larger, bigger, more, (>)
→ equal to, (=)
→ greatest, biggest
→ fewest, smallest
→ tens (10s), ones (1s)
→ how many?, count

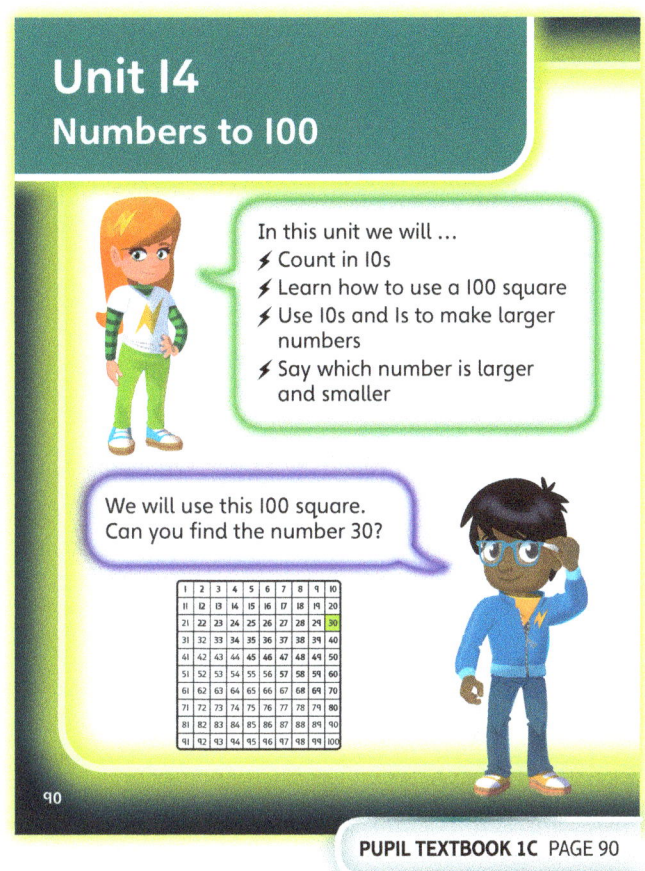

PUPIL TEXTBOOK 1C PAGE 90

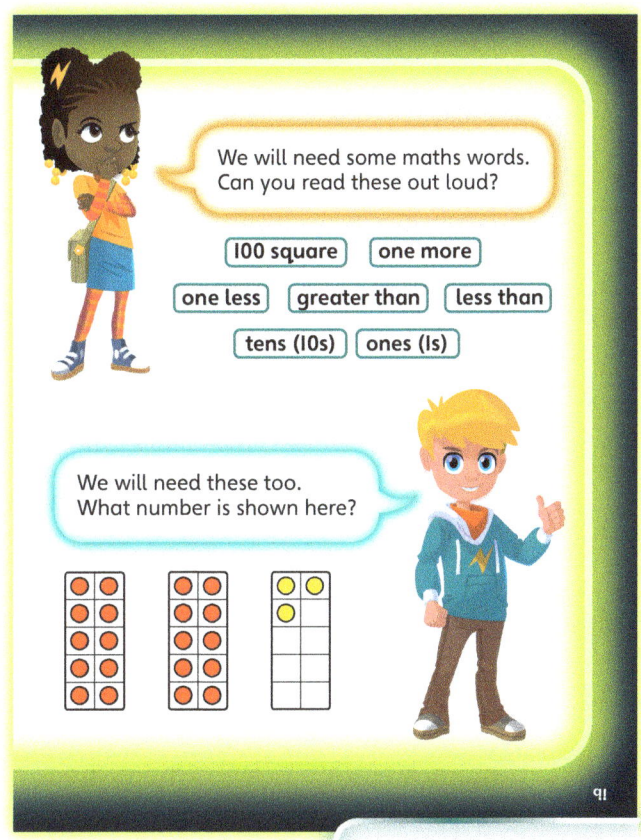

PUPIL TEXTBOOK 1C PAGE 91

Unit 14: Numbers to 100, Lesson 1

Count from 50 to 100

Learning focus
In this lesson, children will develop their ability to count numbers up to 100, supported by number tracks and 100 squares. This lesson focuses on knowing the count of numbers to 100, rather than the structure and representations.

Before you teach
- Can children confidently count to 50?
- Can children count on a number track and complete missing numbers?

NATIONAL CURRICULUM LINKS

Year 1 Number – number and place value

Count, read and write numbers to 100 in numerals; count in multiples of twos, fives and tens.

ASSESSING MASTERY

Children can count on and back to 100 reliably. Children can identify missing numbers in a sequence and can continue a count from any 1- or 2-digit number. Children can count to 100 on a 100 square and can find missing numbers.

COMMON MISCONCEPTIONS

Children may miscount when counting across the next 10: for example, saying 78, 79, 70-10 (seventy-ten). If children are making this mistake, provide a 100 square or number track and count aloud with them. When you reach the potential next 10, ask:
- *Can you read that number? How is it different from the numbers that came before it?*

STRENGTHENING UNDERSTANDING

Use the 100 square to count to 100 with children. Discuss the patterns in the count and what happens when children get to the next 10. Reinforce counting with children daily, if required.

GOING DEEPER

Children may be able to count on in 10s from 0 and recognise the pattern associated with this skill. Deepen their thinking by asking them if they can count in 1s or 10s from a number other than 0. Ask: *Can you count from 13 to 53? Do you notice any patterns when you count in 10s?*

KEY LANGUAGE

In lesson: **100 square**, count, number names from 1 to 100

Other language to be used by the teacher: ones

STRUCTURES AND REPRESENTATIONS

100 square, number track

RESOURCES

Mandatory: large printed 100 squares, counters

Optional: blank 100 squares

 In the eTextbook of this lesson, you will find interactive links to a selection of teaching tools.

Quick recap

Check that children can count to 50 confidently. Ask them to take it in turns to say the next number. Can the class count to 50 accurately within 2 minutes? Encourage children to support one another if they are unsure.

Unit 14: Numbers to 100, Lesson 1

Discover

WAYS OF WORKING Pair work

ASK

- Question 1 a): *Why do you think it is called a 100 square? What number does it start on? What number does it end on? How can you use it to help with your counting?*
- Question 1 b): *What number has your partner covered up?*

IN FOCUS In question 1 a), children learn to use a 100 square to help them with their counting. Discuss why they think it is called a 100 square. Ask them to use it to count by pointing. Once they have developed confidence with the 100 square, move onto question 1 b). Here they are asked to cover up a number with a counter and to see if a partner can work out what the covered numbers is. Repeat this activity.

PRACTICAL TIPS Give children 100 squares that are big enough to allow counters to cover the individual numbers.

ANSWERS

Question 1 a): Count across the row and then go back and start the next row. Continue in this way from 1 to 100.

Question 1 b): Children's answers will vary; ensure they identify a hidden number between 1 and 100.

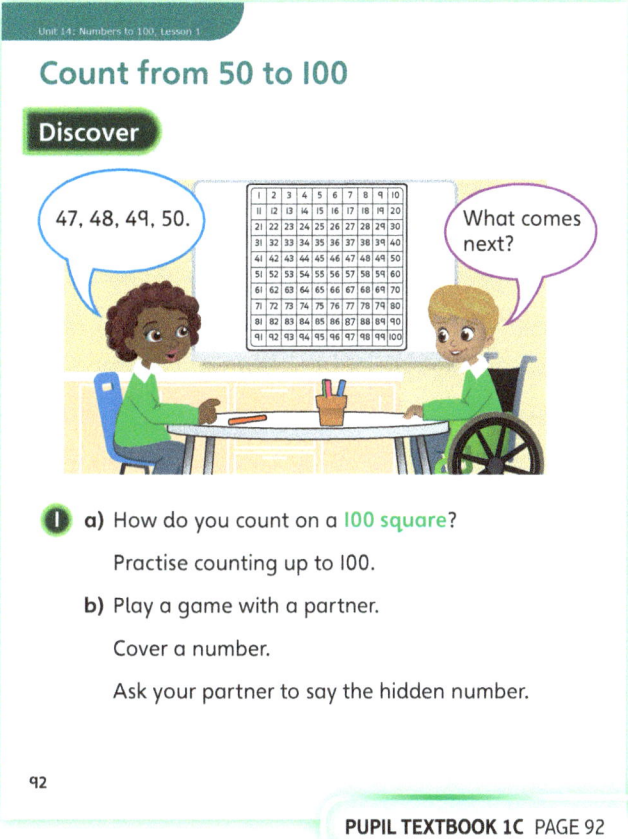

PUPIL TEXTBOOK 1C PAGE 92

Share

WAYS OF WORKING Whole class teacher led

ASK

- Question 1 a): *How can you use the 100 square to help you with counting?*
- Question 1 b): *What number has your partner covered? How do you know? What things help you work out the number? What patterns do you notice?*

IN FOCUS For question 1 a), count as a class, using the 100 square. Ask children to point and count as they go. If they are confident with numbers to 50 already, you might want to focus on the numbers from 50 to 100. Ask children what happens when they reach the end of a line on the 100 square. Then ask them what number their partner has covered. Discuss how they can work it out by looking at the numbers around the covered number. You might want to discuss other patterns they can see on the 100 square.

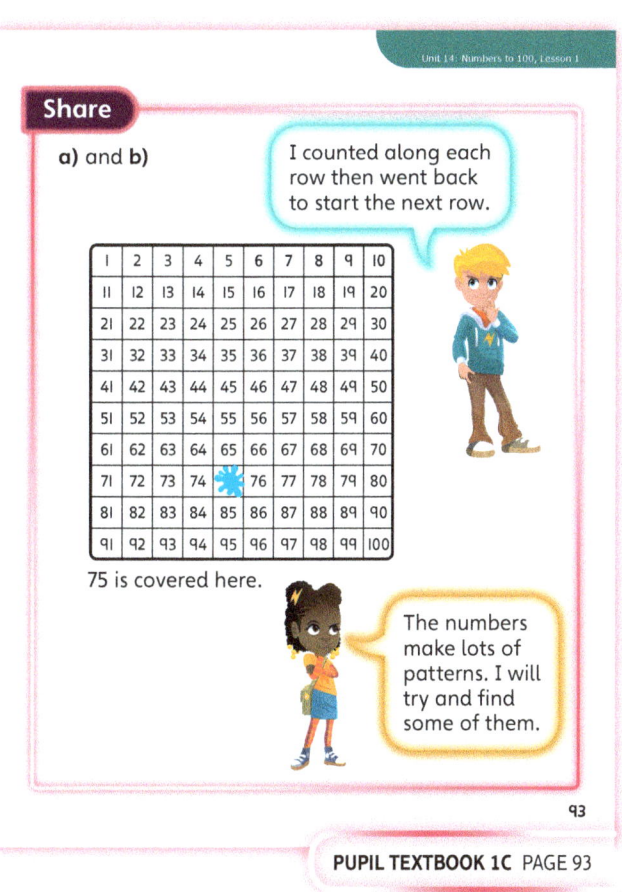

PUPIL TEXTBOOK 1C PAGE 93

131

Unit 14: Numbers to 100, Lesson 1

Think together

WAYS OF WORKING Whole class teacher led (I do, We do, You do)

ASK
- Question ❶: *Have you seen these numbers before? Where would they be on a 100 square?*
- Question ❷: *What numbers are covered? How do you know?*

IN FOCUS Question ❶ uses a number track to help children with their counting. Discuss how each number track is one of the rows from the 100 square. Count aloud and ask children to point as they count. In question ❷, children need to work out the missing numbers. Discuss methods they used to approach this. For example, did they have to count from 1 or did they know some of the numbers without needing to count all the way from 1? For question ❸, you may want to model the game with children before they play it.

STRENGTHEN Use a 100 square to count aloud to 100 with children. Discuss the patterns in the count and what happens when children get to the next 10.

DEEPEN Deepen children's understanding of the 100 square by asking them what patterns they can see. For example, all the numbers in the fifth row start with a 4 (except for 50); all the numbers in the fifth column end with a 5, etc. Discuss how this can help them work out a missing number without having to count on from a nearby number. You could then give children a blank 100 square and ask them to label the numbers 42, 75 and 19.

ASSESSMENT CHECKPOINT Use questions 1 and 2 to check that children can count aloud from 50 to 100 and work out missing numbers in a counting sequence.

ANSWERS

Question ❶ a): Children should accurately count from 51 to 60.

Question ❶ b): Children should accurately count from 61 to 70.

Question ❷: The hidden numbers are 9, 31, 59, 71, 79, 81 and 99.

Question ❸: Children should accurately count on 1, 2 or 3 places each time, from 0 to 100.

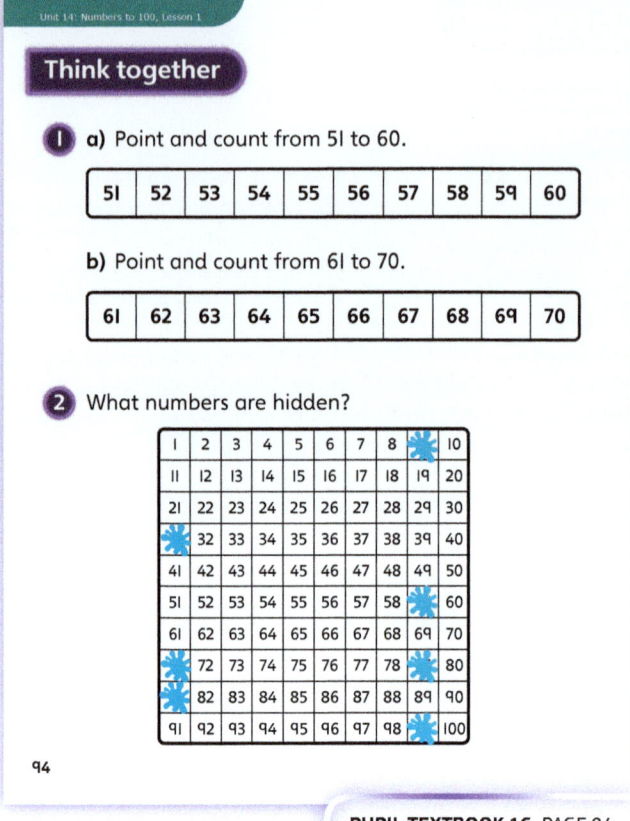

PUPIL TEXTBOOK 1C PAGE 94

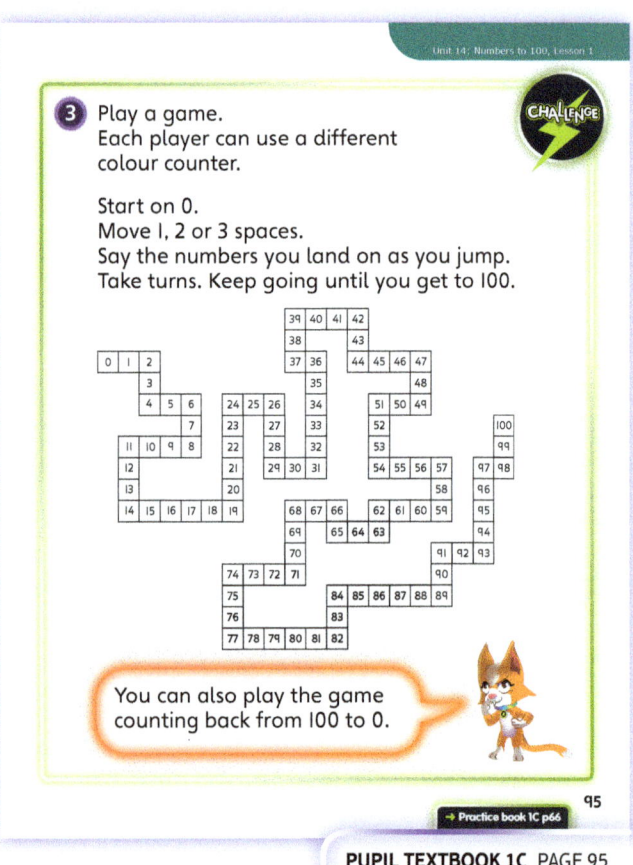

PUPIL TEXTBOOK 1C PAGE 95

Unit 14: Numbers to 100, Lesson 1

Practice

WAYS OF WORKING Independent thinking

IN FOCUS In question 1, children work out the missing numbers on the number tracks. They should recognise that the number tracks are rows from the 100 square. Look for children who understand the counting sequence. In question 2, children find missing numbers on a 100 square. They may need to count to find the missing numbers. Encourage them not to count from 1 each time, but to pick a number close to the missing number. Question 4 uses knowledge of the count to 100. Children will know if they are correct as they will draw a rocket.

STRENGTHEN For question 1, you may want to have 100 squares available for children to help them with their counting. For question 2, provide support for children by starting to count from a number close to the missing number.

DEEPEN Deepen children's understanding of the 100 square by asking them what patterns they see on the grid. For example, all the numbers in the seventh row start with a 6 (except for 70); all the numbers in the eighth column end with an 8, etc. Discuss how this can help them work out the missing numbers in question 2 without having to count on from a nearby number. You could then give them a blank 100 square and ask them to write the numbers 65, 28 and 11.

THINK DIFFERENTLY In question 3 a), children have to count on from a number that is not at the start of a row on the 100 square. In question 3 b), children have to count back instead of counting on. Children who are struggling may want to fill the numbers from right to left.

ASSESSMENT CHECKPOINT Questions 1 and 2 will help you see how confident children are with counting from 1 to 100.

ANSWERS Answers for the **Practice** part of the lesson can be found in the *Power Maths* online subscription.

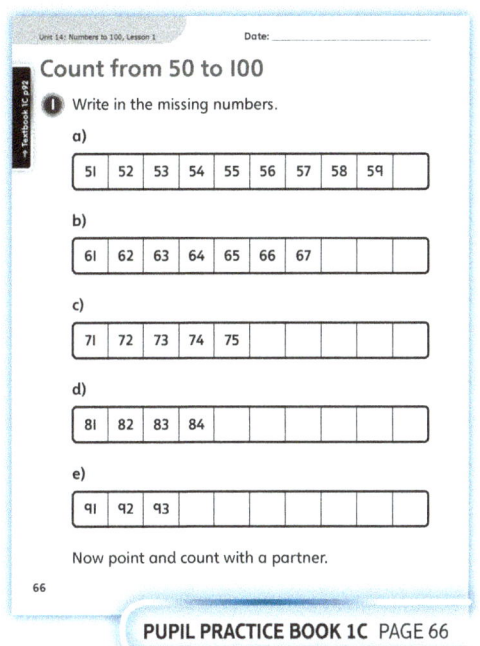

PUPIL PRACTICE BOOK 1C PAGE 66

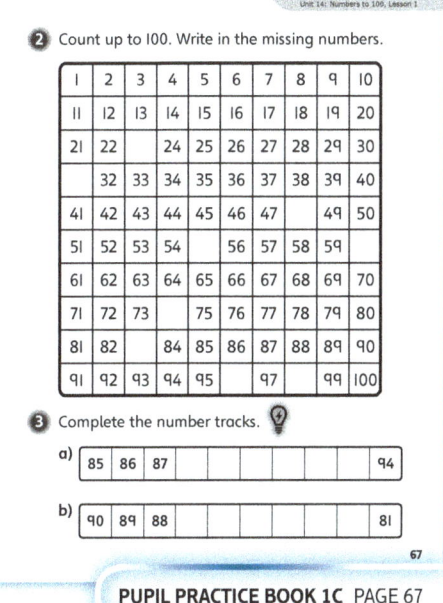

PUPIL PRACTICE BOOK 1C PAGE 67

Reflect

WAYS OF WORKING Whole class

IN FOCUS Discuss with children whether they find counting on or counting back easier. Most will reflect that counting on is easier. This question provides an opportunity for children to practise counting on and back as a class. Count on from 50 to 60 and then back to 50.

ASSESSMENT CHECKPOINT Check whether children can count on from 50, with and without support from number tracks and a 100 square. Check if they can count back from 80 to 70, without the support of a number track. Being able to confidently count on and back deepens children's understanding.

ANSWERS Answers for the **Reflect** part of the lesson can be found in the *Power Maths* online subscription.

PUPIL PRACTICE BOOK 1C PAGE 68

After the lesson

- Can children confidently count from 50 to 100?
- Can children find missing numbers on a number track and a 100 square?

133

Unit 14: Numbers to 100, Lesson 2

10s to 100

Learning focus
In this lesson, children count on in 10s, from 0 to 100. They recognise that these numbers all end in a 0.

Before you teach
- Can children count in 1s from 50 to 100?
- Do children know where different numbers are on a 100 square?
- Can children count in 10s from 0 to 50?

NATIONAL CURRICULUM LINKS

Year 1 Number – number and place value

Count, read and write numbers to 100 in numerals; count in multiples of twos, fives and tens.

ASSESSING MASTERY

Children can count on confidently in 10s, from 0 to 100. They know that all the numbers they say end in 0 and the majority have '-ty' at the end of the word. Children can count in 10s from any 10 and they begin to be able to count back in 10s, too.

COMMON MISCONCEPTIONS

Some children may not be confident counting full ten frames and digits in 10s and may resort to counting in 1s. In order to support these children, count aloud in 10s with them, up to 50. Ask:
- What digit changes when we count in 10s?

Some children may also confuse fourTEEN with forTY. Show the difference between the two numbers by making the numbers on ten frames. Discuss the end of the words that are multiples of 10. Ask:
- What is different about these two numbers?
- What do you notice about the end of the words in the 10s count?

STRENGTHENING UNDERSTANDING

Secure the counting sequence with children using a number track or 100 square. Children say the words out loud and count with you. Point to each number as you say it. Next to each multiple of 10, show a representation. You could use a bead string or rekenrek to help them.

GOING DEEPER

Discuss what children notice about the numbers they are saying in the 10s count. Ask: *Where do these numbers appear on the 100 square? What do you notice about the starting number? What do you notice about the second digit?* Use the 100 square to count in 10s from a number that does not end in a 0. Ask: *Can you count in 10s from 13? What do you notice about the numbers?*

KEY LANGUAGE

In lesson: count, 10s, 100 square

Other language to be used by the teacher: count on, count back, multiple

STRUCTURES AND REPRESENTATIONS

100 square

RESOURCES

Optional: large printed 100 squares, counters, ten frames, bead strings, rekenreks, cubes

 In the eTextbook of this lesson, you will find interactive links to a selection of teaching tools.

Quick recap

Display a large 100 square on the board. Ask one child to come up to the board and the rest of the class to close their eyes. The child selected has to cover up one of the numbers before the other children open their eyes and guess the missing number, writing it down on their mini-whiteboards. Repeat for a few rounds. You could challenge children to cover more than one number.

Unit 14: Numbers to 100, Lesson 2

Discover

WAYS OF WORKING Pair work

ASK

- Question 1 a): *What did we call the grid in the previous lesson? What patterns do you remember about the grid? What numbers are covered?*
- Question 1 b): *How can you show 10 with your fingers or using cubes?*

IN FOCUS Children should recognise the 100 square from the previous lesson, where they used it to count in 1s. In question 1 a), ask them, in pairs, to work out the missing numbers. They may need to count in 1s across every row. Some children may know the numbers immediately. If so, ask them to discuss what they notice about the numbers. Once children understand that these are the 10s, they should try making 10s and counting in 10s with other children on their table.

PRACTICAL TIPS You could make a large 100 square on the floor in the hall or in the playground and allow children to move along it and up and down it, counting as they go.

ANSWERS

Question 1 a): All of the 10s are covered: 10, 20, 30, 40, 50, 60, 70, 80, 90 and 100.

Question 1 b): Show 10 on two hands. Count aloud: 10, 20, 30, 40, 50, 60, 70, 80, 90, 100.

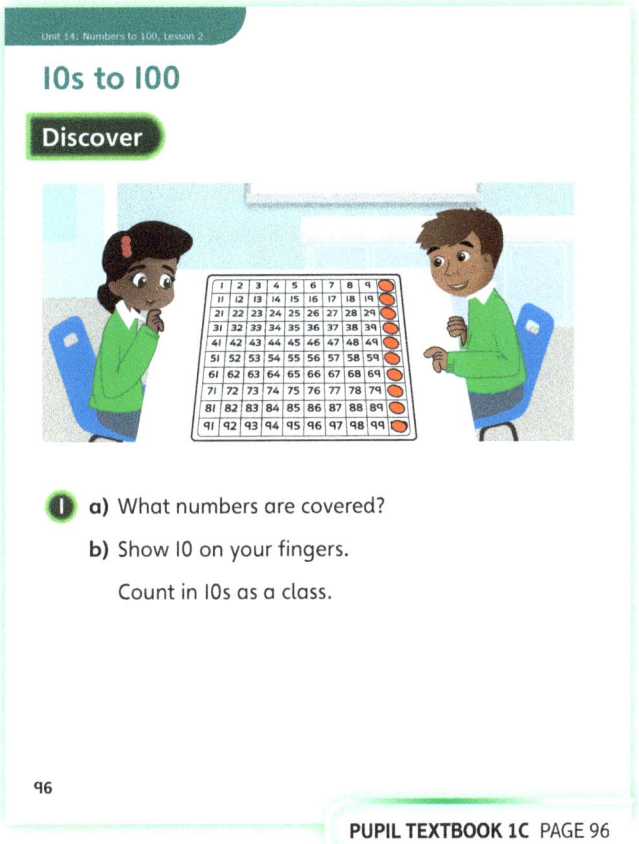

PUPIL TEXTBOOK 1C PAGE 96

Share

WAYS OF WORKING Whole class teacher led

ASK

- Question 1 a): *What are the missing numbers? How did you work them out? What do you notice about all the numbers that were covered?*
- Question 1 b): *How can you show me 10 on your fingers or using cubes? Can you count in 10s from 0 to 100?*

IN FOCUS In question 1 a), discuss how children could have worked out the missing numbers. Stress the importance of counting on, not counting back, and also stress that it is better not to count in 1s along every row to work these out. Reinforce the 10, 20, 30 … counting sequence by counting aloud with the class. In question 1 b), reinforce this further: ask children to hold up 10 of their fingers or arrange 10 cubes and then go around the class pointing to the number on the 100 square with children counting out loud. Repeat the count several times to help children. Stress the '-ty' on the ends of the words.

PUPIL TEXTBOOK 1C PAGE 97

Unit 14: Numbers to 100, Lesson 2

Think together

WAYS OF WORKING Whole class teacher led (I do, We do, You do)

ASK

- Question ❶: *Why is it easier to count in 10s than 1s? Why can we count in 10s?*
- Question ❷: *Can you see each 10? How do you know each ten frame shows 10? Can you continue the count?*

IN FOCUS In question ❶, discuss the difference between counting in 1s and counting in 10s. Explain that counting in 10s is easier and quicker than counting in 1s. Show the two methods and the time it takes to get the answer. In question ❷, children use ten frames to support counting from 0 in 10s. Encourage children, in pairs, to count aloud and point at each ten frame as they go. In question ❸, children are given groups of other 10s, including in question ❸ b) where the 10 sweets in each bag are not countable. Some children may struggle with this concept at this stage, but point out the '10' on each bag to help them.

STRENGTHEN In order to support children, count with them from 0 to 50 in 10s. Children need to know the 10s sequence confidently and should be encouraged to count aloud regularly.

In question ❸ b), if children want to count in 1s because they cannot see the sets of 10 sweets in the bags, then ask them to make each bag using counters on a ten frame.

DEEPEN Ask children to count in 10s using a 100 square from a non-multiple of 10. For example, can children use the 100 square to count in 10s from 17? What numbers do they say? What do they notice about the numbers? Where are those numbers on the 100 square?

ASSESSMENT CHECKPOINT Use the questions to determine whether children can count confidently and accurately from 0 to 100 in 10s.

ANSWERS

Question ❶: The children are showing 30.

Question ❷: Children should count: 10, 20, 30, 40, 50.

Question ❸ a): There are 70 flowers altogether.

Question ❸ b): There are 100 sweets altogether.

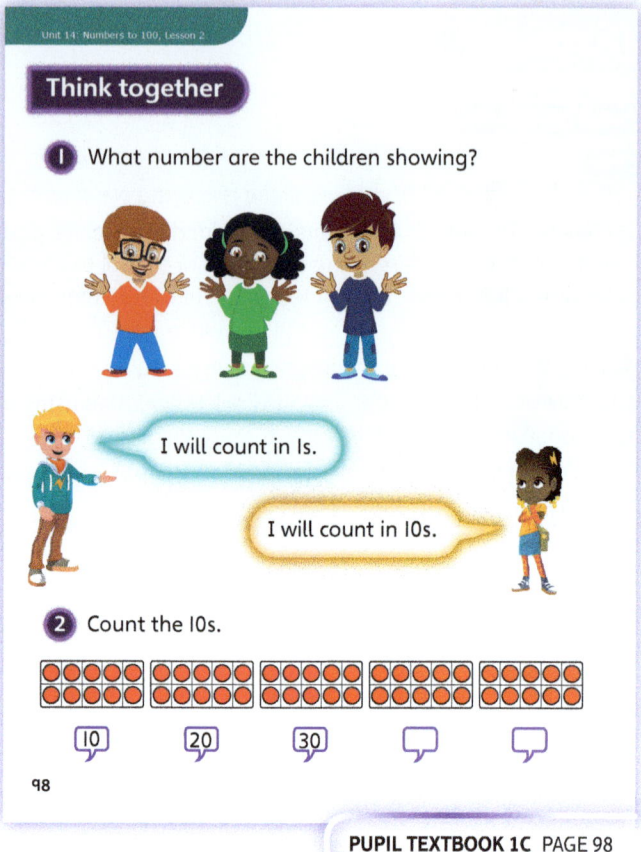

PUPIL TEXTBOOK 1C PAGE 98

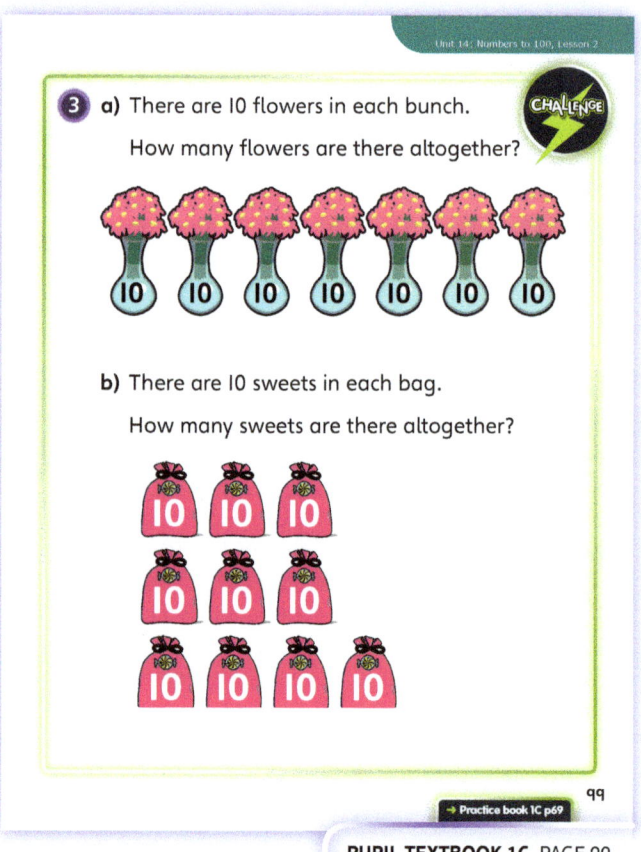

PUPIL TEXTBOOK 1C PAGE 99

Unit 14: Numbers to 100, Lesson 2

Practice

WAYS OF WORKING Independent thinking

IN FOCUS In question ❶, children have to complete the 100 square with the multiples of 10 that they have been reinforcing during the main lesson. In question ❷, they need to focus on the 10s on a number track and complete the count. In questions ❸ and ❹, children count different representations of 10s, again reinforcing the count in 10s from 0 to 100. Encourage them to say the numbers out loud as they count. Count slowly to make sure they do not count too fast and miss out one of the 10s.

STRENGTHEN To support children, provide them with a full 100 square, so that they can see all the 10s and use that to help them count in 10s. In question ❺, if children want to count in 1s because they cannot see all the 10s, then ask them to make each bag or box using counters on a ten frame.

DEEPEN Ask children to count in 10s using a 100 square from a non-multiple of 10. For example, can children use the 100 square to count in 10s from 23? What numbers do they say? What do they notice about the numbers? Ask them to shade in the numbers.

ASSESSMENT CHECKPOINT If children have mastered counting in 10s, then they should be efficient and quick at completing the early questions. This shows that they know the sequence well and they can recognise the 10s.

ANSWERS Answers for the **Practice** part of the lesson can be found in the *Power Maths* online subscription.

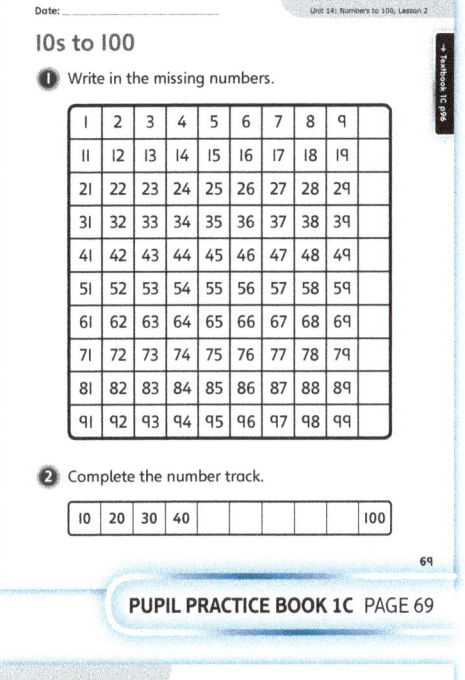

PUPIL PRACTICE BOOK 1C PAGE 69

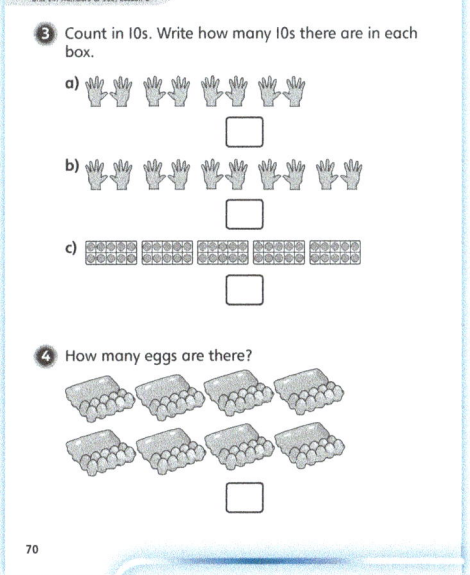

PUPIL PRACTICE BOOK 1C PAGE 70

Reflect

WAYS OF WORKING Group work

IN FOCUS As a group, children count in 10s, starting from 0. As they count, encourage children to show 10 on their fingers or using cubes. They should count together, supporting each other.

ASSESSMENT CHECKPOINT Check that children say the 10s correctly and do not miss any out.

ANSWERS Answers for the **Reflect** part of the lesson can be found in the *Power Maths* online subscription.

After the lesson

- Do children know the count in 10s from 0 to 100?
- Can children count different representations of 10 up to 100?
- Do children know where the 10s lie on a 100 square?

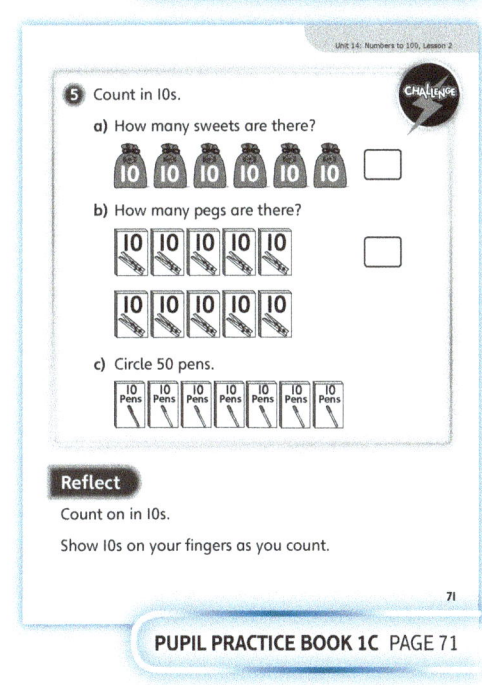

PUPIL PRACTICE BOOK 1C PAGE 71

Unit 14: Numbers to 100, Lesson 3

Partition into 10s and 1s

Learning focus
In this lesson, children will consolidate their understanding of counting numbers in 10s and 1s and will learn to partition numbers.

Before you teach
- How will you make sure children have sufficient opportunities to work with concrete and pictorial representations of 100 squares and number tracks?
- What resources will you provide to help children demonstrate their thinking?

NATIONAL CURRICULUM LINKS

Year 1 Number – number and place value

Identify and represent numbers using objects and pictorial representations including the number line, and use the language of: equal to, more than, less than (fewer), most, least.

Year 2 Number – number and place value

Recognise the place value of each digit in a 2-digit number (tens, ones).

ASSESSING MASTERY

Children can use different representations of numbers to fluently partition them into their 10s and 1s. They can use number squares and ten frames to clearly and efficiently record their partitioning.

COMMON MISCONCEPTIONS

When recording their partitioning, children may record the 10s as the full 10s number, rather than the number of 10s. For example, they may write 36 as 30 tens and 6 ones, rather than 3 tens and 6 ones. Ask:
- *What does the number 36 look like on a 100 square? Can you make the number with cubes? How many groups of 10 are there?*

STRENGTHENING UNDERSTANDING

Strengthen understanding by providing children with practical opportunities to group amounts into 10s and 1s. For example, can they sort classroom resources into groups of 10? How many extra are left after grouping? So, how many do they have altogether?

GOING DEEPER

Deepen children's ability to partition numbers by asking them to partition a number in a different way. Children can work systematically to find all the different ways to partition a given number.

KEY LANGUAGE

In lesson: partition, how many?, how many more?, count, tens, ones, exchange

Other language to be used by the teacher: column, digits

STRUCTURES AND REPRESENTATIONS

100 square, ten frames

RESOURCES

Mandatory: ten frames, counters, blank 100 squares

Optional: bead strings, interlocking cubes, 100 square

 In the eTextbook of this lesson, you will find interactive links to a selection of teaching tools.

Quick recap

Ask children to make 10 on a ten frame using counters. Then ask them to count in 10s on their table. What number did they get to? How many 10s did they make? Ensure children are confident counting in 10s.

Unit 14: Numbers to 100, Lesson 3

Discover

WAYS OF WORKING Pair work

ASK

- Question ① a): *What data is the child recording?*
- Question ① a): *Do you recognise anything in the picture from any other lessons? Can you use that to count the sunny days more quickly?*
- Question ① b): *How many more days until there have been 50 days of sun?*

IN FOCUS Use question ① a) to recap children's learning from the previous lesson. They should recognise the grid as a 100 square. Discuss how they can use this to help them count more quickly, linking the picture to counting in 10s and 1s in question ① b).

PRACTICAL TIPS Recreate the scenario by giving groups of children blank 100 squares and counters. Ask them to fill in the number of counters that match the number of suns. This will help them see how many 10s there are and how many 1s are left over.

ANSWERS

Question ① a): Children should count in 1s. There have been 31 days of sun.

Question ① b): Children should count in 10s and then 1 extra: 10, 20, 30, 31. There have been 31 days of sun.

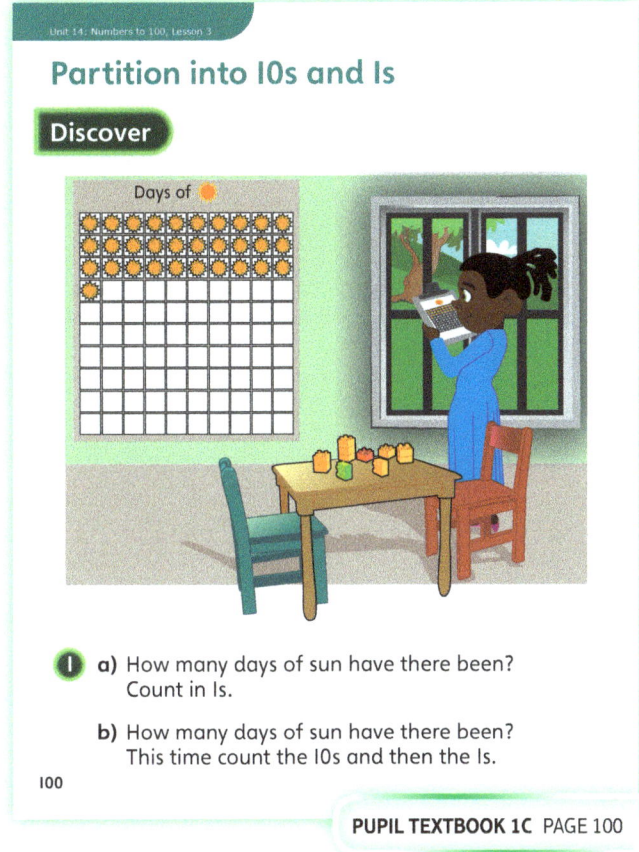

PUPIL TEXTBOOK 1C PAGE 100

Share

WAYS OF WORKING Whole class teacher led

ASK

- Question ① a): *How has Dexter counted the suns? What did he count in?*
- Question ① b): *How has Flo counted the suns? What did she count in first? How do you know there are 10 suns in each row?*

IN FOCUS For question ① a), count in 1s with children, pointing as you count. Count in 1s going across and then move to the next line, like you do on a 100 square. It is important that you take time counting so that children can contrast this with counting in 10s and 1s. In Question ① b), count in 10s and then the extra 1. Discuss with children why it is better to count in 10s and then 1s as opposed to just in 1s. Children should see it is more efficient to count in 10s.

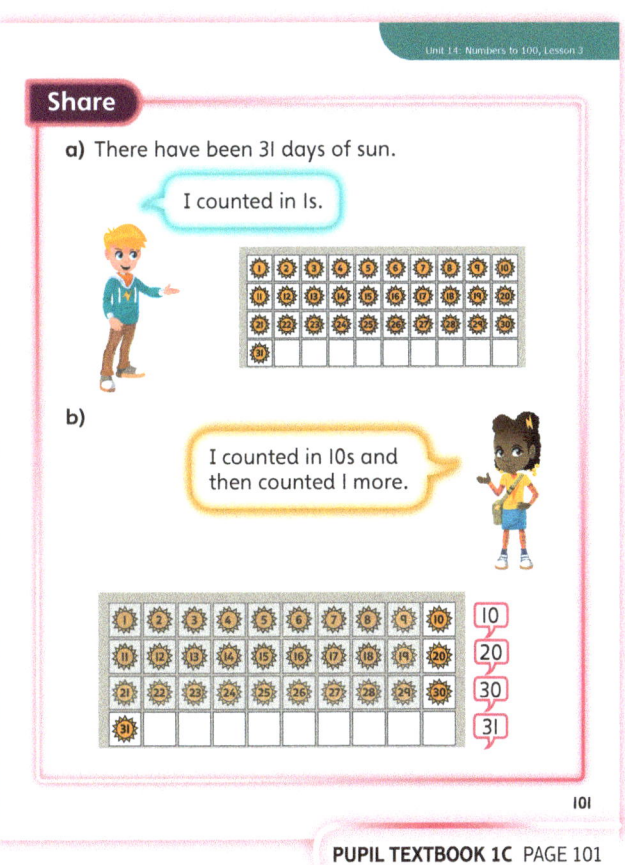

PUPIL TEXTBOOK 1C PAGE 101

139

Unit 14: Numbers to 100, Lesson 3

Think together

WAYS OF WORKING Whole class teacher led (I do, We do, You do)

ASK

- Question ❶: *How will you count how many days of rain there have been? How can you use a 100 square to help you?*
- Question ❷: *How many counters are there in each full ten frame? How can you count the counters?*
- Question ❸: *Why do you think the stars have been circled? Why does this make it easier to count them?*

IN FOCUS Question ❶ is an opportunity to discuss how it is easier to count in 10s and 1s rather than just in 1s. Children should see that it is quicker and more efficient to count in 10s first. In question ❷, children are given a ten frame representation. They again practise counting the 10s first and then the 1s. In question ❸, children should count the stars that are grouped in 10s first and then the 1s. Discuss why, if they group objects into 10s, it makes it easier to count them.

STRENGTHEN Support children counting in 10s then 1s by first ensuring that children are confident counting in 10s. As children count the 10s, encourage them to take it slowly and point as they count. This ensures a correspondence of words. Discuss with children why it is easier to count in 10s and 1s rather than just in 1s.

DEEPEN Ask children to make different numbers on ten frames – for example, 52. Ask them how many 10s there will be and then ask how many 1s. To deepen understanding, children will start to see that a 2-digit number is made up of some 10s and some 1s.

ASSESSMENT CHECKPOINT Use these questions to check whether children can count efficiently and accurately by counting the 10s first and then the 1s, instead of just counting the 1s.

ANSWERS

Question ❶: There have been 62 rainy days.
Question ❷: Children should count: 10, 20, 30, 40, 50, 51, 52.
Question ❸ a): There are three 10s. There are four extra 1s.
Question ❸ b): There are 34 stars altogether.

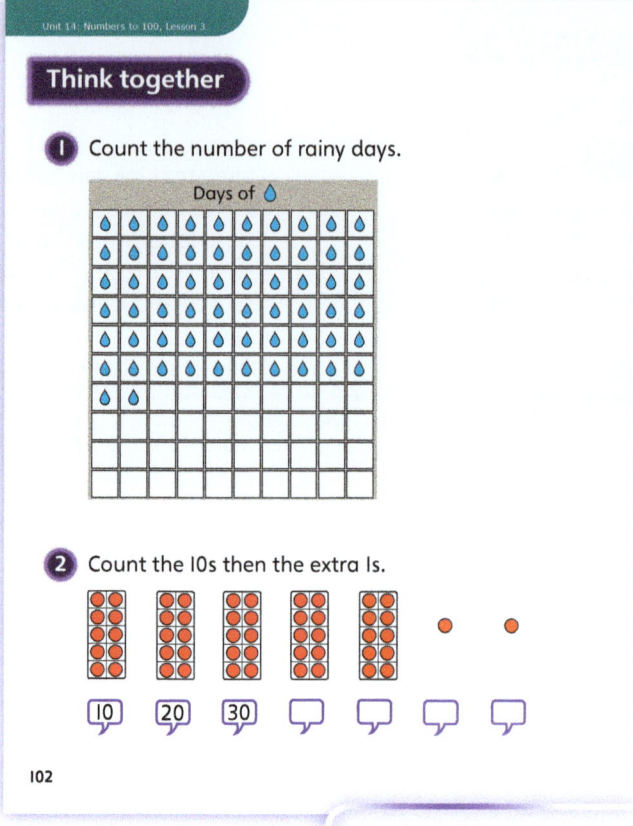

PUPIL TEXTBOOK 1C PAGE 102

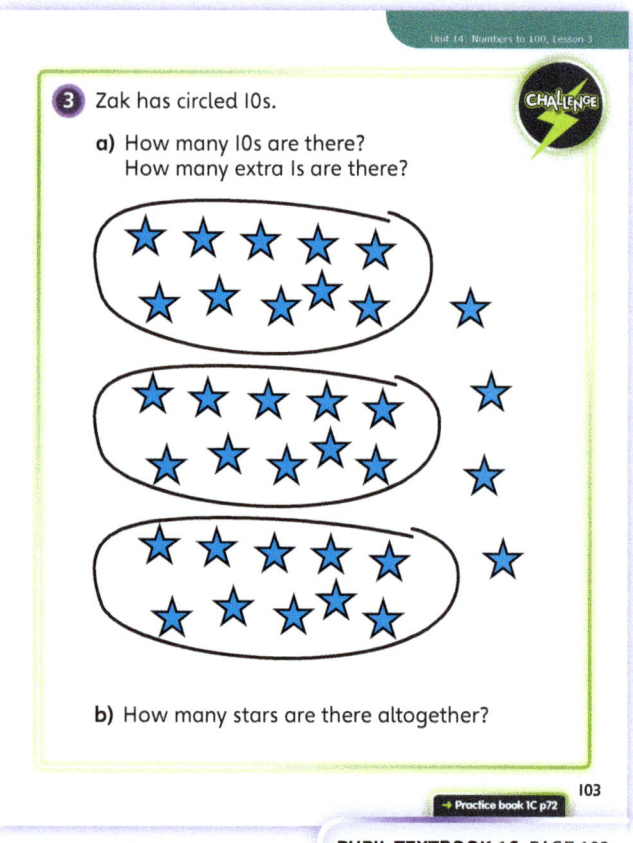

PUPIL TEXTBOOK 1C PAGE 103

Unit 14: Numbers to 100, Lesson 3

Practice

WAYS OF WORKING Independent thinking

IN FOCUS Question ❶ recaps the previous lesson's teaching of using 100 squares to help children count in 10s. Look for children counting in 10s first and then 1s, rather than only counting in 1s. In question ❷, children count in 10s and 1s using ten frame representations. Question ❸ reinforces the same concept using packs of 10 pens and single pens. In question ❹, children have to make their own 10s before counting.

STRENGTHEN In question ❶, children may not realise straight away that there are 10 counters in each row as the rows do not have numbers like the 100 squares they have previously seen. You may want to count the first 10 to check. Once you have counted the first 10, explain to children that they can then count in 10s and 1s. It is important that children can count confidently in 10s first.

DEEPEN Ask children to make different numbers on ten frames – for example, 37. Ask them how many 10s it will be and then how many 1s. To deepen understanding, children will start to see that a 2-digit number is made up of some 10s and some 1s.

ASSESSMENT CHECKPOINT Use these questions to assess children's confidence in counting in 10s and 1s. Check whether they can count efficiently and accurately by counting the 10s first and then the 1s, instead of just counting the 1s.

ANSWERS Answers for the **Practice** part of the lesson can be found in the *Power Maths* online subscription.

Reflect

WAYS OF WORKING Pair work

IN FOCUS Children may want to make the numbers using counters on ten frames to help them with this question. They should discuss with a partner how many counters there are in total. They should discuss that there are some full 10s and then also two 1s. Look for children who still rely on only counting the 1s.

ASSESSMENT CHECKPOINT Check that children have worked out how many counters there are in total by first counting the 10s and then the 1s.

ANSWERS Answers for the **Reflect** part of the lesson can be found in the *Power Maths* online subscription.

After the lesson

- Are children confident when explaining the use of 100 squares and ten frames in partitioning numbers?
- Do children know that it is more efficient to count in 10s first and then in 1s, rather than only counting the 1s?

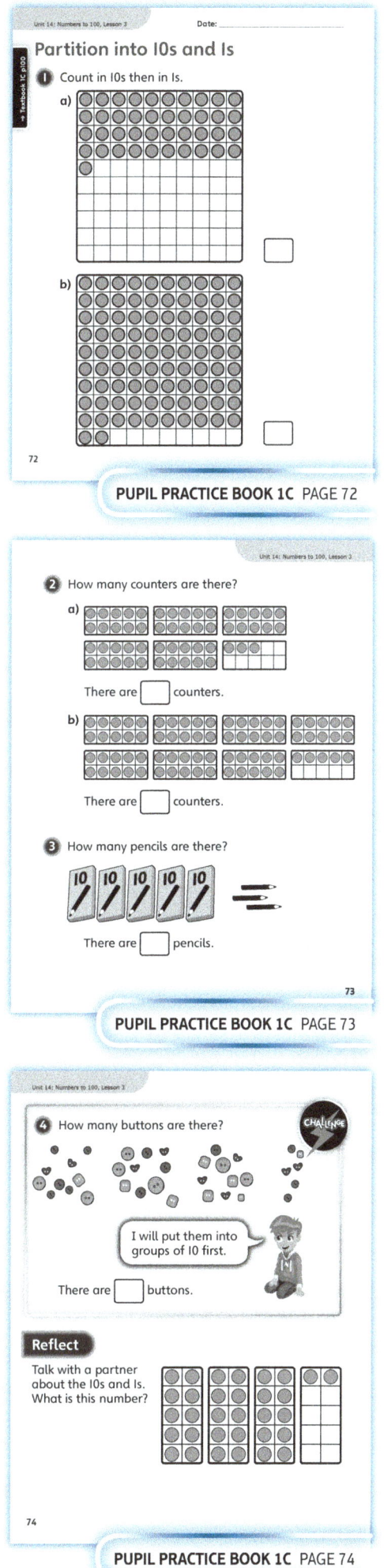

PUPIL PRACTICE BOOK 1C PAGE 72

PUPIL PRACTICE BOOK 1C PAGE 73

PUPIL PRACTICE BOOK 1C PAGE 74

141

Unit 14: Numbers to 100, Lesson 4

Number line to 100

Learning focus
In this lesson, children will explore the number line to 100. They will find missing numbers and place numbers in their correct position.

Before you teach
- Do children have an understanding of the number line from 0 to 20?
- Do children know how to find missing numbers on a number line?

NATIONAL CURRICULUM LINKS

Year 1 Number – number and place value

Identify and represent numbers using objects and pictorial representations including the number line, and use the language of: equal to, more than, less than (fewer), most, least.

ASSESSING MASTERY

Children can find missing numbers on a number line. They can also work out what an arrow is pointing to and determine where a number should be placed. Children will be able to do this for number lines that go up in 1s and 10s.

COMMON MISCONCEPTIONS

Children often think that number lines always go up in 1s. Show them a number line from 0 to 100 that goes up in 10s. Ask:
- *Can you count on in 1s from the start number? Is the end number what you were expecting it to be?*

STRENGTHENING UNDERSTANDING

For children struggling with this concept or who lack confidence, ask them to check their answers by counting from the start of the number line. Point and count aloud as they go and check that all the numbers match up and that they arrive at the correct number at the end of the number line.

GOING DEEPER

Children who are confident with number lines that go up in 1s and 10s could explore number lines that go up by different amounts, such as 2s and 5s. For example, ask children what a number line from 0 to 50 with 10 intervals goes up in. Can they mark on all the numbers? Can they draw an arrow to show where 43 might be?

KEY LANGUAGE

In lesson: number line, value, marked, interval, start, end, 1s, 10s

Other language to be used by the teacher: approximate

STRUCTURES AND REPRESENTATIONS

Number lines

RESOURCES

Optional: dry wipe number lines, large number line displayed in classroom

 In the eTextbook of this lesson, you will find interactive links to a selection of teaching tools.

Quick recap

On the board, draw a number line from 10 to 20, with 1s marked and labelled. Draw three arrows to numbers on the number line and ask children what numbers the arrows are pointing to. Repeat for other number lines: for example, from 30 to 40 and 26 to 36, etc.

Unit 14: Numbers to 100, Lesson 4

Discover

WAYS OF WORKING Pair work

ASK

- Question 1 a): *Have you seen this before? Do you remember what we call it? What do you think the number line is going up in? How can you check?*
- Question 1 b): *Look at the 20. What number do you think comes next?*

IN FOCUS In question 1 a), children focus on reminding themselves about the number lines covered previously in year 1. They have much less information than previously. Support children as necessary through questions to help them work out the values of the shapes. Encourage them to check by counting on from the start number to the end number, to ensure the numbers match up.

PRACTICAL TIPS Displaying a large number line in the classroom might be useful for this activity.

ANSWERS

Question 1 a): The triangle marks the number 3.
The star marks the number 12.

Question 1 b): 21 comes next after 20.

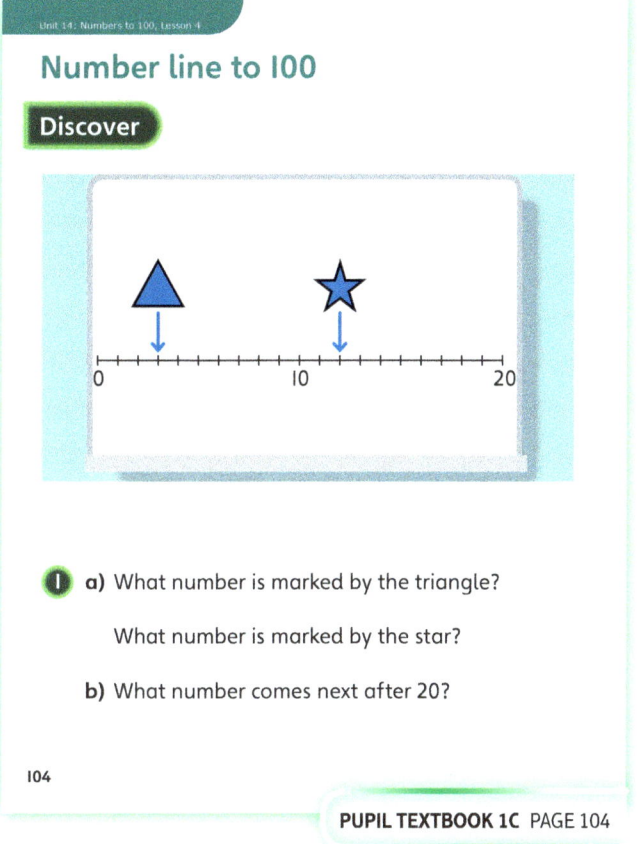

PUPIL TEXTBOOK 1C PAGE 104

Share

WAYS OF WORKING Whole class teacher led

ASK

- Question 1 a): *What does the number line go up in? How do you know? How can you check?*
- Question 1 b): *What is the next number after 20? Where do you think it would be placed on the number line?*

IN FOCUS In question 1 a), children explain that the number line goes up in 1s and they can then count to work out the values that the shapes are pointing to. As a class, count out loud to check that the number line does go up in 1s. Point to the number line as you count. This will provide children with a method that they can use to check their thinking. In question 1 b), it is likely that children will know the next number is 21 as they know the count, but they may be unsure where they put it if they are asked to continue the number line.

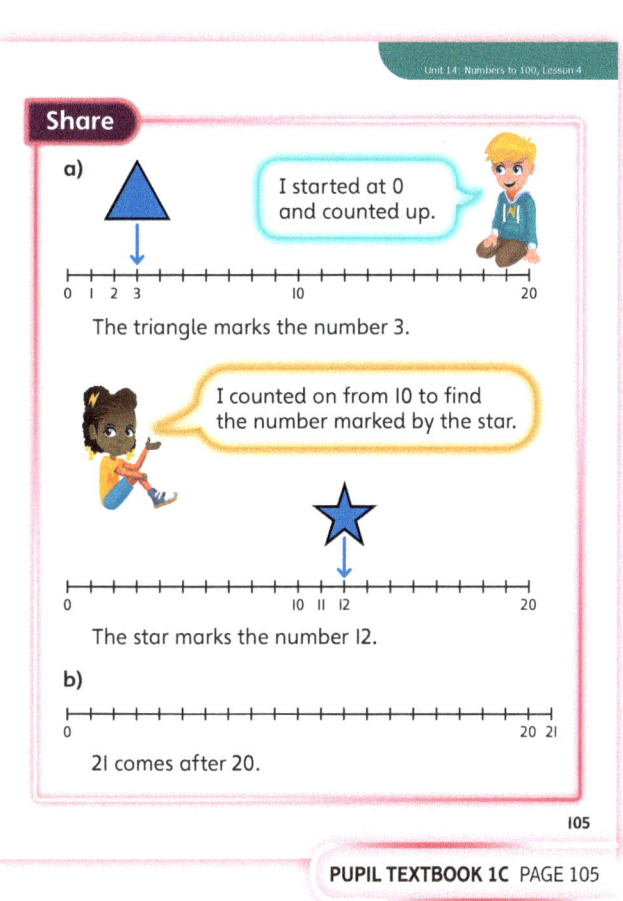

PUPIL TEXTBOOK 1C PAGE 105

143

Think together

WAYS OF WORKING Whole class teacher led (I do, We do, You do)

ASK

- Question ①: *What number does the number line start with? What number does it end with? What do you think it goes up in? How can you check?*
- Question ②: *Do these remind you of number tracks? What are the missing numbers? How did you work it out?*
- Question ③ a): *What is the same and what is different about the two number lines?*

IN FOCUS Question ① asks children to find the location of two numbers on the number line. Discuss how they can work out what the number line goes up in and explain how they can check. Count aloud with children, so they know how to place the two numbers. In question ②, children continue to count on from the given numbers to find the missing numbers. You might want to explain how these are very similar to the number tracks they used in previous lessons. In question ② c), children are given a number line that goes up in 10s. Count along it with children so they can see the difference, as it is the first time they are likely to have seen this.

STRENGTHEN For children struggling with this concept or who lack confidence, ask them to check their answers by counting from the start of the number line. They should point and count out loud as they go and check that all the numbers match up and that they arrive at the correct number at the end of the number line.

DEEPEN Ask children more questions like question ③ b), where children are given an open number line and have to find the approximate position of a number, given a start and end point.

ASSESSMENT CHECKPOINT Use questions ① and ② a) and b) to check whether children are confident at placing numbers on number lines that go up in 1s and can also check whether their answers are correct.

ANSWERS

Question ①: Children should point to 21 and 29 on a number line.

Question ② a): 33, 34, 35, 36, 37, 38, 39

Question ② b): 68, 69, 70, 74, 75

Question ② c): 40, 50, 60, 70, 80, 90, 100

Question ③ a): Can children adjust to the different start and end points of the number lines to locate 45 accurately? It is the midpoint on the top number line and half-way between 40 and 50 on the bottom number line.

Question ③ b): Children should point to 93 on the number line.

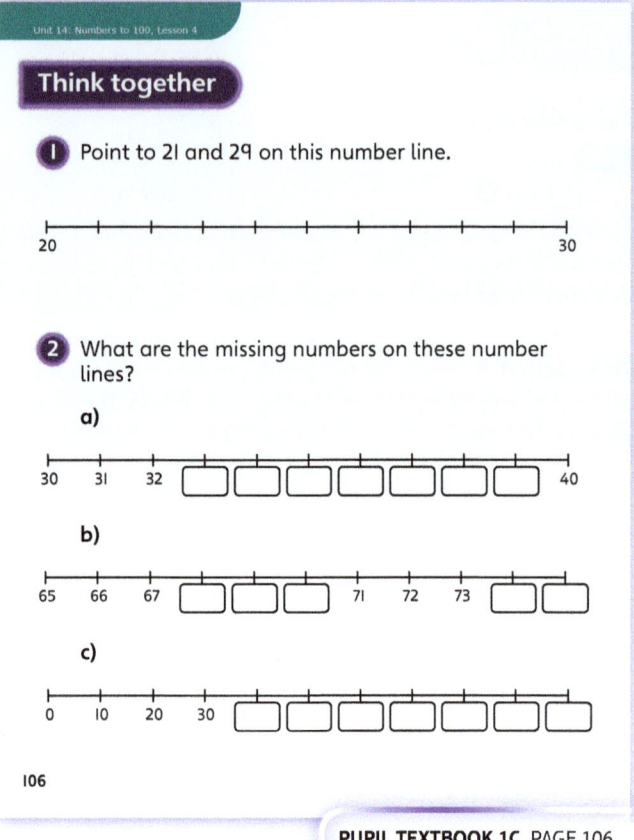

PUPIL TEXTBOOK 1C PAGE 106

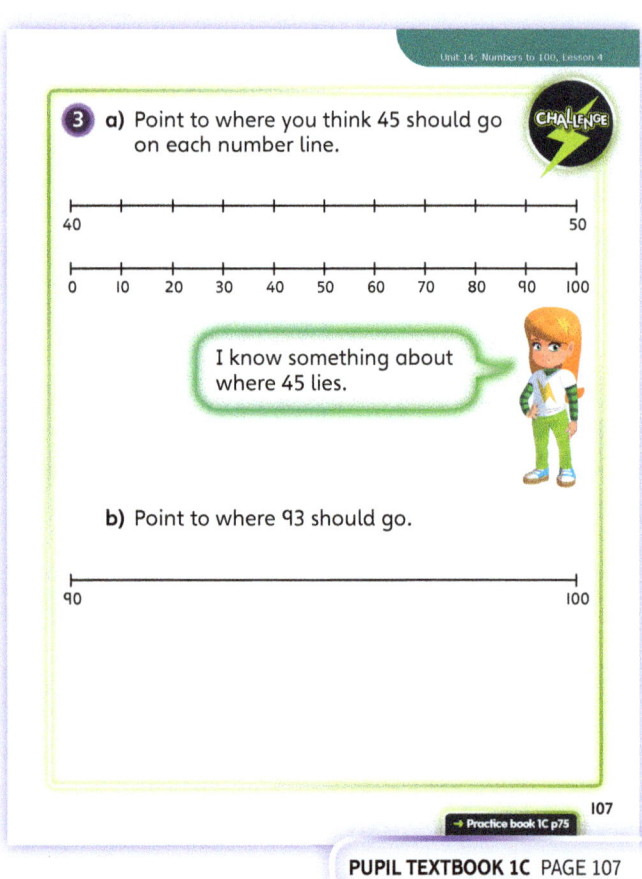

PUPIL TEXTBOOK 1C PAGE 107

Unit 14: Numbers to 100, Lesson 4

Practice

WAYS OF WORKING Independent thinking

IN FOCUS Question ❶ asks children to complete a number line similar to a number track. Question ❷ asks them to find on the number line the specific numbers that the arrows are pointing to. They should point and count and then write in the relevant number when they get to the missing box. For question ❸, children have to match up numbers to their positions on the number lines. In question ❹, children meet a number line that goes up in 10s and one that does not start on a 10, providing a little more challenge and requiring a different way of thinking.

STRENGTHEN For questions ❷ and ❸, encourage children to count from the start number to the end number. Ask them to check whether the number lines go up in 1s or 10s.

DEEPEN Ask children more questions similar to question ❺, where children are given an open number line and have to find the approximate position of a given set of numbers.

ASSESSMENT CHECKPOINT Use questions ❶ to ❹ to check that children can confidently and accurately place numbers on number lines that go up in 1s and 10s, and that they can check if they are correct.

ANSWERS Answers for the **Practice** part of the lesson can be found in the *Power Maths* online subscription.

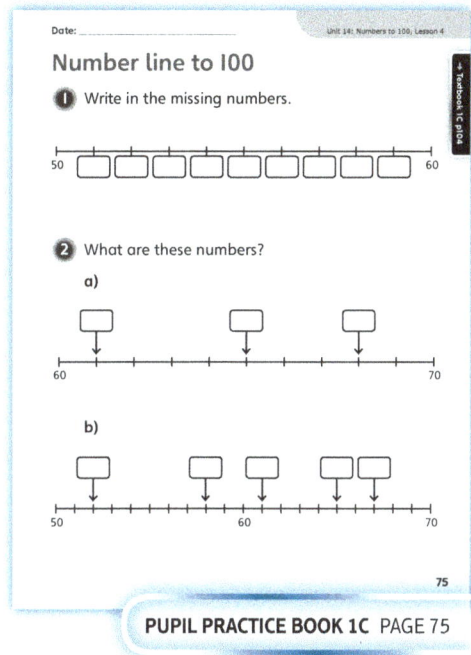

PUPIL PRACTICE BOOK 1C PAGE 75

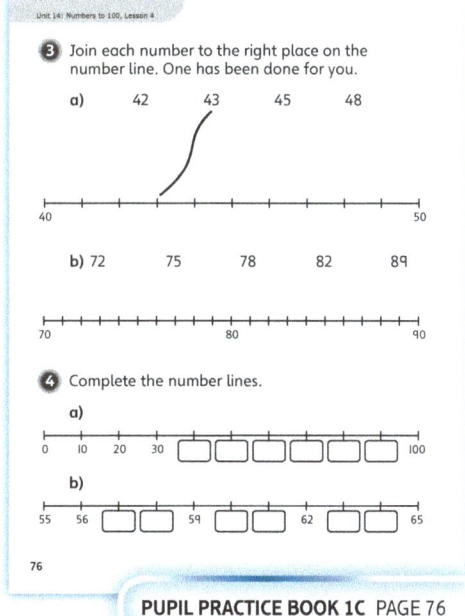

PUPIL PRACTICE BOOK 1C PAGE 76

Reflect

WAYS OF WORKING Pair work

IN FOCUS This is quite an open question and children may draw different number lines. If children are struggling, give them a starting point of 20 to help them. For those children who want to deepen their understanding, they could start at 0 and go up in 10s. Share the different number lines that children come up with.

ASSESSMENT CHECKPOINT Look for the strategy that children use. For example, do they mark on all the points between 20 and 30 or do they estimate where 22 is?

ANSWERS Answers for the **Reflect** part of the lesson can be found in the *Power Maths* online subscription.

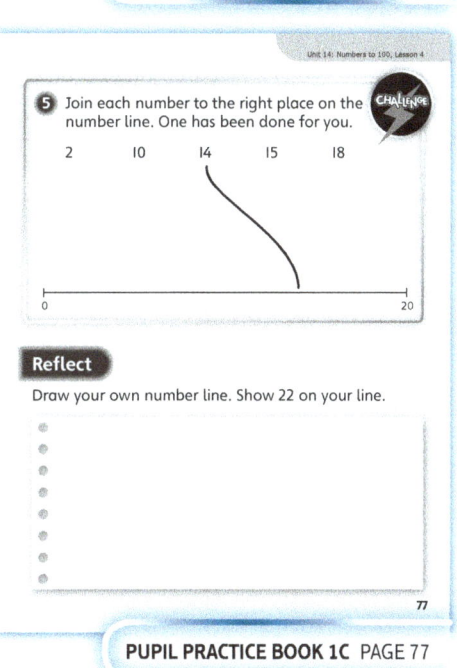

PUPIL PRACTICE BOOK 1C PAGE 77

After the lesson

- Can children find missing numbers on a number line that goes up in 1s?
- Can children find missing numbers on a number line that goes up in 10s?

Unit 14: Numbers to 100, Lesson 5

One more and one less

Learning focus
In this lesson, children will find one more and one less than any number between 1 and 100.

Before you teach
- Can children find one more and one less than numbers up to 50?
- Can children complete the boxes to find one less and one more than a given number in the middle?

NATIONAL CURRICULUM LINKS

Year 1 Number – number and place value

Given a number, identify one more and one less.

ASSESSING MASTERY

Children can find one more and one less than any number from 1 to 100. They are increasingly confident with doing this without the support of mathematical equipment, instead they use their knowledge of the number count.

COMMON MISCONCEPTIONS

Children may struggle with completing statements such as 1 more than ☐ is 73. Often, children see the phrase 'one more' and a number, in this case 73, and assume they have to find 1 more than that number. Ask:
- Have you checked your answer? Is 73 one more than this number?

STRENGTHENING UNDERSTANDING

To support children finding one more and one less than small numbers, use towers of cubes or counters on ten frames and add one counter and remove one counter.

For larger numbers, consider using number tracks or number lines to support children's understanding. For example, when finding one more and one less than 85, children should understand that the number 85 is in the middle; and that the number 84, which is the number to the left, is one less than 85, and the number 86, which is the number to the right, is one more than 85. Show how the number 85 is sandwiched between the two numbers.

GOING DEEPER

Some children may begin to use a number track or number line to find 2 or 3 more than a given number. Ask: *How could you find 10 more than a number using a 100 square?* Questions like this will help children think a little more deeply.

KEY LANGUAGE

In lesson: one more, one less, number

STRUCTURES AND REPRESENTATIONS

Number lines, number tracks, ten frames

RESOURCES

Mandatory: counters, large printed ten frames

Optional: bead string, blank number lines, number tracks, 100 squares, cubes, 36 countable items such as shells or other items found in the classroom or outside environment

 In the eTextbook of this lesson, you will find interactive links to a selection of teaching tools.

Quick recap

On the board, draw three boxes in a row. Draw arrows from the middle box pointing to the other boxes and write 1 less and 1 more above the arrows. Write a number in the middle box that is less than 20 and ask children if they can work out the missing numbers. Discuss how they know the answers. Repeat with a number that is less than 50.

Unit 14: Numbers to 100, Lesson 5

Discover

WAYS OF WORKING Pair work

ASK

- Question 1 a): *How can we count the shells? How many are there in each of the long lines? Do we need to count them one by one?*
- Question 1 b): *How many shells are there now? Do we have to count again from the start?*

IN FOCUS In question 1 a), children first need to count how many shells there are at the moment. Guide children to check that there are 10 shells in each long line – that means counting in 10s then 1s will be quicker and more efficient than counting only in 1s. Children may want to represent the number of shells using counters on ten frames or using cubes. Question 1 b) asks children to work out what happens if one more shell is found. They should not be counting again from 0. Encourage them to look at what has changed between the two questions.

PRACTICAL TIPS Replicate the scenario using shells or other items from a classroom or outdoor environment.

ANSWERS

Question 1 a): There are 35 shells.

Question 1 b): One more than 35 is 36. There are 36 shells now.

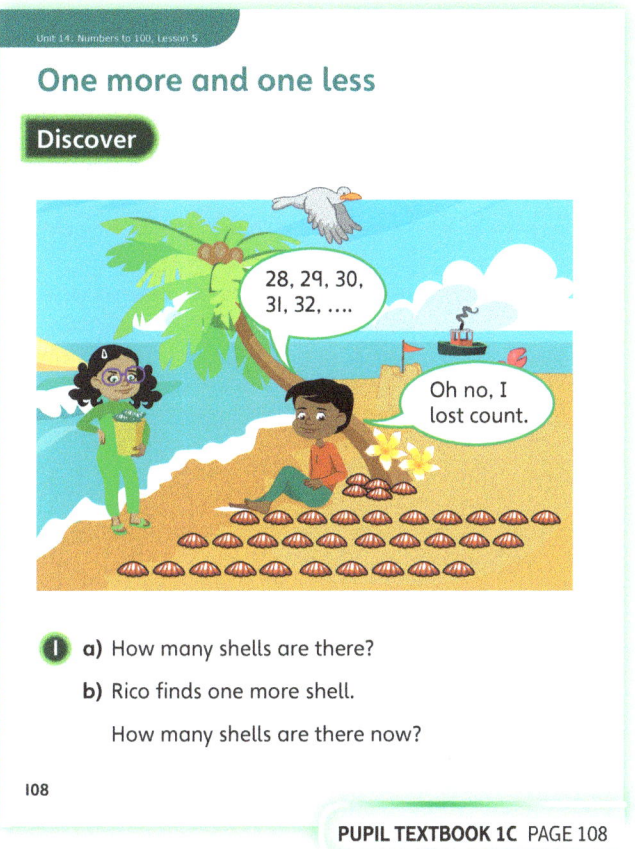

PUPIL TEXTBOOK 1C PAGE 108

Share

WAYS OF WORKING Whole class teacher led

ASK

- Question 1 a): *How did you count the shells? What did you need to check first? Why did you need to check this? Can anyone see there are 5 single shells without counting?*
- Question 1 b): *Did you have to start from 0 again? What has changed? How many shells are there now? What is one more than 35?*

IN FOCUS Discuss with children the methods they used to get the answers. In question 1 a), discuss that they needed to check that there were 10 shells in one line, to make certain, and that then they could count in 10s and then 1s. Go through the count in 10s aloud and then the 1s. Some children may see there are 5 individual shells by subitising, without needing to count them. Children should refer to counters and ten frames to represent the shells in front of them. In question 1 b), children should place one more counter on the ten frame. Explain there are now 36 shells, but we do not need to count from 0 as 36 is one more than 35.

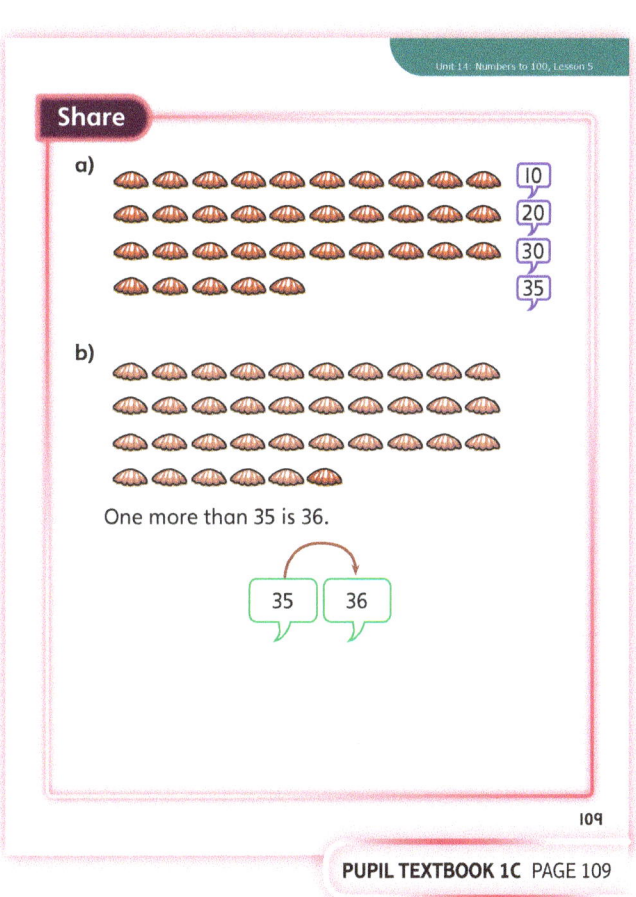

PUPIL TEXTBOOK 1C PAGE 109

147

Unit 14: Numbers to 100, Lesson 5

Think together

WAYS OF WORKING Whole class teacher led (I do, We do, You do)

ASK

- Question ❶: *What numbers are shown? What is one more than each number? How did you work it out?*
- Question ❷: *How many flowers are there in each bunch? How many flowers are there in total? What is one less than 84?*
- Question ❸: *Have you seen these diagrams before? How can you work out one less and one more without having to make the numbers?*

IN FOCUS In question ❶, children practise more examples of finding one more, using mathematical equipment to support them if necessary. Question ❷ provides a context for finding one less than a number. Question ❸ provides more abstract questions related to finding one more and one less. Children should recognise the diagrams from earlier in the year. They use the counting sequence to help them work out one more and one less without having to make the numbers. Throughout all the questions, it is important to encourage children to use full sentences for their answers – for example, one less than ▢ is ▢.

STRENGTHEN To support children finding one more and one less than small numbers, use towers of cubes or counters on ten frames and add one counter and remove one counter. For larger numbers, consider using number tracks or number lines to support children's understanding.

DEEPEN Ask children more questions similar to question ❸, this time where the middle box is not given and one of the other boxes, one more or one less, is given.

ASSESSMENT CHECKPOINT Use these questions to determine whether children can find one more and one less than a number to 100. They should be getting confident with finding one more and one less without the support of diagrams or mathematical equipment.

ANSWERS

Question ❶ a): One more than 42 counters is 43 counters.

Question ❶ b): One more than 56 eggs is 57 eggs.

Question ❷: One less than 84 flowers is 83 flowers.
Max has 83 flowers now.
One less than 83 flowers is 82 flowers.
Max has 82 flowers now.

Question ❸ a): 56, 57, 58

Question ❸ b): 70, 71, 72

Question ❸ c): 49, 50, 51
66, 67, 68

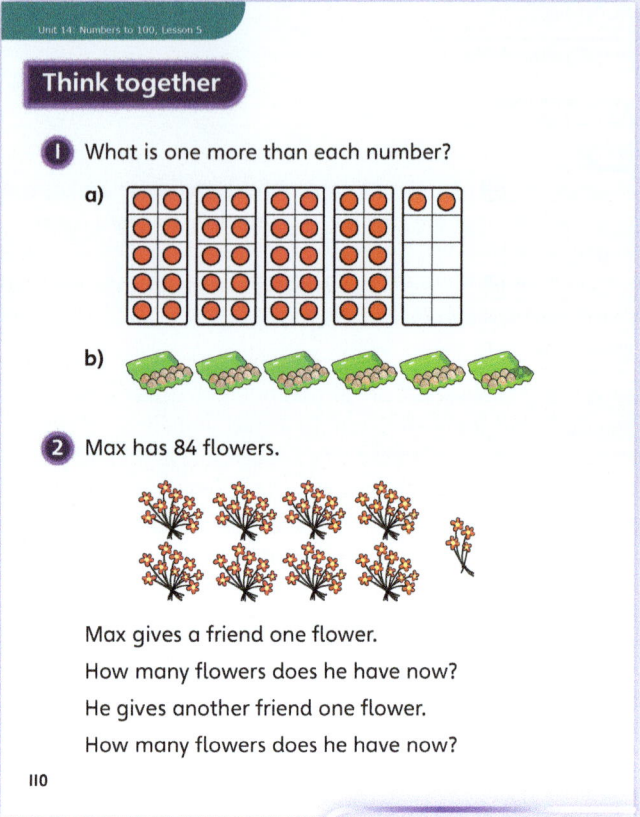

PUPIL TEXTBOOK 1C PAGE 110

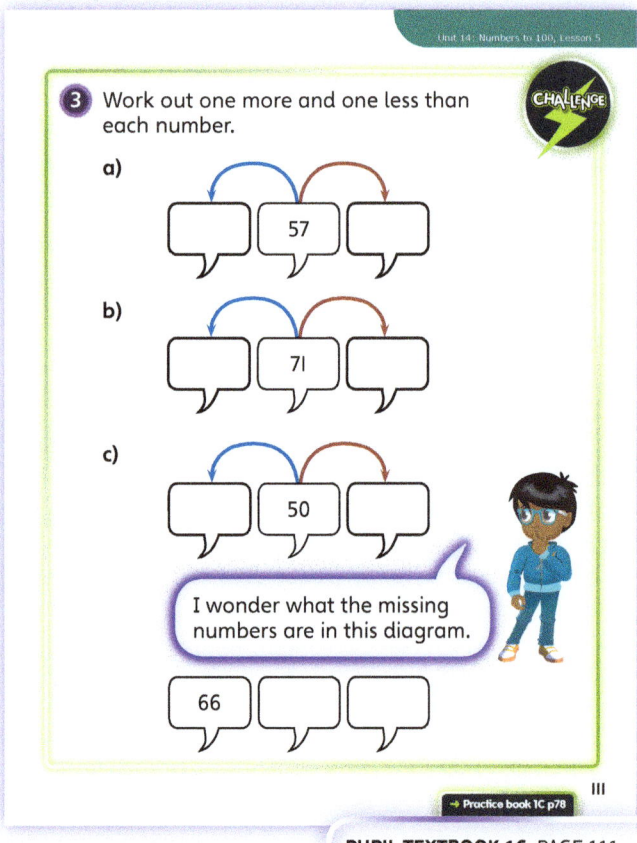

PUPIL TEXTBOOK 1C PAGE 111

Unit 14: Numbers to 100, Lesson 5

Practice

WAYS OF WORKING Independent thinking

IN FOCUS In question ❶, children find one more than numbers that are represented by ten frames and by pictures of concrete objects. In question ❷, children find one less than a number presented on ten frames. In question ❸, children find one more and then one less using their knowledge of the counting sequences. Encourage children throughout to say the answers using full sentences, rather than just giving a number. In question ❺, the idea of 'one more' or 'one less' is formalised in number sentences, using '+1' or '–1'.

STRENGTHEN To support children finding one more and one less than small numbers, use towers of cubes or counters on ten frames and add one counter and remove one counter. For larger numbers, consider using number tracks or number lines to support children's understanding.

DEEPEN Ask children more questions that are similar to question ❹. You could also ask children to find 2 more than a given number or 10 more than a given number. Ask them to explain how they did it.

THINK DIFFERENTLY In question ❹, children are not given the middle number and instead are told either the number that is one less or one more and they first have to work out the original number.

ASSESSMENT CHECKPOINT Questions ❶, ❷ and ❸ will help you determine if children are confident finding one more and one less than given numbers. Use question ❺ to determine whether children can relate the concepts of 'one more' or 'one less' to adding 1 or subtracting 1.

ANSWERS Answers for the **Practice** part of the lesson can be found in the *Power Maths* online subscription.

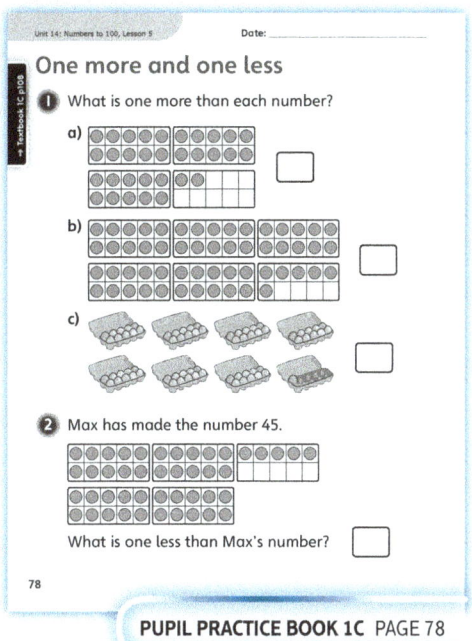

PUPIL PRACTICE BOOK 1C PAGE 78

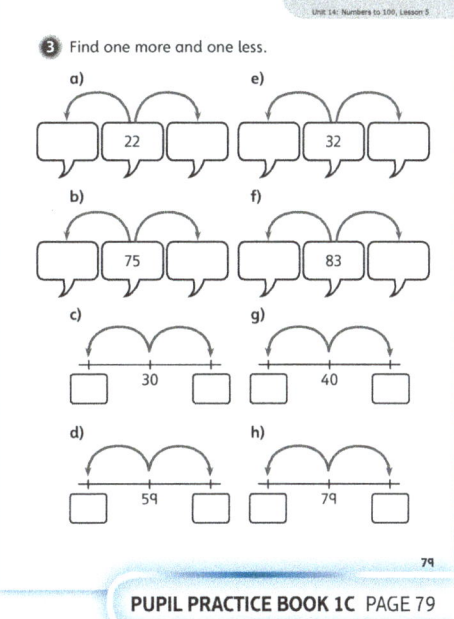

PUPIL PRACTICE BOOK 1C PAGE 79

Reflect

WAYS OF WORKING Pair work

IN FOCUS Explain the game to children by playing a couple of rounds with one child. Then let children play the game in pairs. If children are struggling, you could provide 100 squares to help them pick numbers and work out one more and one less.

ASSESSMENT CHECKPOINT Check that children can confidently and quickly say one more and one less than a given number.

ANSWERS Answers for the **Reflect** part of the lesson can be found in the *Power Maths* online subscription.

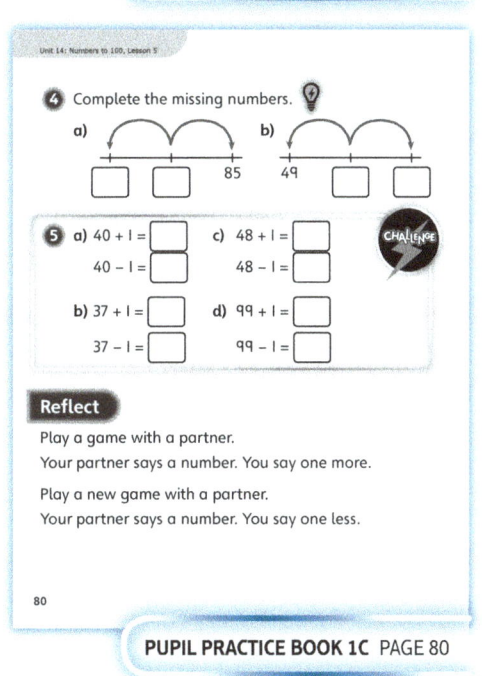

PUPIL PRACTICE BOOK 1C PAGE 80

After the lesson

- Can children find one more and one less than a given number below 100?
- Do children know how finding one more and one less relates to the counting sequence?

149

Unit 14: Numbers to 100, Lesson 6

Compare numbers

Learning focus
In this lesson, children will further consolidate their understanding of how to compare numbers up to 100. They will use the language of comparison accurately.

Before you teach
- Can children compare numbers from 0 to 50?
- Can children find a total by counting in 10s and 1s?
- Do children understand the language and signs for comparison?

NATIONAL CURRICULUM LINKS
Year 1 Number – number and place value

Identify and represent numbers using objects and pictorial representations including the number line, and use the language of: equal to, more than, less than (fewer), most, least.

ASSESSING MASTERY
Children can compare two 2-digit numbers using their knowledge of 10s and 1s and can write or complete simple comparison sentences. At this stage, they may use representations to support their comparison, including looking at the position of a number on a number track or a 100 square.

COMMON MISCONCEPTIONS
As the representations of numbers become more abstract, children may compare the 1s before the 10s and so make incorrect statements such as '19 > 21'. Ask:
- *Can you make the two numbers using resources? Which is greater? Where are these numbers on a number track or number line? Does this help you decide which is greater?*

STRENGTHENING UNDERSTANDING
Give children the opportunity to practise comparing numbers through games or role play. For example, children could play games in pairs and score points. Record the scores using concrete resources (counters, cubes) and also using abstract numerals. Discuss who has won the game. Ask: *How do you know who scored the most? What can you tell me about their score compared to the other player's score?*

Children could role-play a scenario in which some objects have been shared out in two unequal groups. The objects being counted could be put in a bag with the number written on the outside. Ask: *Who has the most? Who has the fewest? How do you know? How could you check?*

GOING DEEPER
Move on to more abstract comparisons of numbers within 100. For example, can children compare 36 and 49 without having pictorial or concrete representations? Can they reason why 49 is greater because it has more 10s? What generalisations and rules can they come up with to compare numbers?

KEY LANGUAGE
In lesson: =, compare, most, tens, ones, greater than, the same, more, less than, digit, between

Other language to be used by the teacher: place value, bigger, equal to, smaller

STRUCTURES AND REPRESENTATIONS
Ten frames, number tracks, 100 squares

RESOURCES
Optional: counters, large printed ten frames, printable blank number tracks, large printed 100 squares, leaves

 In the eTextbook of this lesson, you will find interactive links to a selection of teaching tools.

Quick recap
Check that children can compare numbers within 20 using < and > signs. Write two numbers up to 20 on the board and ask what sign should come between them. Ensure that children understand the different reasons why the different signs are used. For example, because it is further along a number track or because they can see more counters when they make numbers on a ten frame.

150

Discover

WAYS OF WORKING Pair work

ASK

- Question 1 a): *How many leaves has each child collected? How can we represent the number of leaves?*
- Question 1 b): *Who has collected more leaves? How do you know?*

IN FOCUS In question 1 a), ask children in pairs to make the number of leaves using counters on ten frames or using cubes. They could make one number each and pretend to be like the children in the image. Question 1 b) then asks children to discuss who has more leaves. They should discuss who might have the most leaves (counters) by comparing the representations that they have made. They could line up the counters or cubes to see which line is longest.

PRACTICAL TIPS Collect leaves in the playground and then ask children to compare the number of leaves they have found.

DEEPEN You could challenge children to put their comparison for each part into a sentence. Ask: *What mathematical vocabulary will you use?*

ANSWERS

Question 1 a): Children should make 35 and 39 on ten frames or with cubes.

Question 1 b): Seth collected more leaves (39). Children should be able to see this on the ten frames or by comparing the cubes.

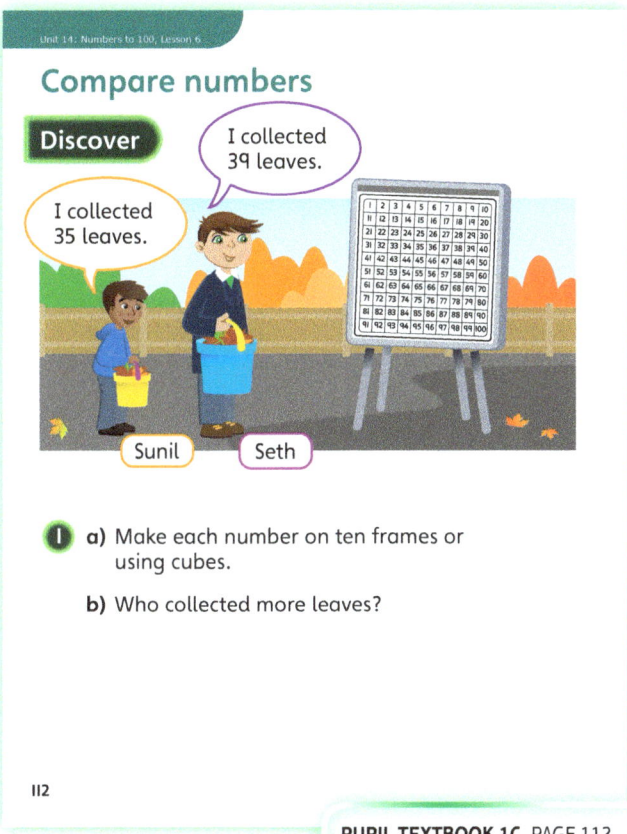

PUPIL TEXTBOOK 1C PAGE 112

Share

WAYS OF WORKING Whole class teacher led

ASK

- Question 1 a): *Can you see the 35 and 39 leaves? Where is the 30 and where are the 1s?*
- Question 1 b): *How do the diagrams show that Seth has collected more leaves than Sunil?*

IN FOCUS In question 1 a), show the ten frame representations of 35 and 39. Discuss from previous lessons where the 3 tens are in each number and then where the 1s are. Ask children if they can see if there are 5 and 9 ones without having to count them. How do they know? For question 1 b), discuss how children know that Seth has collected more leaves than Sunil. Children may know that the numbers increase as they move to the right on a number track or number line.

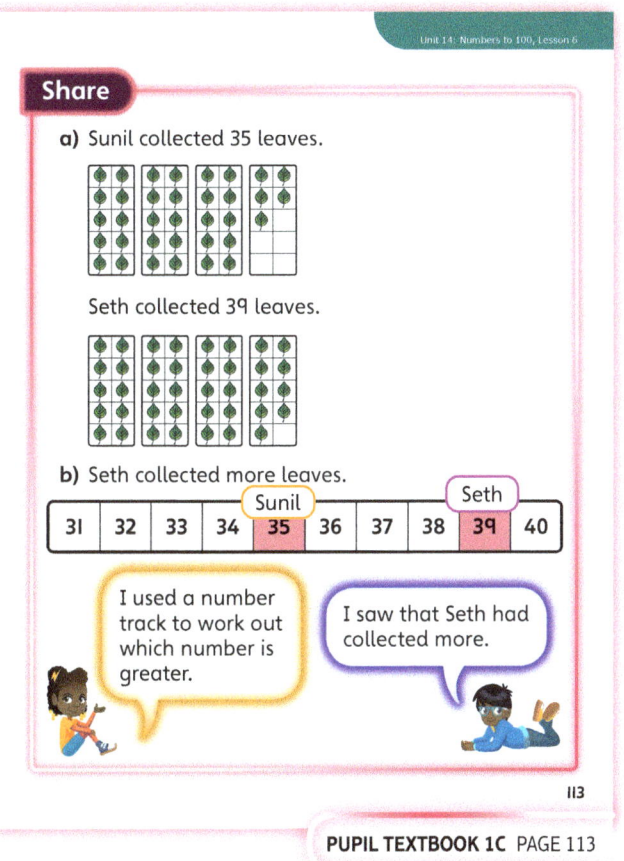

PUPIL TEXTBOOK 1C PAGE 113

Unit 14: Numbers to 100, Lesson 6

Think together

WAYS OF WORKING Whole class teacher led (I do, We do, You do)

ASK
- Question ❶: *What numbers have been made? Which number is greater? Where are the numbers on the number track? How can you use this to compare the numbers?*
- Question ❷: *Where will numbers greater than 65 be on the number track? Is 63 greater than 65? Why not?*
- Question ❸: *Which number is greater? How do you know?*

IN FOCUS In question ❶, children compare numbers using the representations of ten frames and a number track. This example is very similar to the **Discover** questions they have just completed, where they can see that the last ten frame has more counters. Explain how they can use the number track to help them see which number is greater and they do not always need to make the numbers. The further right on an increasing number track or number line, the greater the number. This will then help with question ❷ as children work out which numbers are greater than 65. In question ❸, children compare numbers with a differing number of 10s, using representations and also their position on a 100 square.

STRENGTHEN Encourage children to use counters or cubes to create the numbers. Get them to look at the position of numbers on a number track. If children are still finding the comparisons tricky, they could line up the counters or cubes and see which line is longer or shorter.

DEEPEN Ask children to compare numbers using < and > signs and find numbers that can fit into statements, such as 35 < ☐ < 42.

ASSESSMENT CHECKPOINT Ask children to explain their answers to questions ❶ and ❷ so you can assess the confidence of their explanations. They should be fluent in explaining their method and the use of the vocabulary. Some children may be able to use the comparison signs correctly.

ANSWERS

Question ❶: 44 is smaller than 47.

Question ❷: Children could choose any number from 66, 67, 68, 69 or 70.

Question ❸ a): A: 38 is greater than 29 – it has more 10s.

Question ❸ b): 29 < 38

The 100 square shows us that although 29 has more 1s than 38, 38 has more 10s and so is the greater number. The further down the 100 square the number is, the greater the number of 10s it has, so the greater the number is.

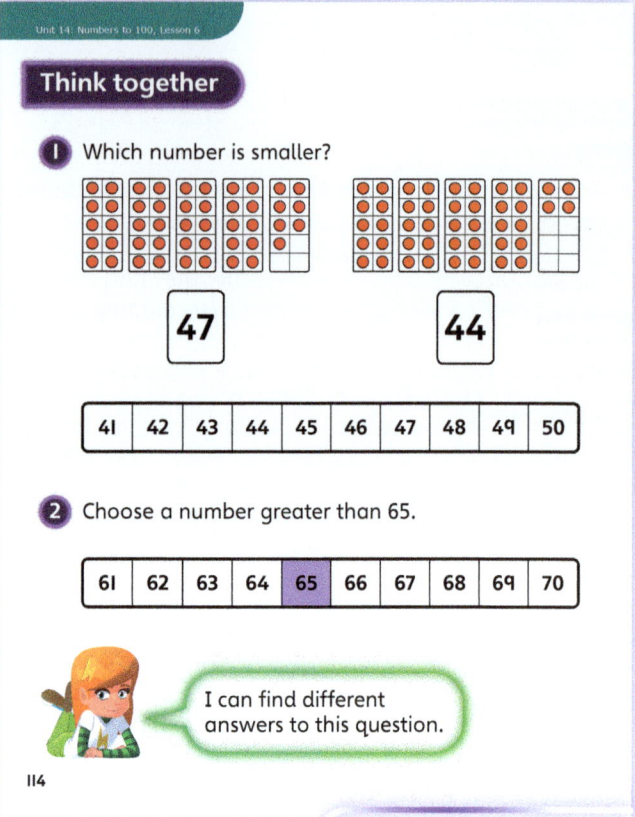

PUPIL TEXTBOOK 1C PAGE 114

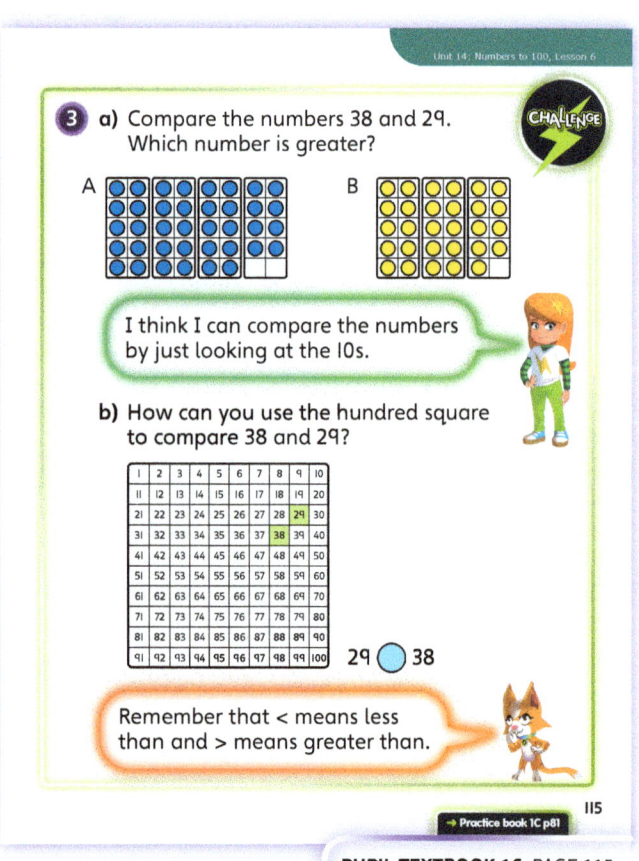

PUPIL TEXTBOOK 1C PAGE 115

Unit 14: Numbers to 100, Lesson 6

Practice

WAYS OF WORKING Independent thinking

IN FOCUS Question 2 asks children to compare numbers using their position on a number track. Children complete the sentences using the correct comparison language that they have been using in the main lesson. In question 3, children compare numbers using the < and > signs and use the 100 square for support.

STRENGTHEN If children are struggling with the abstract representations, encourage them to make the numbers from cubes or counters. Discuss how the further right a number is on an increasing number track or number line, the greater the number.

DEEPEN Extend question 4 by asking children how many solutions they can find for each number sentence. Ask: *Are there any digits that would work in all three number sentences?* (No) *Are there any digits that will not work in any of the number sentences?* (No) Encourage children to use their understanding of all representations used so far to represent their ideas clearly and efficiently.

ASSESSMENT CHECKPOINT Use questions 1 and 2 to check that children can correctly compare numbers to 100 using correct language. Use question 3 to check that children can compare numbers using comparison signs.

ANSWERS Answers for the **Practice** part of the lesson can be found in the *Power Maths* online subscription.

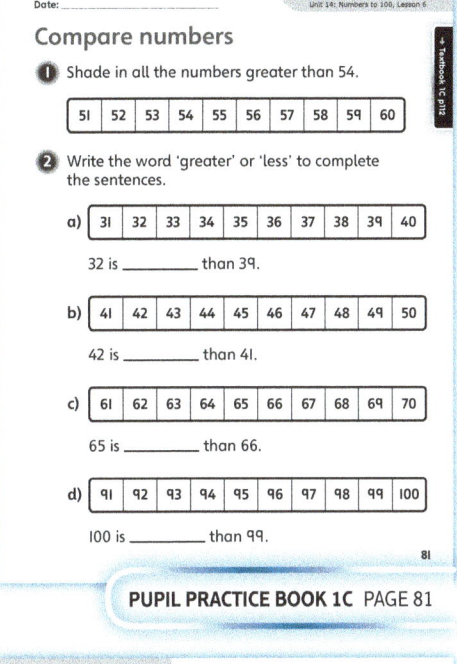

PUPIL PRACTICE BOOK 1C PAGE 81

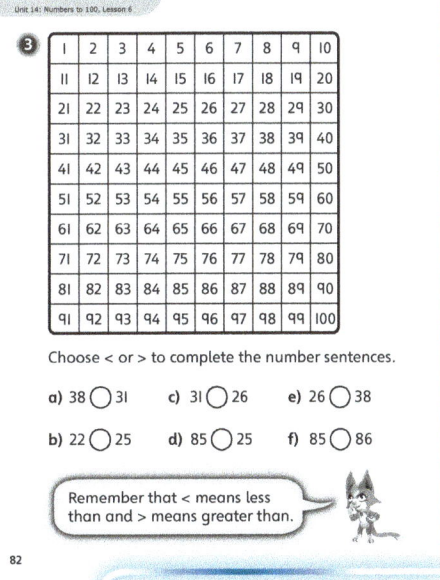

PUPIL PRACTICE BOOK 1C PAGE 82

Reflect

WAYS OF WORKING Independent thinking

IN FOCUS Children work out possible numbers that could go into the boxes. It is important children are confident with the comparison signs, so use words to help them if necessary. The 100 square can be used to help them, although some children may want to make the numbers using mathematical equipment.

ASSESSMENT CHECKPOINT Assess whether children can confidently justify their ideas using the correct vocabulary and comparison signs.

ANSWERS Answers for the **Reflect** part of the lesson can be found in the *Power Maths* online subscription.

After the lesson

- Can children compare two 2-digit numbers, supported or unsupported by mathematical equipment?
- Can they use the language and signs of comparison confidently and accurately?

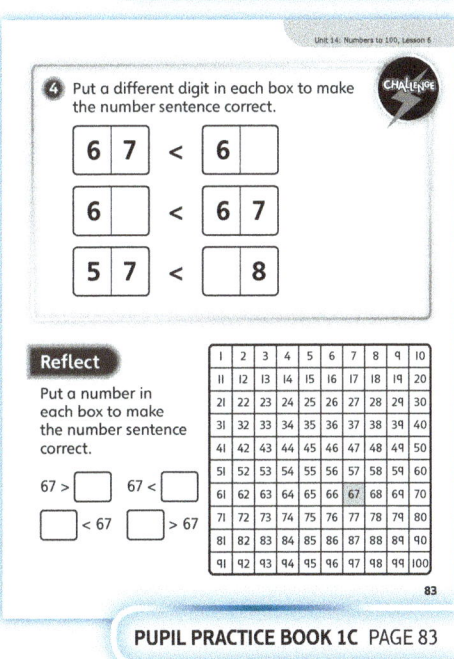

PUPIL PRACTICE BOOK 1C PAGE 83

153

Unit 14: Numbers to 100

End of unit check

Don't forget the unit assessment grid in your *Power Maths* online subscription.

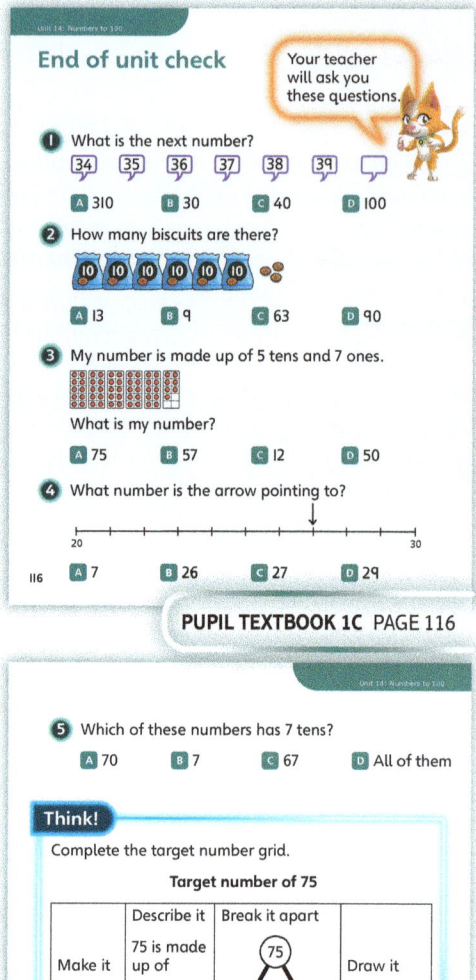

PUPIL TEXTBOOK 1C PAGE 116

PUPIL TEXTBOOK 1C PAGE 117

WAYS OF WORKING Group work adult led

IN FOCUS
- Question ① assesses whether children know what happens when they count across a 10.
- Questions ② and ③ assess whether children can count in 10s and 1s. They are given pictorial representations to support them. Look for children counting the 10s first and then the 1s.
- Question ④ checks that children know the position of a 2-digit number on a number line.
- Question ⑤ assesses whether children know the difference between 10s and 1s.

Think!

WAYS OF WORKING Pair work or small groups

IN FOCUS This question requires children to manipulate the number 75 using concrete, pictorial and abstract representations. Children need to recognise the place value of the digits within the number and represent the parts in different ways.

Draw children's attention to the vocabulary at the bottom of the page.

Encourage children to think through or discuss the structure of the number 75 before writing their answer in **My journal**.

ANSWERS AND COMMENTARY Children who have mastered the concepts of this unit will be able to count confidently in 10s and 1s and recognise how many 10s and 1s there are in any 2-digit number. They will begin to know that a 2-digit number is made up of some 10s and 1s. They will begin to compare 2-digit numbers using the < and > signs.

Q	A	WRONG ANSWERS AND MISCONCEPTIONS	STRENGTHENING UNDERSTANDING
1	C	A suggests that children do not know what happens when they cross the next 10.	If children are struggling with the concept of place value, ask them to make a number using a concrete resource and tell you how many full groups of 10 there are, and how many 'extra' 1s there are. If children struggle to compare numbers, ask them to make two numbers using counters on ten frames. Can they now see which number is greater?
2	C	B suggests that the child has counted all items in 1s, rather than the bags of biscuits in 10s.	
3	B	A suggests children have mixed up the 10s and 1s. D suggests children have just counted the 10s and not included the 1s.	
4	C	A suggests children have just counted along and have not noticed the starting point. B suggests that children could have possibly miscounted along the number line.	
5	A	Any other answer suggests that children are not confident with recognising place value of 10s and 1s.	

154

Unit 14: Numbers to 100

My journal

WAYS OF WORKING Independent thinking

ANSWERS AND COMMENTARY

Children should choose a clear, concrete method to represent 75 (for example, they could use counters on a ten frame or use a bead string). They should write that '75 is made up of 7 tens and 5 ones' and complete the part-whole model to show 75 = 70 + 5 (with the parts in either order). Their drawing should clearly represent 75.

Assess how confidently children approach the task. Those who find it difficult may not yet have a secure understanding of place value and will need intervention support before this topic is taught again in Year 2, Unit 1.

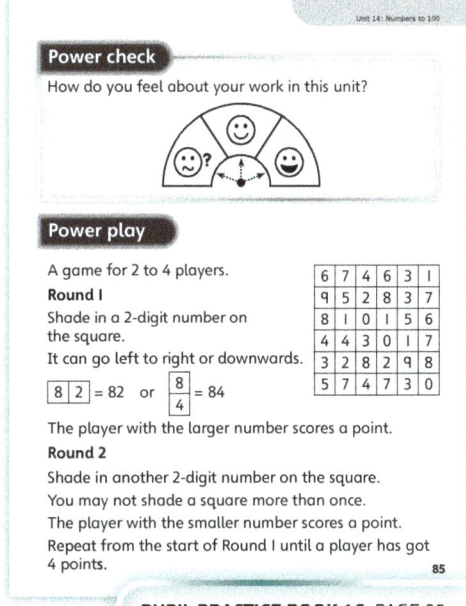

PUPIL PRACTICE BOOK 1C PAGE 84

Power check

WAYS OF WORKING Independent thinking

ASK

- Do you think you could explain what you have learnt about place value to someone else?
- How confident do you feel about comparing and ordering numbers?

Power play

WAYS OF WORKING Pair work or small groups

IN FOCUS Use this **Power play** to see if children can compare two 2-digit numbers or (if playing in groups of three or four) order 2-digit numbers in ascending and descending order. Children who have mastered the concepts involved may be able to spot advantageous combinations of digits that they can choose to help them win the game.

ANSWERS AND COMMENTARY If children are able to play the game confidently and compare and order the numbers fluently, it would suggest that they are confident with place value and comparing numbers. If they are still struggling, they may need to be offered the opportunity to go back to some of the strengthening activities for these two concepts.

PUPIL PRACTICE BOOK 1C PAGE 85

After the unit

- How could the concepts covered in this unit be taken further in other areas of the curriculum? Comparing and ordering numbers can be implemented easily into PE through comparing scores. How else could these concepts be practised?
- How confidently were children able to apply their understanding of this unit to their reasoning? Were they able to use the vocabulary fluently?

Strengthen and **Deepen** activities for this unit can be found in the *Power Maths* online subscription.

Unit 15
Money

Mastery Expert tip! 'My class really enjoyed this work; we talked a lot about what the coins and notes are worth, and what you could buy with them. I like to use real coins (rather than plastic ones) wherever possible, but this is not essential.'

Don't forget to watch the Unit 15 video!

WHY THIS UNIT IS IMPORTANT

This unit focuses on recognising coins and banknotes, and understanding their relative and absolute values. This work has obvious practical significance, in that it is clearly important that children develop familiarity with money in a range of everyday settings. Less obvious is the importance of money as a context for developing fundamental ideas about measurement; the value of a coin or note depends on both the numerical value assigned to it and the unit (pounds or pence) that is involved. There is also a degree of abstraction involved: notes and coins are compared according to an assigned value, rather than any inherent property such as size or weight.

WHERE THIS UNIT FITS

→ Unit 14: Numbers to 100
→ **Unit 15: Money**
→ Unit 16: Time

This unit stands alone but draws on the key skills of reading and writing numbers, counting and addition.

Before they start this unit, it is expected that children:
- can read, write and understand whole numbers to 100
- know that money is used to buy things and that it is measured in pounds and pence/pennies
- can count in 1s, 2s, 5s and 10s (with or without number lines)
- can compare values using the signs <, > and = .

ASSESSING MASTERY

Children who have mastered this unit will recognise real coins and banknotes and know their value. They will be able to find the total value of a small set of coins or banknotes (counting in 2s, 5s and 10s where appropriate), and to compare the value of two sets of coins or notes. They will know that there are a limited number of denominations for coins and notes.

COMMON MISCONCEPTIONS	STRENGTHENING UNDERSTANDING	GOING DEEPER
Children may think that larger coins (physically) are worth more than smaller ones.	Emphasise the importance of the unit when naming coins and notes, making sure children say 'a five pence piece', and not just 'a five'.	Ask children to investigate combinations of coins and notes that make particular totals. Ask: *What is the smallest number of coins you could use to make 83 pence? What is the smallest number of notes and coins that you could use to make 83 pounds?*
Children may think that larger sets (sets with more items) of coins or notes must be worth more than sets with fewer items.	Compare the number of items and their value explicitly. Ask: *Which pile has got more coins? Which pile is worth more?*	

Unit 15: Money

UNIT STARTER PAGES

Use these pages to introduce the unit focus to children. How many coins do they recognise? You can use the characters to explore different ways of working too. Practise counting in 2s, 5s and 10s.

STRUCTURES AND REPRESENTATIONS

Real coins and banknotes:

Number line: Number lines help children count. They allow children to identify the starting point, the number counted on and the end point.

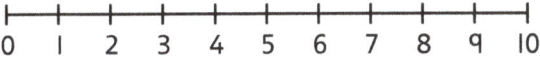

KEY LANGUAGE

There is some key language that children will need to know as part of the learning in this unit.

- pound, penny, pennies, pence
- coins, notes, banknotes
- greater than, less than, equal, total, altogether
- <, >, and =
- value, worth

PUPIL TEXTBOOK 1C PAGE 118

PUPIL TEXTBOOK 1C PAGE 119

Unit 15: Money, Lesson 1

Recognise coins

Learning focus

In this lesson, children will learn to recognise coins and become familiar with their relative values.

Before you teach

- Are children familiar with handling money?
- If you use plastic coins, are they realistic enough to not confuse children?
- Ensure that the coins you use have the numbers, not just the words, on their tails side.

NATIONAL CURRICULUM LINKS

Year 1 Measurement – money

Recognise and know the value of different denominations of coins and notes.

ASSESSING MASTERY

Children can recognise all of the coins (1p, 2p, 5p, 10p, 20p, 50p, £1 and £2) and can order them according to value.

COMMON MISCONCEPTIONS

Children may not be familiar with some of the coins, especially the higher denomination ones. If children are not already familiar with the relative values of the coins, they may attempt to sort them by size rather than value, or they may just compare the numerical values shown on the coins, ignoring the denomination (pounds or pence). Ask:
- Which of these coins is physically larger (50 pence or 1 pound)? Which number is bigger: 50 or 1? Which one is worth the most?

STRENGTHENING UNDERSTANDING

Children who are not familiar with the coins will benefit from practising 'naming and selecting'. Ask: *Can you tell me what this coin is? Can you find the 1 pound coin? Choose a coin that is worth more than 10 pence or less than 50 pence.*

GOING DEEPER

Children who grasp this work quickly could make up amounts to 20p using 1p, 2p, 5p and 10p coins and then amounts to £1 or £2 using silver coins. Ask: *How many different ways can you make 15 pence? How many different ways can you make 1 pound with silver coins?*

KEY LANGUAGE

In lesson: coins, pound, pence, left, right, worth, greater than, less than, value, before, after

Other language to be used by the teacher: amount, most valuable, least valuable, silver, copper, heads, tails

RESOURCES

Mandatory: coins – either real ones (choose coins with numbers not just words), or realistic plastic versions

 In the eTextbook of this lesson, you will find interactive links to a selection of teaching tools.

Quick recap

Ask children what coins they know and how much the coins are worth. You may want to show images of the different coins on the board and ask them which they remember. Although this lesson focuses on recognising coins, finding out what children know already will allow you to build on this.

158

Unit 15: Money, Lesson 1

Discover

WAYS OF WORKING Pair work

ASK

- Question 1 a): *Look at these two trays. What is in them? Do you know the names of any of the coins?*

IN FOCUS Question 1 a) introduces all of the coins in current use. It provides an opportunity to see which coins children already recognise and the key at the bottom of the page can be used to help name any of the coins that are less familiar.

PRACTICAL TIPS Hand out real coins or plastic coins to small groups of children and ask them to recreate the scenario.

ANSWERS

Question 1 a): The coins in the tray to start with are:

 1 pence coin   20 pence coin

2 pence coin   50 pence coin

5 pence coin   1 pound coin

10 pence coin    2 pound coin

Question 1 b): The 5 pence coin has been removed.

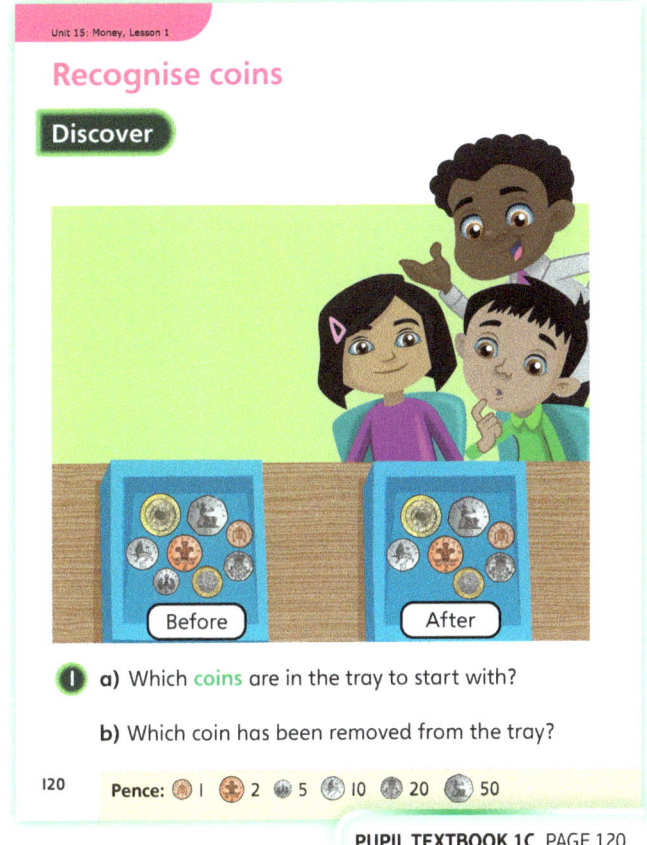

PUPIL TEXTBOOK 1C PAGE 120

Share

WAYS OF WORKING Whole class teacher led

ASK

- Question 1 a): *How do the colours of the coins help you to see which ones are worth more or less? What colour are the most valuable coins? What about the least valuable ones? What can you say about the silver coins?*
- Question 1 a): *Is there a 3 pence coin? Is there a 40 pence coin?*

IN FOCUS In question 1 a), discuss the difference between a 1 pence coin and a 1 pound coin. Ensure that children understand the relationship between pounds and pence, and that 1 pound is worth more (is more valuable) than 1 pence. It is also important that children understand that there are a set number of coins and, for example, there are no 3 pence or 4 pence coins. The numbers are found on the tails side, but some coins only have the words, no numbers. Explain that the side with the Queen's head or King's head is called 'heads' and the other side is called 'tails', and this is where they will usually find the value of the coin. Some newer or special edition coins may only have the value in words.

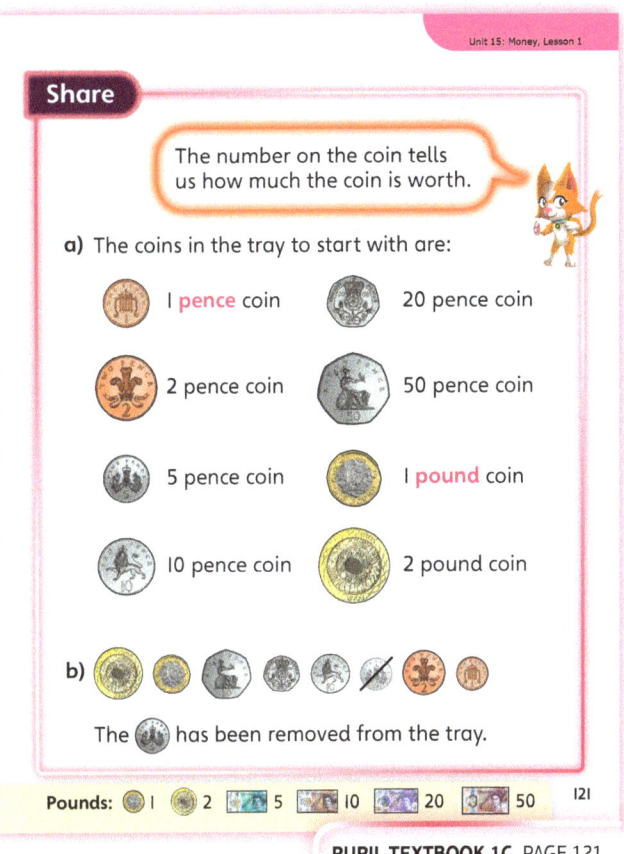

PUPIL TEXTBOOK 1C PAGE 121

Think together

WAYS OF WORKING Whole class teacher led (I do, We do, You do)

ASK

- Question 1: *Which coins do you recognise?*
- Question 2: *What do you notice about the way that the coins have been put on the table? Why is it easier to compare the coins when they are this way up? Which coins on the table are worth the most? Which are worth the least?*

IN FOCUS In question 1, children point to the different coins. You may want to discuss what coins children have seen in real life. Ask them to tell a partner which coins they have used before, and what they bought with them. In question 2, you may want to ask children to make their own table on a mini-whiteboard, so they can record their answers. They may want to start with the smallest (in denomination) first. Question 3 combines most of the key learning points for this lesson with children having to identify coins from their images and decide which are worth more. Look out for the misconception in question 3 b) where children think the biggest coins represent the greatest value.

STRENGTHEN Children who are having difficulty recognising coins will benefit from further practice with real or plastic coins. Make sure that children turn them 'tails up' so that they can compare them to the pictures in the Textbook. For question 2, ask children to select pairs of the same coin. This will help them work out how many of each coin there are.

DEEPEN Children who grasp this work quickly could be encouraged to think about other possible coin descriptions for question 3, especially, for example, which coin has the least or greatest value in a set of 3 coins.

ASSESSMENT CHECKPOINT Questions 1 and 2 provide a good opportunity to check that children recognise and name the coins. Make up further questions like this to check individual children's understanding.

ANSWERS

Question 1: Children should point to:

a)

b)

c)

Question 2: 2 × 1p, 1 × 2p, 3 × 10p, 4 × 20p, 2 × £1

Question 3 a): 5 pence is worth five 1 pence coins.
10 pence is worth ten 1 pence coins.
20 pence is worth twenty 1 pence coins.
Altogether, 35 1 pence coins = 20p + 10p + 5p.

Question 3 b): 20p has the greater value.

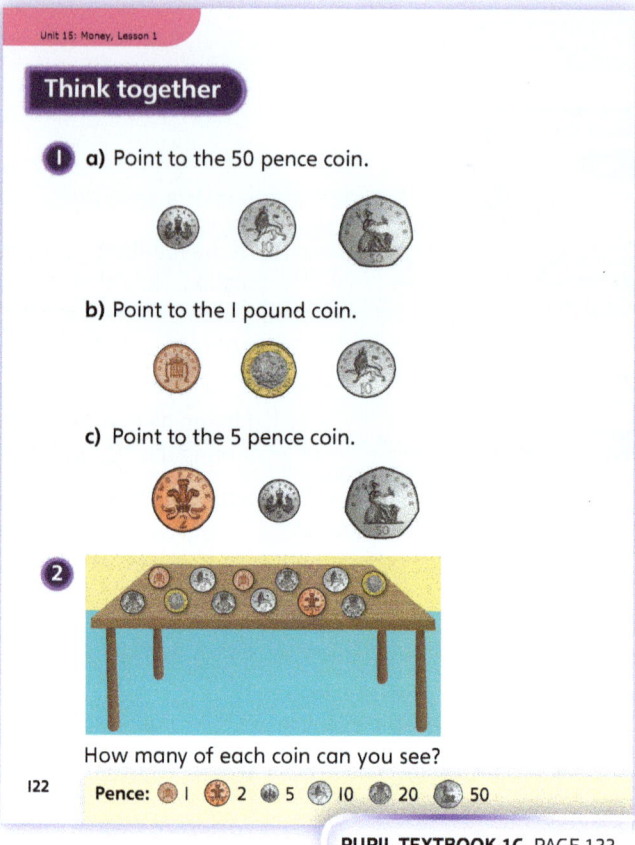

PUPIL TEXTBOOK 1C PAGE 122

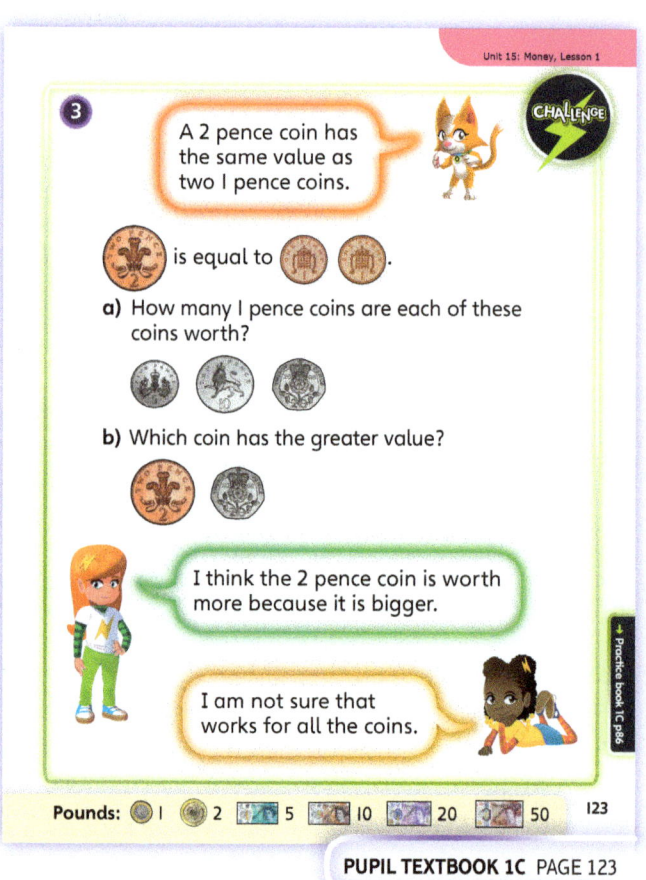

PUPIL TEXTBOOK 1C PAGE 123

Unit 15: Money, Lesson 1

Practice

WAYS OF WORKING Independent thinking

IN FOCUS Questions ❶ and ❷ assess whether children can recognise the different coins. Ask them initially to try to answer these questions without looking back at their Textbook. If they are unsure, children should look at the coins in detail and use the value on the coin to help them. Question ❸ asks children to work out how many of each type of coin there are.

STRENGTHEN Children who are having difficulty recognising coins will benefit from further practice with real or plastic coins. Make sure that children turn them 'tails up' so that they can compare them to the pictures in the Practice Book. For question ❹, ask children to exchange the coin for the number of 1p coins it represents. Children can then work out whether they have more than 10 coins or less than 10 coins.

DEEPEN Children could start to look at which coins can be added together to make the value of another coin. For example, two 5 pence coins are the same as one 10 pence coin. Ask: *What other coins add together to make 10 pence?*

THINK DIFFERENTLY Question ❹ requires children to work out whether the value of a coin is less than or greater than 10 pence.

ASSESSMENT CHECKPOINT Use questions ❶ and ❷ to check whether children recognise different coins. Use question ❹ to check that all children are confident with the relative values of the coins in relation to 10 pence. Check that they use the terms 'more than' and 'less than' correctly in this context. This example allows you to check the basic idea without the added complication of dealing with the 1 pound and 2 pound coins.

ANSWERS Answers for the **Practice** part of the lesson can be found in the *Power Maths* online subscription.

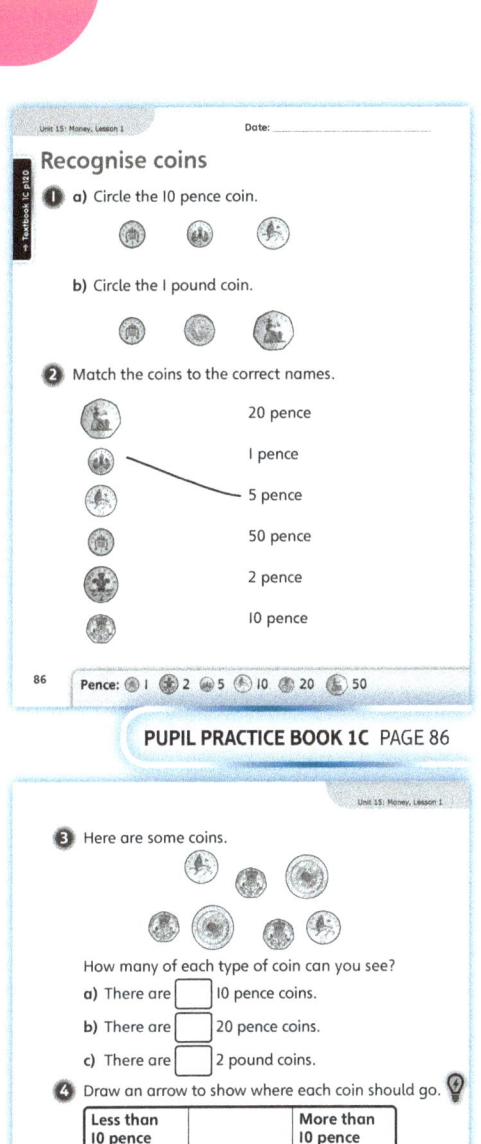

PUPIL PRACTICE BOOK 1C PAGE 86

PUPIL PRACTICE BOOK 1C PAGE 87

Reflect

WAYS OF WORKING Pair work

IN FOCUS Encourage children to use a logical series of questions, rather than just guessing. Children could use different kinds of questions (for example, about shape or colour) to reinforce their coin recognition as well as their understanding of the values of the coins.

ASSESSMENT CHECKPOINT Ask: *Can you always work out what coin your partner chose within three questions?* It can be done: for example, start by asking: *Is it silver?* If it is, ask: *Is it round?* Finally ask about value. If the coin is not silver, ask: *Is it copper?* Then ask about value.

ANSWERS Answers for the **Reflect** part of the lesson can be found in the *Power Maths* online subscription.

After the lesson

- Are children able to recognise all of the coins? What opportunities can you provide for extra practice where needed (for example, consider shopping games)?
- Do children have a good understanding of the 'real-world' value of the coins – that is, what the coins could actually buy?
- What links can be made with other areas of maths, such as addition and subtraction?

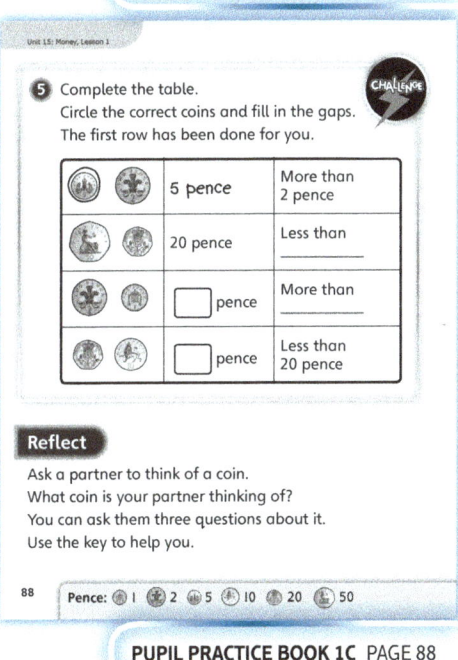

PUPIL PRACTICE BOOK 1C PAGE 88

161

Unit 15: Money, Lesson 2

Recognise notes

Learning focus
In this lesson, children will learn to recognise and compare banknotes.

Before you teach
- Have children had experience dealing with banknotes (perhaps in the context of 'play money' in board games)?
- Can children see the connection between the value of some of the coins and the value of the banknotes (5, 10, 20, and 50)?

NATIONAL CURRICULUM LINKS

Year 1 Measurement – money
Recognise and know the value of different denominations of coins and notes.

ASSESSING MASTERY

Children can recognise all of the different banknotes and can arrange them in order of value. Children can compare two banknotes using the words 'less than' or 'greater than'. Children know that there are limited denominations available (£5, £10, £20 and £50) and that each banknote has its own colour and size.

COMMON MISCONCEPTIONS

Children may not recognise all the banknotes. Show children a selection of banknotes. Ask:
- *Which banknote is this? What is its value?*

STRENGTHENING UNDERSTANDING

Children who lack a secure grasp of place value may have difficulty when comparing the value of the £5 note with some of the others. It may be useful to refer to the purchasing power of the £5 note in relation to the other notes, to establish that it is the least valuable of them.

GOING DEEPER

Deepen understanding by encouraging children to attempt some harder problems with notes. Ask: *What would one hundred pounds look like? How many different ways could you count out one hundred pounds in notes?*

KEY LANGUAGE

In lesson: notes, banknotes, pound, money, greatest, smallest, least, most, count, >, <

Other language used by the teacher: compare, greater than, less than, worth, value

STRUCTURES AND REPRESENTATIONS

Currency – banknotes and coins

RESOURCES

Mandatory: images of banknotes (real and pretend)

 In the eTextbook of this lesson, you will find interactive links to a selection of teaching tools.

Quick recap

Show children some coins on the board. Ask them if they can recognise the coins. You might want to give children multiple choice options (for example, show a £1 coin and ask if it is 1 pence or 1 pound, etc.).

Discover

WAYS OF WORKING Pair work

ASK
- Question 1 a): *Can you see the sponsorship money? Are there any coins?*
- Question 1 a): *Is it a lot of money? How can you tell?*

IN FOCUS Question 1 a) should be used to draw out the idea that we are dealing with larger sums of money. Explain that a lot of money has been collected and that it has been sorted out into piles of banknotes – perhaps to make it easier to count.

PRACTICAL TIPS Show children real or pretend banknotes, so they can look at them to see how they are different.

ANSWERS

Question 1 a):  5 pound notes

 10 pound notes

 20 pound notes

 50 pound notes

b):

least ──────────→ greatest

PUPIL TEXTBOOK 1C PAGE 124

Share

WAYS OF WORKING Whole class teacher led

ASK
- Question 1 a): *Which of these notes have you seen in real life?*
- Question 1 a): *Can you see the numbers that tell you how much each note is worth?*

IN FOCUS Recognising real notes is an important part of the learning in this lesson. You might like to have some real notes available to show children, especially so that they can see the designs on the other sides, and get an idea of their actual sizes. Point out that you are showing the 'heads' side of the banknotes – the distinct colour and size of the notes and the clear numerical markings should be sufficient to distinguish them, despite the overall similarity of their designs. Remind children of the connection between the value of the silver coins and the value of the banknotes (5, 10, 20 and 50 only).

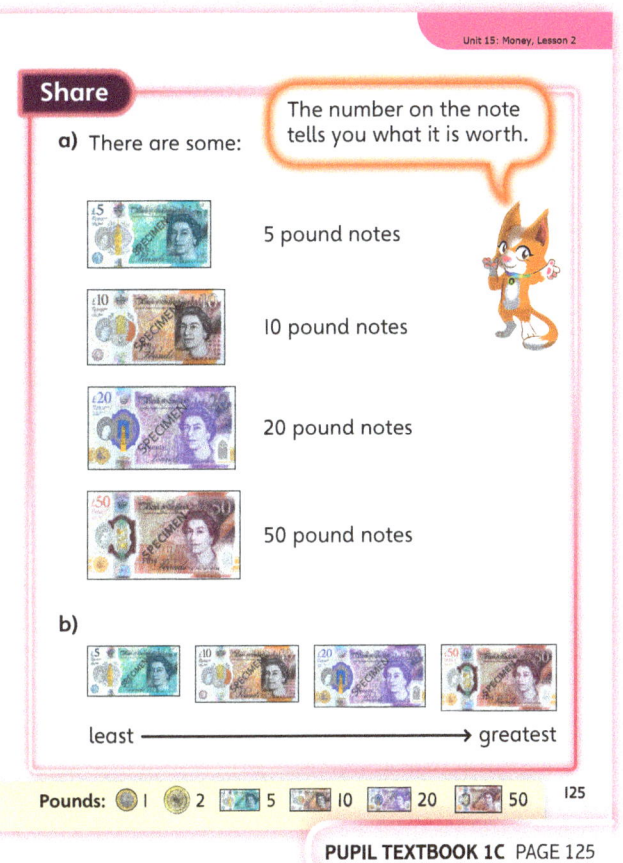

PUPIL TEXTBOOK 1C PAGE 125

163

Unit 15: Money, Lesson 2

Think together

WAYS OF WORKING Whole class teacher led (I do, We do, You do)

ASK
- Question ❶: *This is the same sponsorship money as in the* **Discover** *section – what is different now?*
- Question ❶: *Can you see all of the notes now? What will you have to do to find out how many of each note there are?*
- Question ❸: *What information will you use to decide which note is worth more?*

IN FOCUS Question ❶ asks children to count how many notes of each kind there are on the table. Make sure everyone understands that the 'How many?' amount that is asked for is the number of notes of each denomination, not the total amount of money.

STRENGTHEN Children who find it difficult to determine the value of a banknote should be encouraged to look for the number in the top left corner (with the King or Queen's head uppermost). The designs of the notes are deliberately complicated, but the denomination shown clearly in the top left corner is a consistent feature in all of them.

DEEPEN Question ❸ asks children to compare the values of the notes. Make the comparisons more challenging by comparing images where the denomination is obscured, or asking questions such as: *Which is worth more – a small banknote or a larger one?*

ASSESSMENT CHECKPOINT Question ❷ should be used to check that children have grasped the key learning point of this lesson, by being able to distinguish and recognise each of the notes.

ANSWERS

Question ❶:

Note	How many?
£5	5
£10	4
£20	2
£50	3

Question ❸ a):

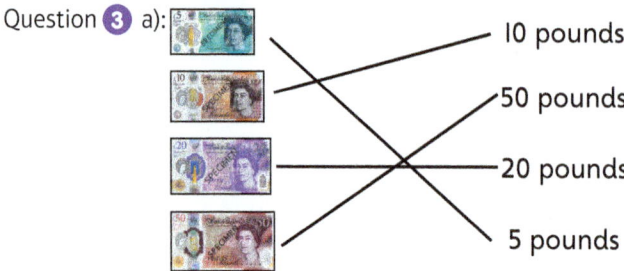

Question ❸ a): £5 < £50
Question ❸ b): £20 > £10
Question ❸ c): £10 < £50

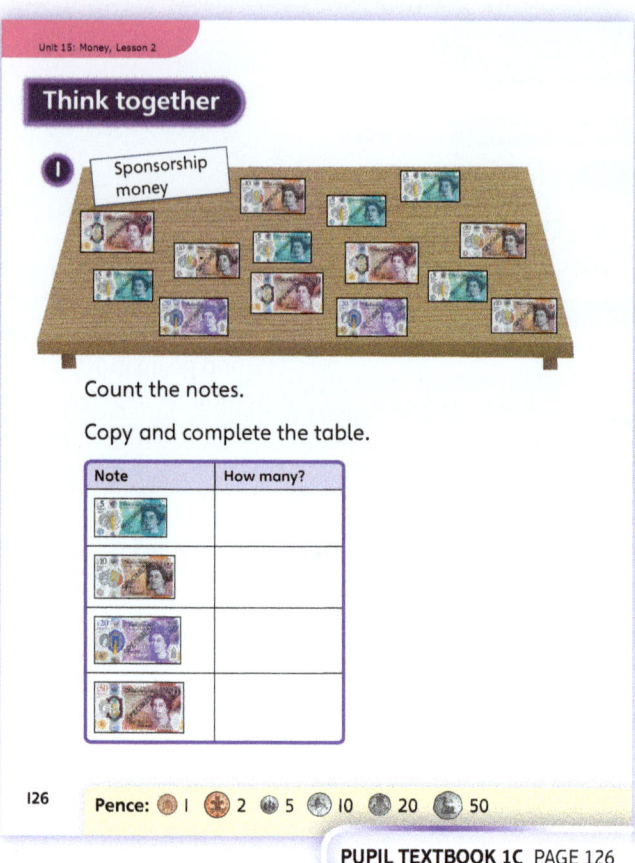

PUPIL TEXTBOOK 1C PAGE 126

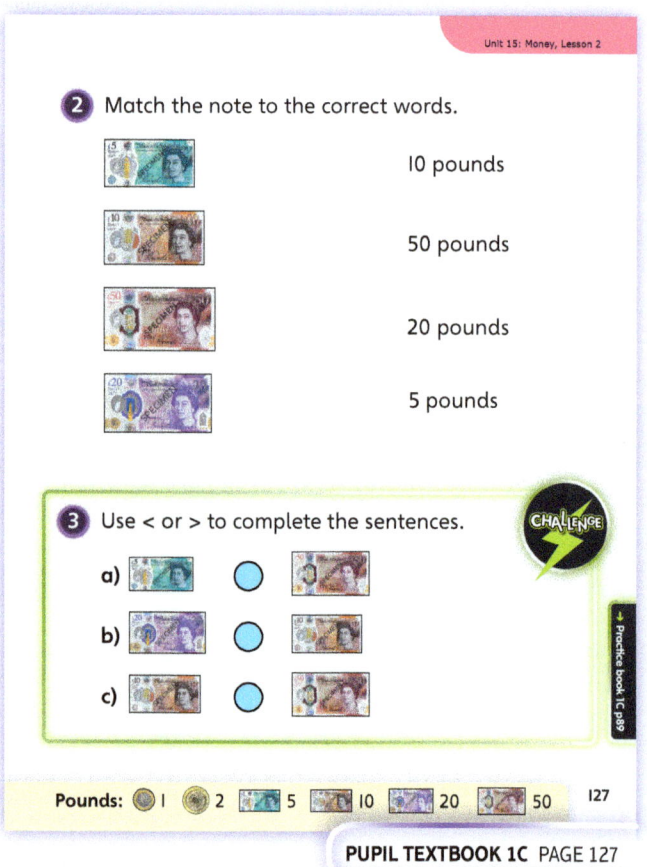

PUPIL TEXTBOOK 1C PAGE 127

Unit 15: Money, Lesson 2

Practice

WAYS OF WORKING Independent thinking

IN FOCUS In questions ❶, ❷ and ❸, children are expected to recognise the banknotes and know their values. Question ❺ requires children to use reasoning to explain why the notes are not ordered from greatest value to least value.

STRENGTHEN Children who still find it difficult to recognise the notes may benefit from further work with colour images made to the correct relative sizes.

DEEPEN Children who find question ❺ easy could be asked to draw the sequence of notes with the appropriate inequality sign between each pair. They should notice that the position of the note that is in the wrong order corresponds to an inequality sign that 'changes direction' relative to the others.

THINK DIFFERENTLY Question ❹ involves comparing 'notes with notes', as well as 'notes with values'. The extra level of abstraction involved provides an interesting additional challenge.

ASSESSMENT CHECKPOINT Question ❸ provides a good opportunity to check that children can recognise all of the notes; even without the availability of colour as a cue, the designs should be sufficiently distinct to make counting the different denominations reasonably straightforward. Some children might like to think about checking strategies – perhaps counting the total number of notes and making sure that the numbers in the boxes make the same total.

ANSWERS Answers for the **Practice** part of the lesson can be found in the *Power Maths* online subscription.

PUPIL PRACTICE BOOK 1C PAGE 89

PUPIL PRACTICE BOOK 1C PAGE 90

Reflect

WAYS OF WORKING Pair work

IN FOCUS The point here is to focus on the denomination of the notes – however realistic the design looks, there is no such thing as a 15 pound or 30 pound note.

ASSESSMENT CHECKPOINT This task provides a check on an important piece of real-world knowledge, rather than anything purely mathematical. Children who fail to circle any of the real notes could be shown relevant images in the book; if they indicate that the 15 pound or 30 pound notes are real, you could ask them to look back and try to find the matching picture.

ANSWERS Answers for the **Reflect** part of the lesson can be found in the *Power Maths* online subscription.

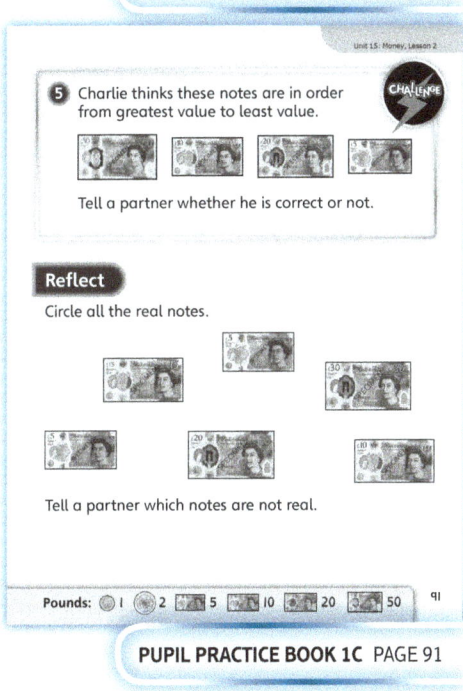

PUPIL PRACTICE BOOK 1C PAGE 91

After the lesson ⏸

- Are children using visual clues to recognise the notes or are they still reading the numerical values printed on the notes?
- Did children make the connection with the relative values of the coins, knowing that there are a limited number of different denominations?

165

Unit 15: Money, Lesson 3

Count in coins

Learning focus

In this lesson, children will find and compare the total value of small sets of coins.

Before you teach

- Do children recognise all of the different coins?
- Is children's understanding of the relative values of the various coins secure?
- Can children count in 2s, 5s and 10s?

NATIONAL CURRICULUM LINKS

Year 1 Measurement – money

Recognise and know the value of different denominations of coins and notes.

ASSESSING MASTERY

Children can recognise coins and use their number skills and knowledge to find the total of a small group of coins of the same kind. They may do this by 'skip counting' in multiples of the coin denomination (for example, counting '5, 10, 15, …' to find the total value of a group of 5 pence coins), or they may use known number facts (for example, 'four 2s are 8' to find the value of four 2 pence coins).

COMMON MISCONCEPTIONS

The most likely confusion in this lesson is likely to arise in moving from counting to calculating – children are expected to compare groups of coins based on their total value, rather than the number of coins in each group. Ask:
- *How many coins are there? What is their total value?*

STRENGTHENING UNDERSTANDING

Use language patterns that emphasise the idea that we are looking for the total value of each for the groups of coins that we are dealing with in this lesson. For example, be careful to ask: *How much is this group worth?* rather than: *How much is this?*

GOING DEEPER

Children who grasp this learning quickly could go on to finding totals of mixed groups of coins. For example, a simple game for pairs could involve one child taking a small collection of mixed coins and challenging a partner to guess the total value, which can then be worked out accurately.

KEY LANGUAGE

In lesson: worth, amount, total, coin, money, altogether, pence, <, >, =, copper

Other language used by the teacher: value, denomination

STRUCTURES AND REPRESENTATIONS

Coins, number lines

RESOURCES

Mandatory: coins – real or realistic plastic

Optional: number lines, 100 squares

 In the eTextbook of this lesson, you will find interactive links to a selection of teaching tools.

Quick recap

In this lesson, it is important that children are confident in counting in 2s, 5s and 10s. Start at 0 and count in 2s first, then 10s and then 5s. Use a number line or manipulatives to support children with their counting.

Unit 15: Money, Lesson 3

Discover

WAYS OF WORKING Pair work

ASK
- Question 1 a): *Where are the coins? How have they been arranged on the table?*
- Question 1 a): *What sorts of coins can you see? Why do you think they have been put into lines like that?*

IN FOCUS Question 1 b) focuses on the key distinction between the number of coins in a group and the total value of the group. More coins in a group does not necessarily mean the total value is greater than that of a group with fewer coins in it. Some children may need number lines or a 100 square to support their counting in 2s, 5s and 10s.

PRACTICAL TIPS Recreate the scenario in the classroom using real or plastic coins.

ANSWERS

Question 1 a): There are five 1 pence coins
= 5 pence altogether.
There are four 2 pence coins
= 8 pence altogether.
There are six 5 pence coins
= 30 pence altogether.
There are four 10 pence coins
= 40 pence altogether.

Question 1 b): The most coins are in the 5 pence line.
The most money is in the 10 pence line.

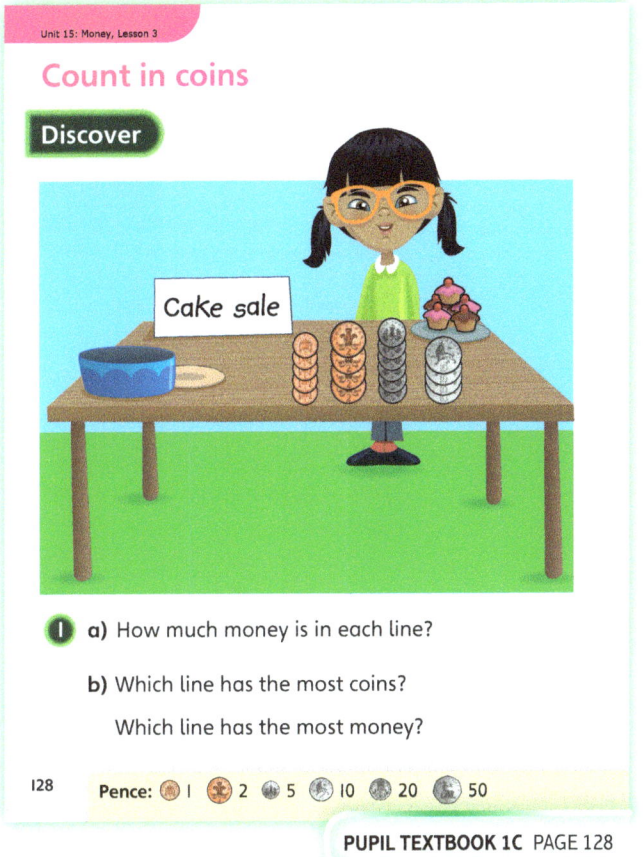

PUPIL TEXTBOOK 1C PAGE 128

Share

WAYS OF WORKING Whole class teacher led

ASK
- Question 1 a): *Dexter counted the coins in each line. Did you start by doing the same thing?*
- Question 1 a): *It is easy to work out the value of the 1 pence pieces. Can you explain why?*
- Question 1 a): *You have to do a bit more work for the other lines of coins. Can you explain how Dexter found the total of the line of 2 pence pieces?*

IN FOCUS Question 1 b) focuses on the distinction between the number of coins in a line (which is a simple count of the coins, disregarding their value) and the total value of the line (which is essentially a multiplication problem, although the actual calculation will most likely be done by repeated addition or 'counting in steps' in this case). This is the key content of the lesson and it is worth spending some time making sure that the distinction between the number of coins and their total value is clearly understood.

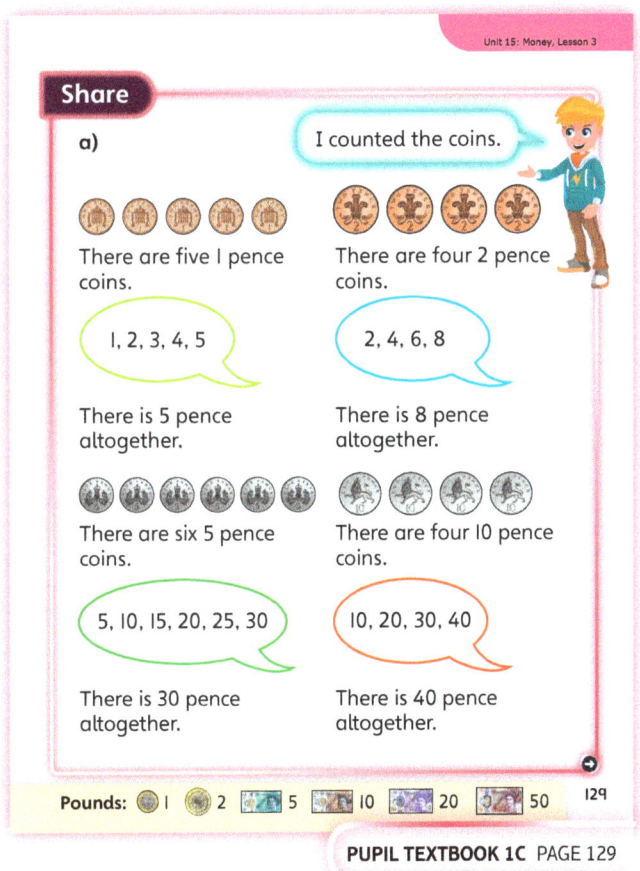

PUPIL TEXTBOOK 1C PAGE 129

Unit 15: Money, Lesson 3

Think together

WAYS OF WORKING Whole class teacher led (I do, We do, You do)

ASK
- Question ❶: *What kind of coins are these? How many of them are there? What is their total value?*

IN FOCUS In question ❷, make sure that children understand that it is the total values of the groups of money on each side that are being compared. Notice how the inequality signs are being used in a slightly more demanding context than previously – children have to compare values calculated on the basis of the images shown, rather than numbers that are immediately apparent in the question.

STRENGTHEN Where children find it difficult to compare the total values of the coins in question ❷, support them in using a number line or 100 square to find each total before working out the correct inequality sign to use. Provide more examples with fewer coins for children to try in pairs. For example: two 5 pence coins and one 20 pence coin; one 5 pence coin and four 1 pence coins (or two or three 2 pence coins).

DEEPEN Question ❸ can easily be extended and will provide a good opportunity to explore number patterns. Ask: *What if Mia had a different amount of money such as 30 pence, or 12 pence? What coins could she have then? Can you make up some more examples of your own? Can you make 11 pence with just 2 pence coins?*

ASSESSMENT CHECKPOINT Use question ❷ c) to check that children have understood the importance of comparing total values of coins rather than the number of coins. There are more 1 pence coins but the 2 pence coins are worth more. Question ❸ tests children's understanding of multiples of 2, 5 and 10.

ANSWERS

Question ❶ a): 12 pence

Question ❶ b): 50 pence

Question ❶ c): 20 pence

Question ❷ a): 3 × 5p > 3 × 2p

Question ❷ b): 5 × 2p = 10p

Question ❷ c): 5 × 1p < 3 × 2p

Question ❸: Mia could have fifteen 1 pence coins or three 5 pence coins.
15 is an odd number so it cannot be made with just 2 pence coins.
15 cannot be made exactly with 10 pence coins.

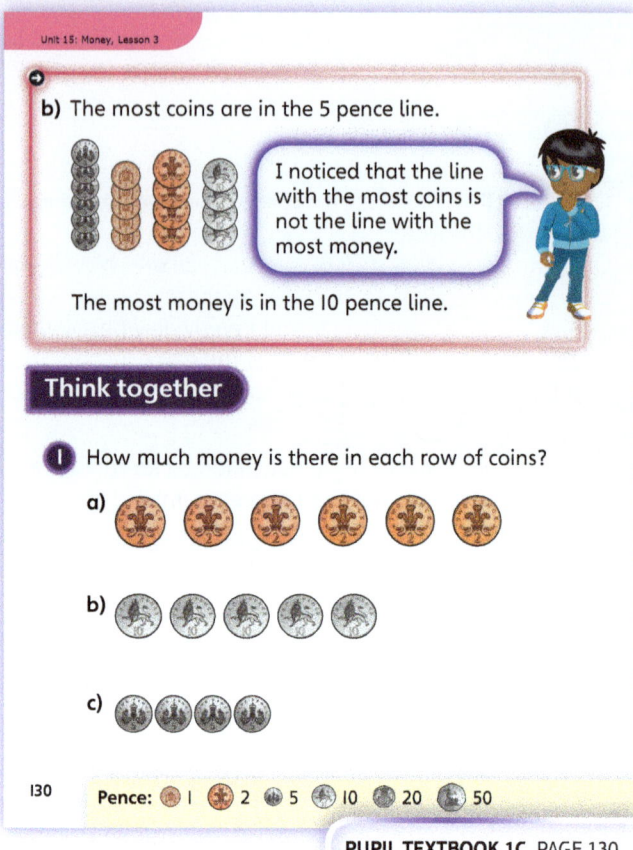

PUPIL TEXTBOOK 1C PAGE 130

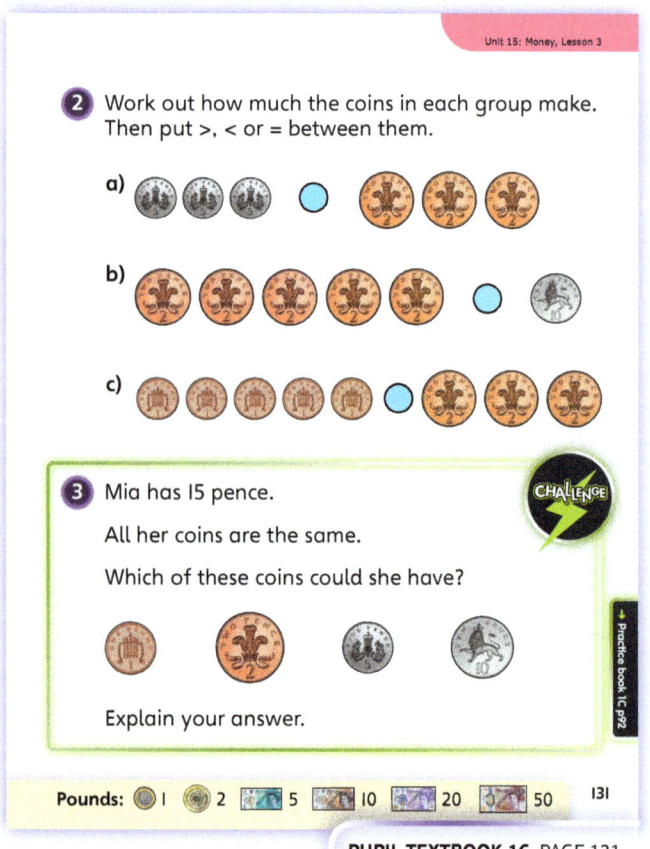

PUPIL TEXTBOOK 1C PAGE 131

Unit 15: Money, Lesson 3

Practice

WAYS OF WORKING Independent thinking

IN FOCUS Questions ❶ to ❹ help reinforce children's understanding of the value of coins, and also give them further practice in counting in 1s, 2s, 5s and 10s. Question ❸ illustrates that, because a 10 pence coin is worth twice the value of a 5 pence coin, children need twice the number of 5 pence coins to make the same total as they do 10 pence coins. In question ❺, emphasize that Lucy has 6 coins all the same and Amy has 3 coins all the same. The key understanding that children should gain is that the value of Amy's coin must be twice the value of Lucy's coin, as Amy has half the number of coins that Lucy has, yet they have the same amount of money.

STRENGTHEN Some children may find the drawing in question ❷ difficult. Provide apparatus if required; children could make the totals using real or plastic coins and then draw around the coins to record their answers.

DEEPEN Question ❺ can be extended easily. Children who have a sound grasp of the material in the lesson could be asked: *If Lucy had 5 copper coins and Amy had 10 copper coins, and they still had the same amount of money, what coins would they have then?*

ASSESSMENT CHECKPOINT Use question ❷ to check that children can make totals using either 1p or 2p coins. Question ❺ tests to see if children can work logically to find a solution. Can they explain why there is only one solution?

ANSWERS Answers for the **Practice** part of the lesson can be found in the *Power Maths* online subscription.

Reflect

WAYS OF WORKING Pair work

IN FOCUS The choice of a total of 20 pence means that any of the coins can be used in this example. Ask: *What happens if the total amount you want to make is different? Which coin can you always use? How could you tell if you could make the new amount with 2 pence coins?*

ASSESSMENT CHECKPOINT Children may choose to draw the coins rather than write them out. Children's drawings may be a little hard to follow, especially with the 1 pence and 2 pence coins. If the drawings look wrong, check whether the child's intention was correct by asking them how they decided what to draw.

ANSWERS Answers for the **Reflect** part of the lesson can be found in the *Power Maths* online subscription.

After the lesson

- Are children confident in finding the totals of small groups of coins?
- What strategies did you observe children using for finding totals?
- What opportunities will there be for children to work with money away from formal lessons?

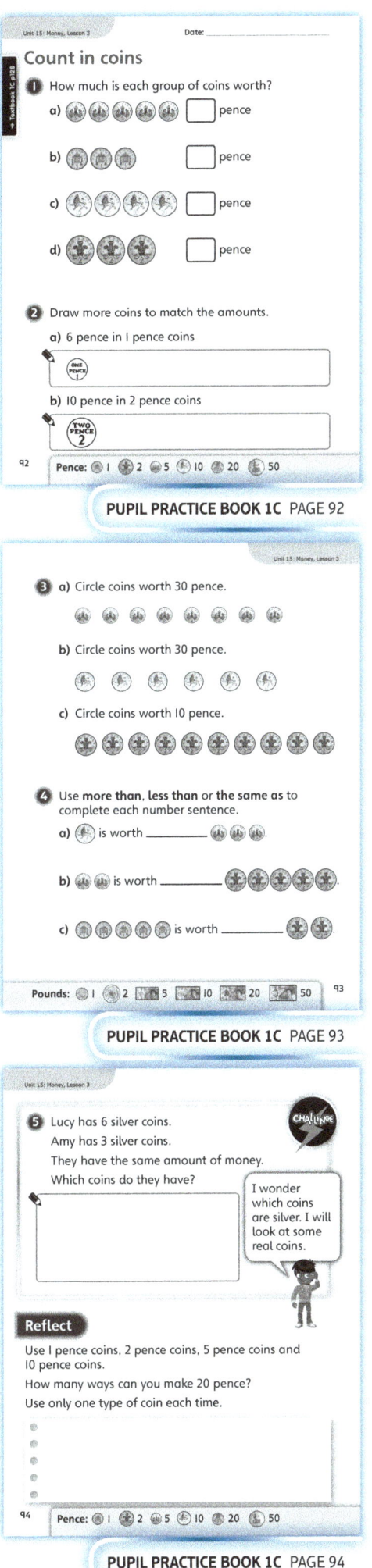

PUPIL PRACTICE BOOK 1C PAGE 92

PUPIL PRACTICE BOOK 1C PAGE 93

PUPIL PRACTICE BOOK 1C PAGE 94

Unit 15: Money

End of unit check

Don't forget the unit assessment grid in your *Power Maths* online subscription.

WAYS OF WORKING Group work adult led

IN FOCUS
- Questions 1 and 2 check that children recognise coins and notes from their value.
- Question 3 is a good opportunity to check that children can recognise the coin shown, that they understand that the total amount of money is required, and that they can count using the pattern of multiples (2 in this case).
- Question 4 tests whether children know the relative worth of a coin in comparison to the values of other coins.
- Question 5 checks that children know there are only certain denominations of notes (and coins) and that a 25 pounds note is not a real note.

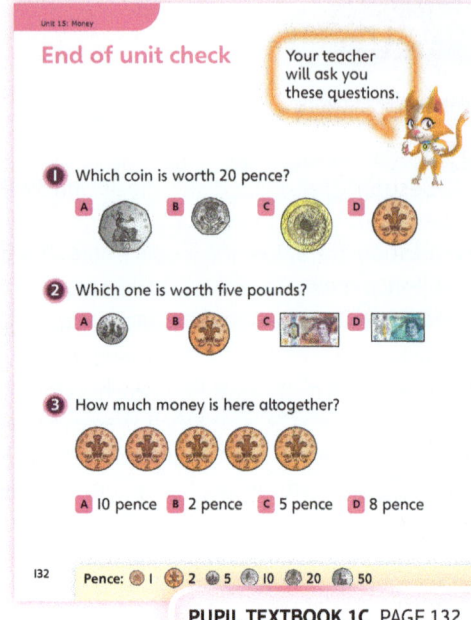

PUPIL TEXTBOOK 1C PAGE 132

Think!

WAYS OF WORKING Pair work or small groups

IN FOCUS All children will benefit from the practice and reinforcement provided by this activity.

Where appropriate, encourage more systematic ways of working and recording. Ask: *How can you be sure that you have found all the ways of doing it? What is the clearest way of writing all this down? How many different ways can you find that use one type of coin? Can you make 20 pence just using 5 pence and 10 pence coins? How many different ways can you make 20 pence using only 5 pence and 1 pence coins; or 5 pence and 2 pence coins; or 10 pence, 2 pence and 1 pence coins?*

Ask children to think about how many different types of coin each combination uses. Ask: *Can you make 20 pence using at least one of each of the 1 pence, 2 pence, 5 pence and 10 pence coins?*

ANSWERS AND COMMENTARY Children who have mastered the concepts of this unit will be able to recognise real coins and banknotes, find the total value of a small group of coins and banknotes, compare the value of two groups of coins or notes and make totals with different combinations of coins.

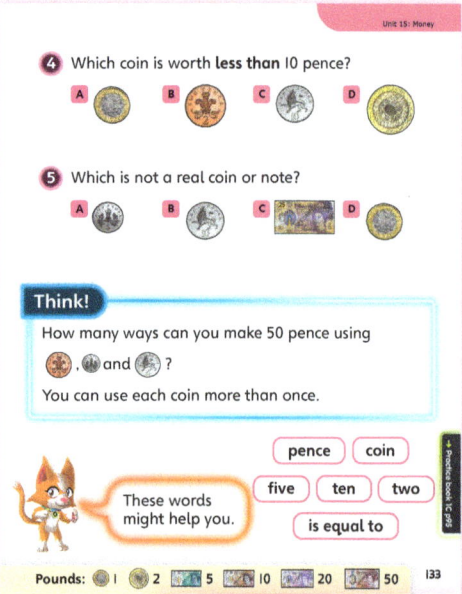

PUPIL TEXTBOOK 1C PAGE 133

Q	A	WRONG ANSWERS AND MISCONCEPTIONS	STRENGTHENING UNDERSTANDING
1	B	Where children claim that A (50 pence) or C (2 pounds) are 'worth 20 pence', check that they did not misinterpret the question as asking whether the coins are worth '20 pence or more'.	Children may be less familiar with the higher value coins (1 pound and 2 pounds); these coins are also 'unusual' in that their values are in pounds rather than pence, and the numerical values are often given in words only (not figures). All of these factors may mean that some children may need a little more practice in dealing with these coins. It is important not to assume that any hesitancy necessarily reflects significant mathematical difficulties.
2	D	Children who answer A may not understand the difference between pounds and pence (or may have simply misread the question). Answer C could signify a lack of understanding of place value.	
3	A	Children who give answer C may be counting the coins rather than understanding that the total amount of money is required.	
4	B	Answer A may reflect a confusion between 'less than' and 'smaller than'.	
5	C	Other answers indicate that children are not sure of the limited denominations of the notes.	

Unit 15: Money

My journal

WAYS OF WORKING Independent thinking

ANSWERS AND COMMENTARY

Children may start by using just one type of coin and will often simply draw a collection of coins to represent their answer. More efficient representations will involve writing simple calculations (5 pence + 5 pence + 5 pence + 5 pence = 20 pence). When using more than one type of coin, some children may not appreciate that 'the same coins in a different order' do not in fact represent a new combination. Some children may move towards a tabular form of recording, such as a table with three columns headed 2 pence, 5 pence and 10 pence, where each row of the table shows how many of each coin are used in each combination totalling 20 pence.

Power check

WAYS OF WORKING Independent thinking

ASK

- Do you think that you know more about money now?
- Can you tell quickly which coin is which, or do you have to read what it says on the coin?
- The coins are copper, silver or gold coloured. Do you know which is which?

Power play

WAYS OF WORKING Pair or small group work

IN FOCUS This activity provides a further opportunity for children to use and practise their understanding of coin properties (size, shape, value, colour), as well as developing mental arithmetic approaches and devising efficient sets of questions.

ANSWERS AND COMMENTARY Look out for children whose questions or answers suggest insecure knowledge of coin properties; for example, if we already know that the selected coins are round, there is no point asking whether they are 20 pence pieces. These children may need further practice at handling coins.

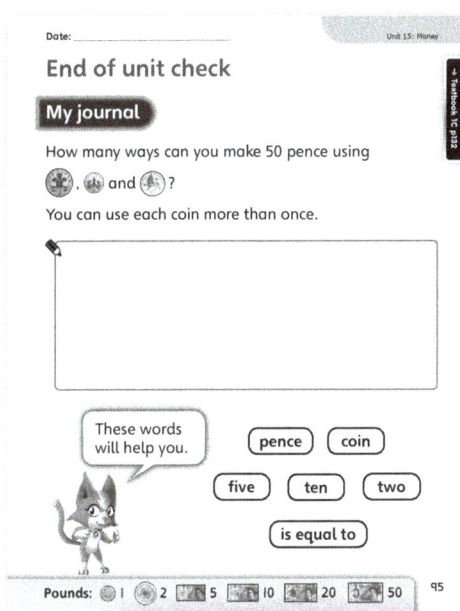

PUPIL PRACTICE BOOK 1C PAGE 95

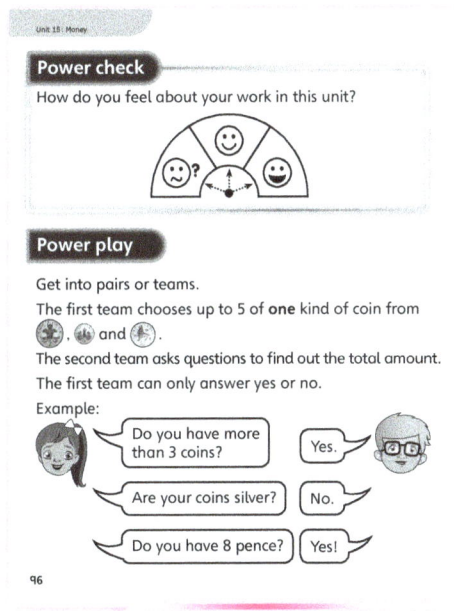

PUPIL PRACTICE BOOK 1C PAGE 96

After the unit

- Are children confident enough to recognise and use all of the different coins?
- Did you see any unexpected approaches to recording or calculation during this unit?
- What further opportunities will children have to use money?

Strengthen and **Deepen** activities for this unit can be found in the *Power Maths* online subscription.

Unit 16
Time

Mastery Expert tip! 'As a teacher, I am always talking about time and this means that there are lots of ways to reinforce the learning from this unit outside of lessons. I use the classroom clock a lot when talking about the events of the day and how long things will take.'

Don't forget to watch the Unit 16 video!

WHY THIS UNIT IS IMPORTANT

This unit introduces children to various aspects of time. Children will develop their ability to tell the time by reading an analogue clock or watch, estimating and comparing durations, and carrying out simple calculations involving time. All of these skills have real practical importance in daily life. Despite the popularity of digital displays, the ability to read time 'at a glance' from an analogue display remains a vital skill for children to learn.

In this unit, children strengthen their understanding of the hands of a clock. Children will develop their understanding of units of measurement of time (hours and minutes). They will use the following vocabulary to develop their understanding of durations of time and the ordering of events in time: 'before', 'after', 'yesterday', 'today', 'tomorrow', 'day', 'week', 'date', 'month', 'year' and 'calendar'.

WHERE THIS UNIT FITS

→ Unit 15: Money
→ **Unit 16: Time**

This unit builds on, and formalises, children's experiences of using various measurements of time in daily life, as well as their prior experience with numbers, calculations and problem solving.

Before they start this unit, it is expected that children:
- recognise a clock face and a calendar, and understand that they are used to tell the time and day or date, respectively
- can carry out simple addition and subtraction calculations
- can use real-world knowledge and experience to sequence events.

ASSESSING MASTERY

Children who have mastered the work in this unit will be able to work confidently within simple situations involving time, including understanding clocks and calendars. They will be able to tell the time to the half hour using an analogue clock or watch, and use a calendar to say what day of the week a particular date falls on. They will be able to use a range of language to order familiar events.

COMMON MISCONCEPTIONS	STRENGTHENING UNDERSTANDING	GOING DEEPER
Some children may find sequencing events hard and may need support in using the appropriate language (such as 'before' and 'after').	Use cards with pictures of events that children can move around, to help them see the events in a different order.	Some children will be ready for more advanced language patterns about sequencing events, involving several factors: 'the blue car arrived after the yellow car, but before the red one'.
Some children may have little knowledge of important facts (like the sequence of months) and some real-world experiences may be misleading: children may have heard adults say things such as 'We had to wait hours for the bus this morning!'.	Make sure that the learning in this unit is related to real-world experiences wherever possible, and use opportunities to talk about times, dates and sequences of events as these arise during the school routine.	Some children will already know the months of the year, days of the week and how to tell the time. The activities in this unit will allow you to find out how secure this knowledge is. Can children explain what they know, and use it to solve problems?

Unit 16: Time

UNIT STARTER PAGES

Use these pages to introduce the unit focus to children. You can use the characters to explore different ways of working.

The rate of progress through this unit will depend to some extent on children's existing knowledge and real-world experiences. You may find it useful to talk through each of the examples here in turn and try to get an idea of children's current understanding.

STRUCTURES AND REPRESENTATIONS

Clocks: Clocks are used regularly to show how to tell the time and measure durations of time. They are used for demonstration purposes and also as the basis of problems to solve. Various clocks should be used, including analogue wall clocks, watches, stopwatches and sand timers.

Calendars: Calendars are used to show particular dates and the passage of time over periods of days, weeks and months within a year. They indicate the day of the week as well as the number of days within a month and the months of the year.

KEY LANGUAGE

There is some key language that children will need to know as part of the learning in this unit:

- before, after
- faster, slower, shorter, longer, earlier, later
- yesterday, today, tomorrow
- day, week, month, year
- Monday, Tuesday, Wednesday, Thursday, Friday, Saturday, Sunday
- calendar, date
- minute hand, hour hand
- o'clock, half past
- minute, hour

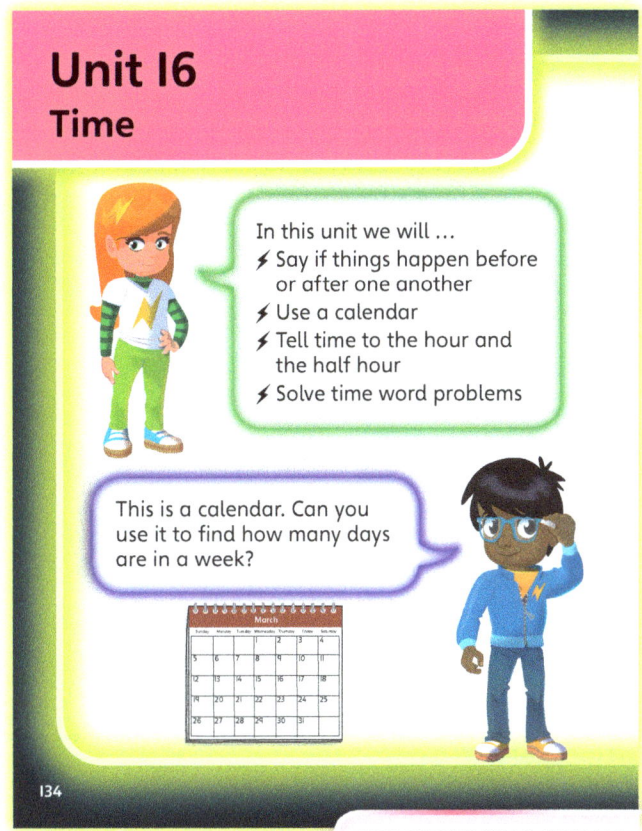

PUPIL TEXTBOOK 1C PAGE 134

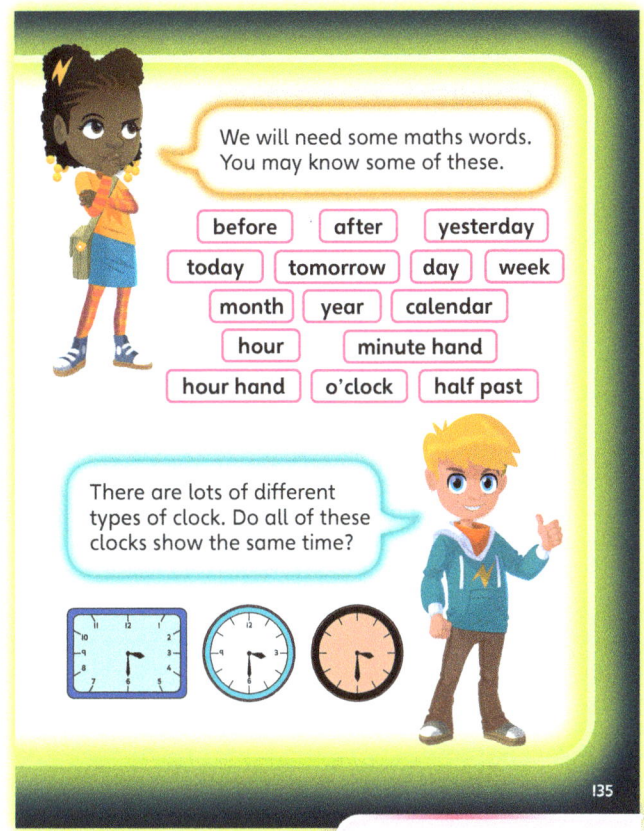

PUPIL TEXTBOOK 1C PAGE 135

Unit 16: Time, Lesson 1

Before and after

Learning focus
In this lesson, children will use a range of language to sequence events in chronological order.

Before you teach
- Are children confident sequencing numbers?
- What resources will you provide for children who are still developing this skill?

NATIONAL CURRICULUM LINKS

Year 1 Measurement – time

Sequence events in chronological order using language (for example, before and after, next, first, today, yesterday, tomorrow, morning, afternoon and evening).

ASSESSING MASTERY

Children can use a range of language, especially 'before' and 'after', but also 'yesterday', 'today', 'tomorrow' to describe the order of a series of events, including events in their own daily routine. Children can make sensible suggestions about things that might have happened before or after a given event.

COMMON MISCONCEPTIONS

Children may have problems with sequencing. Earlier work on number and measurement will probably already have identified children who find sequencing particularly difficult, but it is sensible to use this lesson as a further check. Ask:
- *What will we do before lunch? What will we do after lunch?*

STRENGTHENING UNDERSTANDING

Where children find it difficult to handle the 'information content' of a number of events, make some simple cards to represent the events – arranging a set of cards makes less demand on working memory than trying to sort events 'in your head'. In addition, some children may not know the sequence of days of the week, so this lesson is a good time to check that this knowledge is secure.

GOING DEEPER

The language patterns used in this lesson are very simple, but form the basis for more complex forms that some children may be ready for. For example, ask: *Can you tell me what things you do regularly after lunch and before bedtime?* Make sure that children understand that we are talking about one set of activities that meets both conditions (between lunch and bedtime).

KEY LANGUAGE

In lesson: before, after, first, then, next, yesterday, today, tomorrow

Other language used by the teacher: last, morning, afternoon, evening

STRUCTURES AND REPRESENTATIONS

Timelines (similar in style to number lines, but without numbers at this stage; they are used here solely for sequencing events)

RESOURCES

Optional: cards to represent events (pictorially)

 In the eTextbook of this lesson, you will find interactive links to a selection of teaching tools.

Quick recap
Talk generally about the activities that children did before they came to school this morning. At this stage, do not talk about the order in which they did them and avoid using the terms 'before' and 'after'. A general think about the activities they do to get ready for school will be helpful as they can then compare these to the **Discover** scenario.

Unit 16: Time, Lesson 1

Discover

WAYS OF WORKING Pair work

ASK

- Question 1 a): *What is the first thing that Nina does in the morning?*
- Question 1 a): *Is this a school day? How can you tell?*
- Question 1 b): *What is the last thing that you see Nina doing?*

IN FOCUS This activity features a set of pictures that are organised in 'comic strip' format to represent a series of activities taking place on one school day. The pictures are arranged in chronological order – before, during and after school.

PRACTICAL TIPS It may be helpful to hand out cards with pictures on, representing each activity that Nina does in a day. Children can then put them in order and move them into activities that Nina does before school and activities that Nina does after school.

ANSWERS

Question 1 a): Before school, Nina wakes up, gets dressed, eats her breakfast, brushes her teeth and walks to school.

Question 1 b): After school, Nina says goodbye to her friend, rides her bicycle and reads a book.

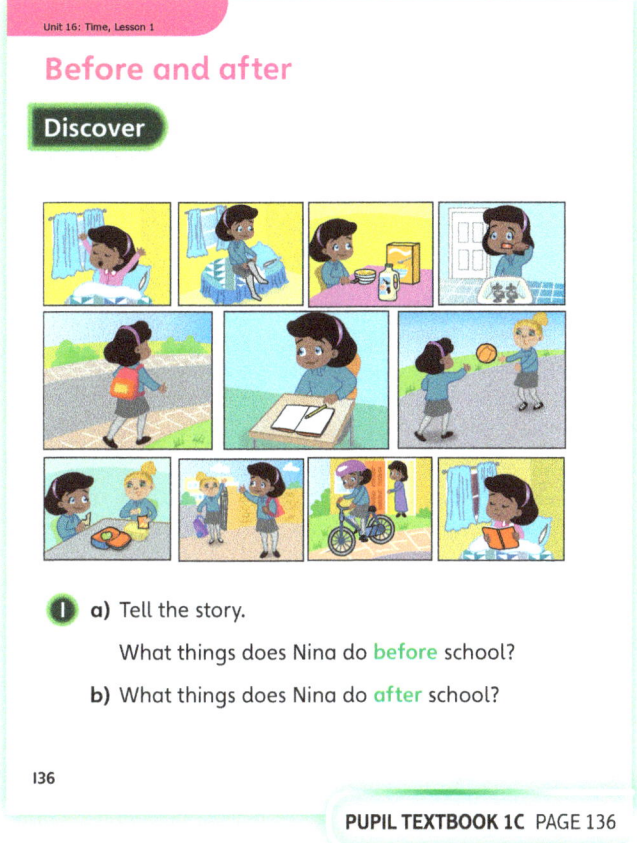

PUPIL TEXTBOOK 1C PAGE 136

Share

WAYS OF WORKING Whole class teacher led

ASK

- Question 1 a): *Look at what Flo says. What things do you do before school?*
- Question 1 b): *What things do you do after school?*

IN FOCUS The fact that the pictures are set out in order means that children could have pointed to the first picture of school and looked at the pictures that came before it, to answer question 1 a). If the pictures were in a random order, we would only have the 'common sense' method that Flo used. This might be ambiguous in some cases – for example, it could be argued that Nina could have gone out on her bike before school. However, the point of this activity is to get children thinking in terms of 'before, during and after', and using the corresponding language.

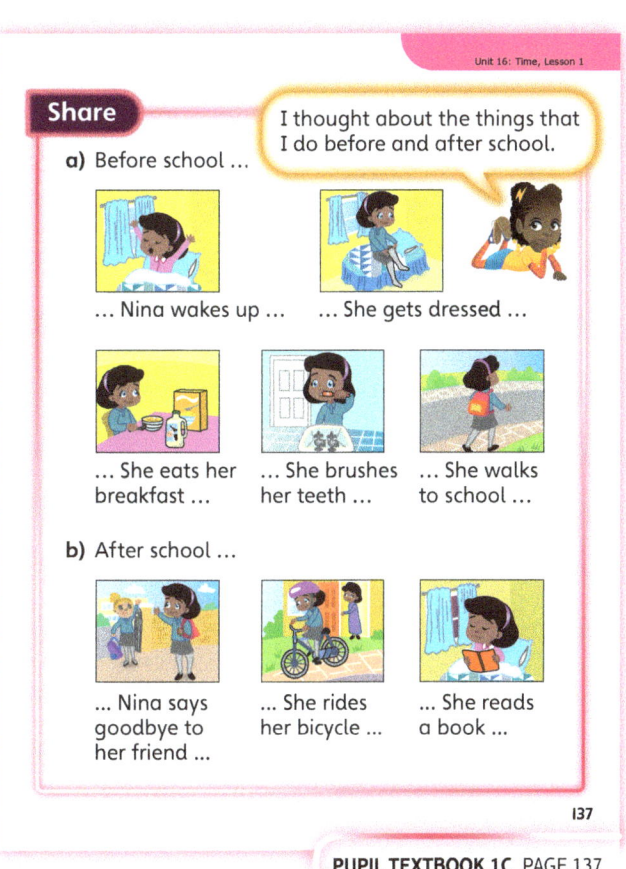

PUPIL TEXTBOOK 1C PAGE 137

175

Think together

WAYS OF WORKING Whole class teacher led (I do, We do, You do)

ASK

Question ②: *What is the child doing? Why are they doing that? What do you think might have happened after they watered the plant? What could have happened before?*

IN FOCUS Question ① asks children to use the language 'before' and 'after' to describe the story. Encourage them to use these words as they go through the story. Question ② asks them to think about what might have happened before and after the child waters the plant. You may relate this to any plants you have in the classroom or around school. Question ③ asks children to put the images in a sequential order. Again, encourage the correct use of the terms 'before' and 'after'.

STRENGTHEN There will be plenty of opportunities in children's daily routine to reinforce the sequence of day names and the use of the key vocabulary related to chronological order. You could make a point of asking questions using this vocabulary. Ask: *What are you doing tomorrow afternoon?*

DEEPEN The basic language patterns used in this lesson can naturally be extended to others that are more precise or longer term. For example, ask: *What did you do yesterday afternoon? What will you do in three day's time?*

ASSESSMENT CHECKPOINT Use questions ① and ② to check the correct use of the key language 'before' and 'after'.

ANSWERS

Question ①: Before: children should point to Joe putting his boots on. After: children should point to Joe having a drink of water.

Question ②: Children's answers will vary, for example: 1. (Before) plants a seed; 2. waters the shoot; 3. (After) the plant grows a big flower.

Question ③: Children's answers will vary, for example: First, the baker takes milk and eggs from the fridge. Then, she puts all the ingredients in a bowl. Before it goes in the oven, she mixes it all together with a spoon. Next, she puts the cake in the oven. After it has cooked, she takes the cake from the oven and decorates it.

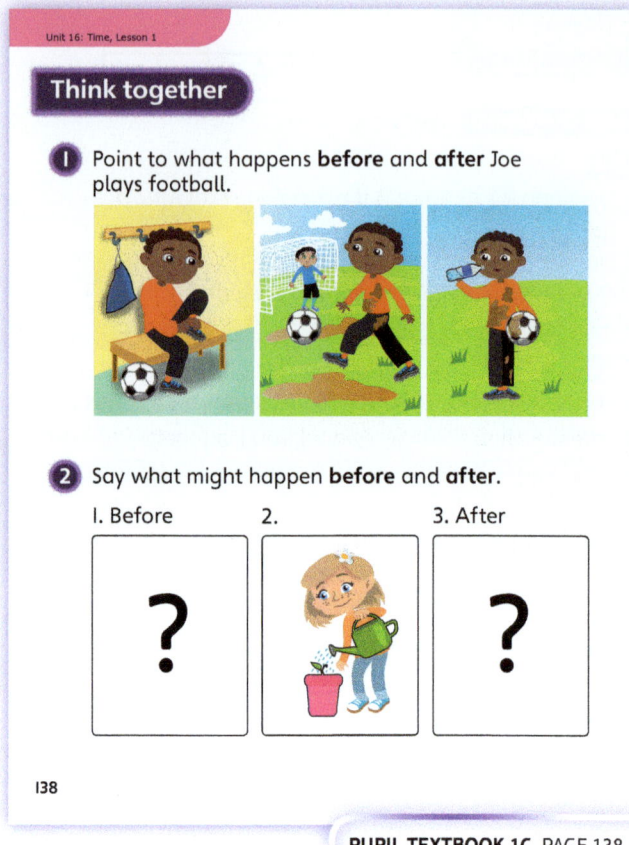

PUPIL TEXTBOOK 1C PAGE 138

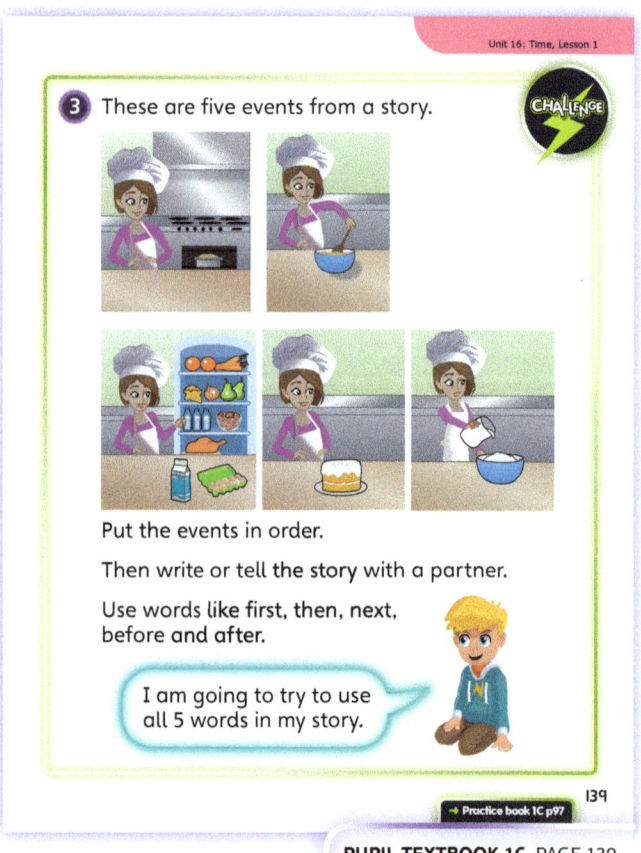

PUPIL TEXTBOOK 1C PAGE 139

176

Unit 16: Time, Lesson 1

Practice

WAYS OF WORKING Independent thinking and pair work

IN FOCUS These questions check that children understand the key terms 'before' and 'after'. As children move through them, they will need to think about the scenarios in each of the questions and what is happening. Encourage them to discuss and talk about the images with a partner. Children may not be used to some of the scenarios, such as walking a dog in question ④, and so it could be an opportunity to discuss this as a whole class.

STRENGTHEN The sequence of questions in this section follows some of the stages that children will need to go through in developing their understanding; first understanding the logical order of events, then labelling these with simple markers like 'before' and 'after', and then using a broader range of relationships and language to describe more complicated situations. To be successful, children really need to understand what is taking place in the images, so spend time talking to them about what they can see. It is important that they can relate to this, so you may want to find an activity similar to something that children have experienced to allow them to use the key language.

DEEPEN Describe a day to children with one of the activities in the wrong order. Can they spot which one it is and put it in the correct order?

THINK DIFFERENTLY Question ③ presents images that are not in the correct order. Children need to use logic to identify the correct order of the activities.

ASSESSMENT CHECKPOINT Questions ① and ② involve using real-life experience to order events in time. Check that children can make sense of all of these situations; they should be straightforward, but some children may not be aware that chicks hatch from eggs and they may not have had the practical experience of posting a letter.

ANSWERS Answers for the **Practice** part of the lesson can be found in the *Power Maths* online subscription.

PUPIL PRACTICE BOOK 1C PAGE 97

PUPIL PRACTICE BOOK 1C PAGE 98

Reflect

WAYS OF WORKING Independent thinking

IN FOCUS This question asks children to think about what they do before and after school. Children are natural story tellers, so encourage imaginative responses.

ASSESSMENT CHECKPOINT Check that children's explanations distinguish between events occurring before and after school.

ANSWERS Answers for the **Reflect** part of the lesson can be found in the *Power Maths* online subscription.

After the lesson

- Did children have difficulty using the basic patterns of 'before' and 'after'?
- Are children able to order events and identify what comes before and after?

PUPIL PRACTICE BOOK 1C PAGE 99

Unit 17: Time, Lesson 2

Days of the week

Learning focus

In this lesson, children will recognise and read the days of the week and know their correct order. They will use words such as 'today' and 'tomorrow'.

Before you teach

- Do children know how to count to 30?
- How many days of the week do children already know? Do they know the days that make up the weekend?

NATIONAL CURRICULUM LINKS

Year 1 Measurement – time

Recognise and use language relating to dates, including days of the week, weeks, months and years.

ASSESSING MASTERY

Children can recognise and read the days of the week and know their correct order. They will use words such as 'today', 'yesterday' and 'tomorrow' confidently. Given two days, they will be able to tell you what day comes between them. At this stage, children may not be able to confidently write down the correct days of the week.

COMMON MISCONCEPTIONS

Children may get the order of the days of the week wrong and confuse different days. Ask:
- *What day is it today? What day comes next? Do you know any rhymes or songs to help you remember the order of the days of the week?*

STRENGTHENING UNDERSTANDING

Provide daily reminders of the current day to help children understand the sequence. You could do this by displaying a poster of the days of the week in the classroom and pointing to the current day each morning, or by writing the day of the week on the whiteboard. Children should then write the day and date in their *Power Maths* Practice Book throughout the year. This will reinforce the correct sequence of days. Discuss words like 'yesterday', 'tomorrow', 'weekend', and so on. Ask children to tell you activities they do on particular days. You could suggest they make a weekly calendar.

GOING DEEPER

Challenge children to answer questions that bridge more than one day. Ask: *If today is Wednesday, what day is two days after today? What day is the day before yesterday? Sam orders a book on Monday and it arrives this Saturday. How many days do they have to wait for the book to arrive?*

KEY LANGUAGE

In lesson: day, week, yesterday, tomorrow, day names, (Monday, Tuesday, Wednesday, Thursday, Friday, Saturday, Sunday), today

STRUCTURES AND REPRESENTATIONS

Timeline for sequencing days in the correct order

RESOURCES

Mandatory: calendar showing a week to a page, so children can see the days of the week in order

Optional: a range of printed and digital calendars

 In the eTextbook of this lesson, you will find interactive links to a selection of teaching tools.

Quick recap

Check if children know the days of the week. Ask: *What days of the week do you know?* Avoid going into too much detail. Essentially, you are just checking at this stage which children are more confident about the days of the week than others.

Unit 16: Time, Lesson 2

Discover

WAYS OF WORKING Pair work

ASK

- Question 1 a): *What day do you think the teacher is writing on the board? Why do you think this?*
- Question 1 b): *If today is Tuesday, what day was it yesterday? What day comes before Tuesday? How do you know? What day will it be tomorrow? What day comes after Tuesday?*

IN FOCUS Question 1 a) is about whether children recognise a day from part of the word being written out. They are then asked to work out the day before and the day after in question 1 b) and use the language of yesterday and tomorrow. Ask children to discuss this in pairs and get them to say their answers as full sentences.

PRACTICAL TIPS Act out the scenario in class using a different day of the week. Start writing the word 'Thursday' on the board and see if children can identify what day of the week this is. In question 1 b), provide a weekly calendar showing the days of the week so children can identify the day before and after Tuesday.

ANSWERS

Question 1 a): Tuesday

Question 1 b): Yesterday was Monday and tomorrow will be Wednesday.

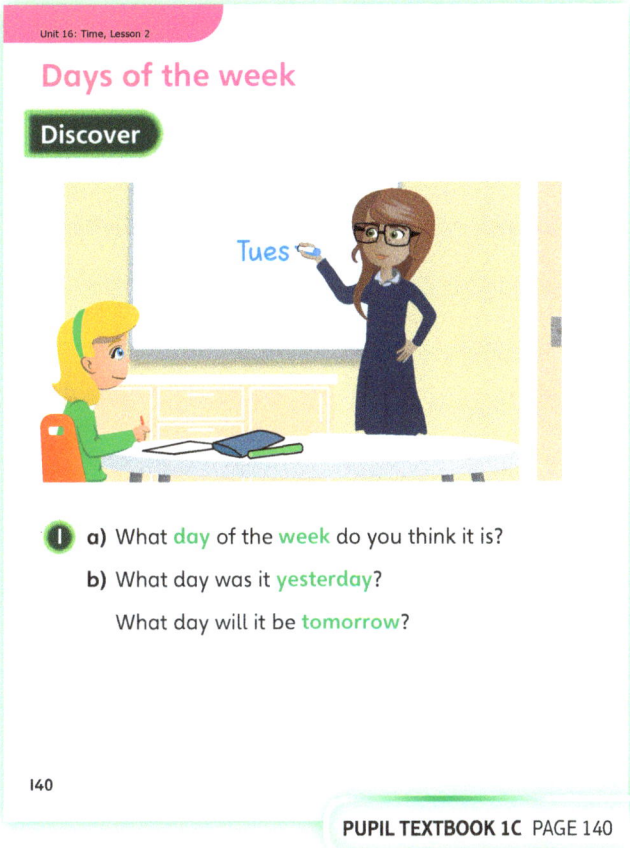

PUPIL TEXTBOOK 1C PAGE 140

Share

WAYS OF WORKING Whole class teacher led

ASK

- Question 1 a): *Could it be any other day than Tuesday? What do you usually do on a Tuesday? Is there a way you can tell what day it is?*
- Question 1 b): *What word can you use to describe the day that comes before today? What word can you use to describe the day that comes after today?*

IN FOCUS Talk through the answers to the questions. Ask children to use full sentences. Discuss other days of the week using the sequence of days given in **Share** and discuss what days come before and after Tuesday (or another day that you may want to choose). You could get children to cover up the days before and after with paper to check. Discuss what happens when you get to Sunday and agree that the sequence starts again.

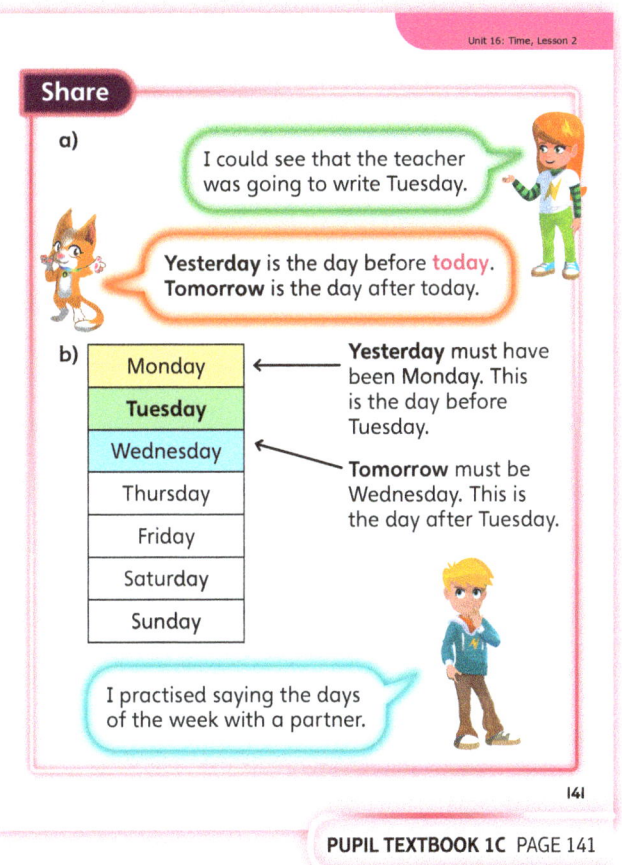

PUPIL TEXTBOOK 1C PAGE 141

Think together

WAYS OF WORKING Whole class teacher led (I do, We do, You do)

ASK

- Question ❶: *What day comes next in each example? How do you remember the days of the week?*
- Question ❷: *What day is it today? Can you find Thursday on the timeline?*
- Question ❸: *Has anyone seen the weather forecast on TV?*

IN FOCUS Question ❶ is checking whether children understand the sequence of the days in the week. Before you start, you may want to use a rhyme or song to help children get the sequence in their head. Say or sing this rhyme to help them with the missing next day if necessary. Question ❷ assesses children's understanding of the words 'yesterday' and 'tomorrow'. Ask them to rehearse their answer with a partner and to use full sentences.
Question ❸ is more of a problem-solving question. The image shows the days of the week on a weather forecast. Children tell the story of the weather in the week using different key words.

STRENGTHEN Provide daily reminders of the current day to help children understand the sequence of the days of the week. Encourage children to write the day and date in their *Power Maths* Practice Book throughout the year. This will reinforce the correct sequence of days of the week. Discuss words like 'yesterday', 'tomorrow', 'weekend'.

DEEPEN For question ❸, ask children to use language such as 'the weather today is sunny and in two days' time, the weather will be …'.

ASSESSMENT CHECKPOINT Use questions ❶, ❷ and ❸ to check whether children understand the sequence of days and can correctly use key terms such as 'tomorrow' and 'yesterday'.

ANSWERS

Question ❶ a): Thursday

Question ❶ b): Sunday

Question ❶ c): Monday

Question ❷ a): He is walking the dog.

Question ❷ b): He played football.

Question ❷ c): He will paint a picture.

Question ❸ a): Before Tuesday, it was sunny on both Sunday and Monday.

Question ❸ b): The day after Wednesday is Thursday. On Thursday, it will be stormy with rain, thunder and lightning.

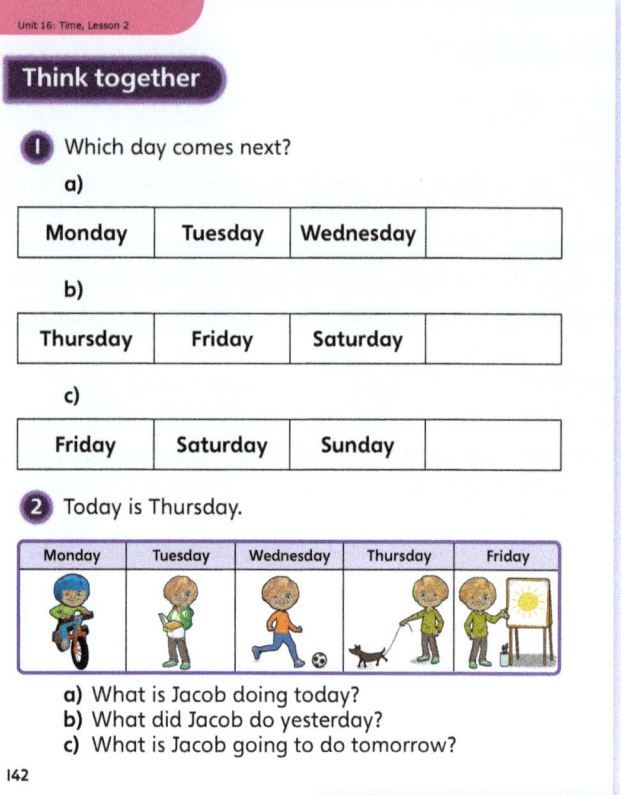

PUPIL TEXTBOOK 1C PAGE 142

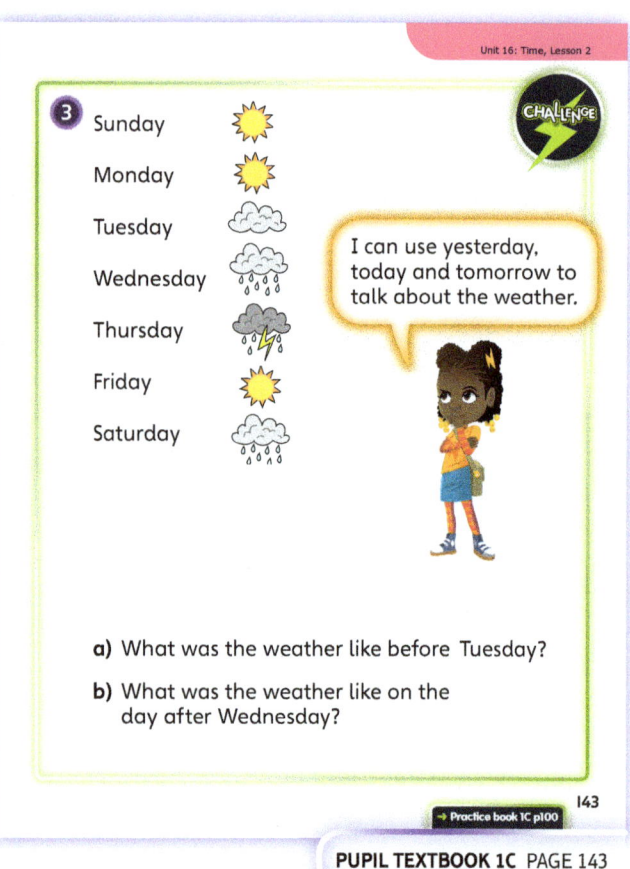

PUPIL TEXTBOOK 1C PAGE 143

Unit 16: Time, Lesson 2

Practice

WAYS OF WORKING Independent thinking

IN FOCUS Question ❶ asks children if they know what day of the week it is and then to identify which day it was yesterday and which day it will be tomorrow. Question ❷ checks whether children know what day comes next after being given a sequence of days. This question is designed to check that children know the correct sequence of days. Question ❸ provides a small wordsearch in which children have to find the days of the week. Make sure the days of the week are displayed in the classroom, to help children with their spellings.

STRENGTHEN Provide daily reminders of the current day to help children understand the sequence. Encourage children to write the day and date in their *Power Maths* Practice Book too throughout the year. This will reinforce the sequence of days that make up the week. Discuss words like 'yesterday', 'tomorrow', 'weekend', and so on.

DEEPEN Challenge children with questions that bridge more than one day, and cross a weekend into the next week. Ask: *If today is Saturday, what day is three days after today? What day is the day before yesterday? Peter calls the dentist on Monday and makes an appointment for Friday. How many days does Peter have to wait until his dentist appointment?*

THINK DIFFERENTLY Question ❹ requires children to know the missing day, when presented with the day before and after it.

ASSESSMENT CHECKPOINT Use question ❹ to check that children understand the days of the week and the sequence in which they come.

ANSWERS Answers for the **Practice** part of the lesson can be found in the *Power Maths* online subscription.

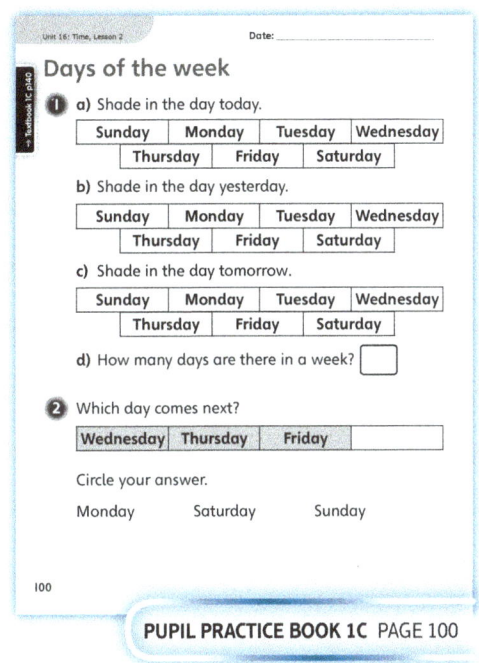

PUPIL PRACTICE BOOK 1C PAGE 100

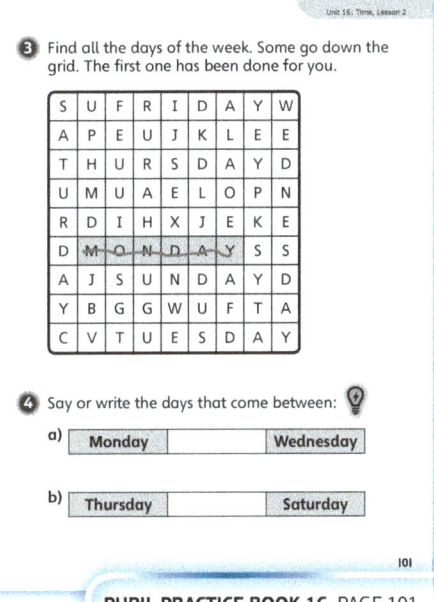

PUPIL PRACTICE BOOK 1C PAGE 101

Reflect

WAYS OF WORKING Pair work

IN FOCUS Working in pairs, children say the days of the week together. Encourage them to use sentences and to tell each other what they do on different days of the week. For example, 'On Wednesday, I go to dance practice.'

ASSESSMENT CHECKPOINT Check that children use the correct language for the days of the week.

ANSWERS Answers for the **Reflect** part of the lesson can be found in the *Power Maths* online subscription.

PUPIL PRACTICE BOOK 1C PAGE 102

After the lesson

- Do children know all the days of the week? Are they confident in saying them in the correct order?
- Are children able to say what day comes before or after a particular day?
- Can children use the terms 'yesterday', 'today' and 'tomorrow' correctly?

Unit 17: Time, Lesson 3

Months of the year

> **Learning focus**
>
> In this lesson, children use a calendar to read and record information related to days and dates.

> **Before you teach**
>
> - Do children know the correct sequence of the days of the week?
> - Is there a calendar in your classroom? If so, how do you use it?
> - Do children know the sequence of month names? How can you strengthen children's knowledge of month names?

NATIONAL CURRICULUM LINKS

Year 1 Measurement – time

Recognise and use language relating to dates, including days of the week, weeks, months and years.

ASSESSING MASTERY

Children can use a calendar to identify and record events on a particular day, and can find the day of the week that a particular date falls on. Children can say a date in the form, for example, 'Sunday the 9th of November', and read and understand dates given in the form 'Sunday 9 November'.

COMMON MISCONCEPTIONS

Some children may have difficulty with the number of sequences that need to come together to produce a calendar – month names, day names and the numerical sequence of dates. They will need time and practice to make sense of this complexity. Ask:

- What can you tell by looking at a calendar?

STRENGTHENING UNDERSTANDING

Build on children's existing understanding and experience, using calendars to record key events such as leisure activities, birthdays and holidays. Help children to reflect on the structure of the calendar, in particular how each row represents a week, usually starting on a Sunday. You may find it useful to have a large calendar in the classroom and use it to record the dates of key events.

GOING DEEPER

There is a lot more to learn about the calendar, and some children will be ready to go further. Encourage children to memorise the sequence of month names, understanding that months have different numbers of days, eventually learning the number of days in each month ('Thirty days hath September …') and understanding leap years.

KEY LANGUAGE

In lesson: day, week, date, **calendar**, **month**, **year**, day names (Sunday, Monday, …), month names (January, February, …), day number in the month, day of the week, today, yesterday, tomorrow

STRUCTURES AND REPRESENTATIONS

Timeline for sequencing months of the year

RESOURCES

Mandatory: a range of printed and digital calendars showing both one month to a page and also one year to a page

 In the eTextbook of this lesson, you will find interactive links to a selection of teaching tools.

> **Quick recap**
>
> Ask children if they can tell you the date of their birthday and if they can write it down. What day of the month were they born? What month were they born in? What year were they born? Tally the number of children born in each month on the whiteboard.

Unit 16: Time, Lesson 3

Discover

WAYS OF WORKING Pair work

ASK

- Question 1 a): *Which month does the calendar show? What are the letters written at the top of the calendar, underneath the word 'November'?*
- Question 1 a): *What day of the week is Meg's birthday?*

IN FOCUS Question 1 a) involves reading a calendar, and identifying the date and day of Meg's birthday. The point here is to make sure that children are aware that the calendar is more than just an unstructured list of days – it incorporates information about which month we are in and which day of the week each date falls on.

PRACTICAL TIPS Allow children to physically handle calendars, find where Meg's birthday would be, and count how many months there are in a year.

ANSWERS

Question 1 a): Meg's birthday is Sunday 19 November.

Question 1 b): There are 12 months in a year.

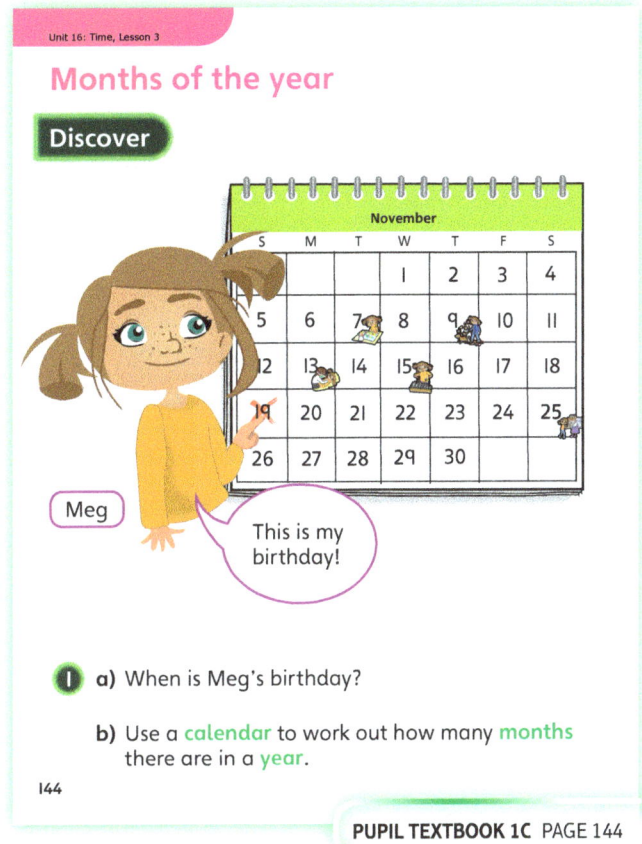

PUPIL TEXTBOOK 1C PAGE 144

Share

WAYS OF WORKING Whole class teacher led

ASK

- Questions 1 a) and b): *In these calendars, we have only one letter for each day of the week. Do you still know which is which? Why are there 2 Ts and 2 Ss?*
- Question 1 a): *Once you have found the date that you are looking for, how do you find out what day of the week it is?*
- Question 1 b): *Where would you look on a calendar to find out what month is shown?*

IN FOCUS In question 1 a), ensure that children are clear about the difference between the idea of a date (which is a number), and the day of the week (which is a day name). Talk about how these are used in everyday life. For example, you may use day names when you are talking about events that are happening soon or events that recur. Ask: *What are you doing on Thursday? What evening does the club meet?* For events planned further in the future, we tend to use dates: *Remember, Sports Day is the 10th July.*

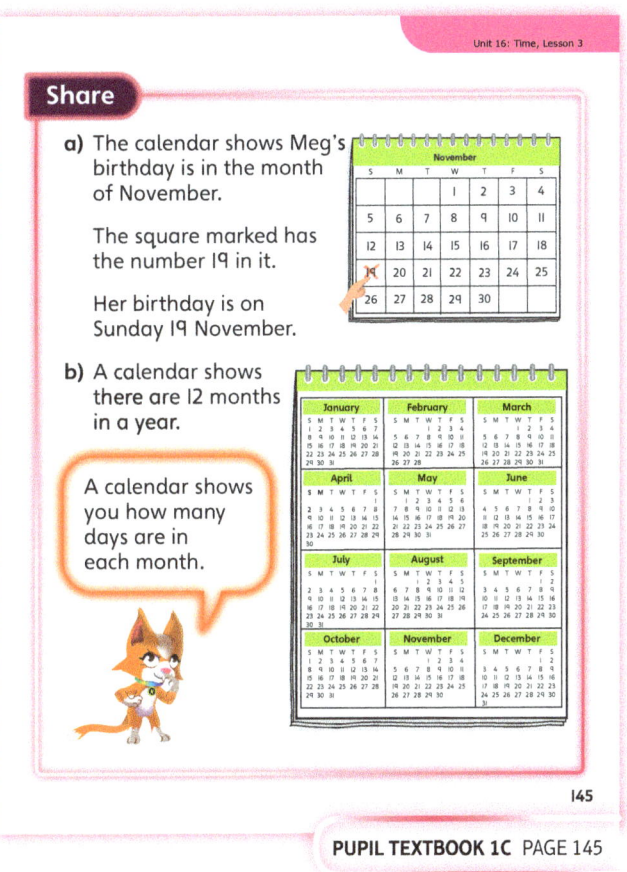

PUPIL TEXTBOOK 1C PAGE 145

183

Unit 16: Time, Lesson 3

Think together

WAYS OF WORKING Whole class teacher led (I do, We do, You do)

ASK

- Question ❶: *What month is Charlie looking at? Can you see where he has marked his birthday?*
- Question ❶: *It says 'W' above Charlie's birthday. What day of the week does this mean?*
- Question ❶: *Can you say Charlie's birthday in full – with the date, the day of the week and the name of the month?*

IN FOCUS Question ❸ shows a full calendar for a complete year. You may wish to use a 'real' calendar for the current year instead. Ensure that children understand the order of the months – 'across first, then down'. You could ask a range of further questions here. For example, ask: *What day of the week is your birthday? How many months start on a Monday?*

STRENGTHEN In Question ❸, some children may find the density of information in the full calendar overwhelming. Try working with one 'layer' of information at a time – perhaps just finding a month at first, then a date within a month, and then finding the day of the week for a date.

DEEPEN Children with a secure understanding of the work in this section could start to look at the structure of the calendar in more depth. Ask: *Which months have 31 days? What happens to February in a leap year? If I know today's date, how can I use the calendar to find the date in a week's time?*

ASSESSMENT CHECKPOINT Check that children are familiar with the sequence of month names, even if they have not yet memorised the complete sequence, and check that they can read and locate the months in the calendar. If you are confident that they can find months reliably, check that children can read up from a date to a day of the week, and that they can distinguish between the days with the same initial letters (Tuesday and Thursday, Saturday and Sunday).

ANSWERS

Question ❶: Charlie's birthday is Wednesday 20 April.

Question ❷: Children should point to Thursday 8 July.

Question ❸ a): There are 12 months in a year.

Question ❸ b): The months are: January, February, March, April, May, June, July, August, September, October, November, December.

Question ❸ c): April, June, September and November have 30 days.

Question ❸ d): Children's answers will vary. Check they can accurately locate their birthday on a calendar.

PUPIL TEXTBOOK 1C PAGE 146

PUPIL TEXTBOOK 1C PAGE 147

Unit 16: Time, Lesson 3

Practice

WAYS OF WORKING Independent thinking

IN FOCUS Question ❶ checks that children know which month it is now and which months come before and after. The layout of the grid may help children to really understand the order of the different months. Make sure they go along the rows, rather than down the columns. Questions ❷ and ❸ further check this understanding, with children knowing which months follow and which come in between given months. It is fine at this stage to provide children with calendars or get them to look back in the Textbook if they are unsure.

STRENGTHEN Using calendars is a skill that children need to develop over time. For those struggling with the sequence, consider suggesting to parents and carers that they discuss the 'family calendar' with their child, regularly recording and talking about upcoming events, and discussing what dates and days of the week they fall on.

DEEPEN Children who are confident with calendars could be challenged to carry out more complicated date calculations, including those that go beyond the information shown. For example, given the calendar showing August in question ❹, can children work out what day of the week 1 September falls on? What about 1 October?

THINK DIFFERENTLY In question ❹ b), children are given a date and are asked to correctly mark it on the calendar. They need to demonstrate that they are able to read a calendar correctly. Children may need additional support with this.

ASSESSMENT CHECKPOINT Use questions ❶, ❷, ❸ and ❺ to assess whether children understand the correct order of the months. Use question ❹ to check they can read a calendar accurately.

ANSWERS Answers for the **Practice** part of the appear can be found in the *Power Maths* online subscription.

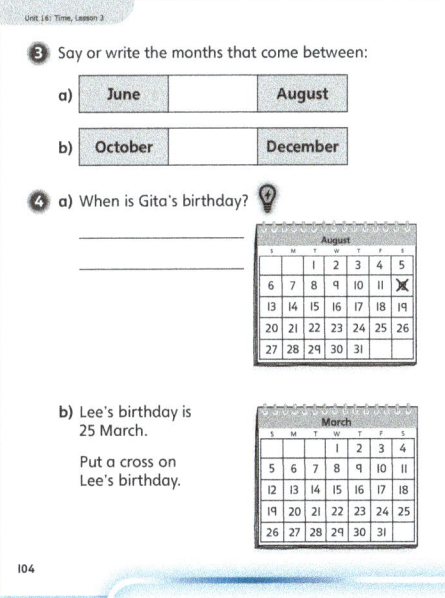

PUPIL PRACTICE BOOK 1C PAGE 103

PUPIL PRACTICE BOOK 1C PAGE 104

PUPIL PRACTICE BOOK 1C PAGE 105

Reflect

WAYS OF WORKING Pair work

IN FOCUS This question allows children to reflect on what they have learnt about calendars. Encourage children to work in pairs and to say the names of the months together, in the correct order.

ASSESSMENT CHECKPOINT Responses here will indicate whether children know the months in the year and whether they are able to say them in the correct order. Did they miss any? Counting from 1 to 12 will help children to make sure they say them all.

ANSWERS Answers for the **Reflect** part of the lesson can be found in the *Power Maths* online subscription.

After the lesson

- Will children have opportunities to continue to use calendars as part of their regular school routine?
- Are children familiar with the sequence of month names? Have they committed these to memory?

Unit 17: Time, Lesson 4

Tell the time to the hour

Learning focus
In this lesson, children will use an analogue clock face to tell the time to the nearest hour ('something o'clock').

Before you teach
- Can any children already tell the time?
- Do any children wear a watch? Digital or analogue?
- Do you have a classroom clock? Do you use it as part of your classroom routine?

NATIONAL CURRICULUM LINKS

Year 1 Measurement – time

Tell the time to the hour and half past the hour and draw the hands on a clock face to show these times.

ASSESSING MASTERY

Children can tell the time using an analogue clock set to a 'whole hour' and they know that all such times are 'something o'clock'. Children can understand the importance of distinguishing between the hour hand and the minute hand, they know that the hour hand is the shorter one, and they are starting to relate the time shown on a clock to times of the day and familiar events.

COMMON MISCONCEPTIONS

Children may encounter many potential problems in starting to tell the time: they may confuse the minute and hour hands, or find it difcult to understand that there are two distinct scales (hours and minutes) arranged in a circular fashion around the clock face. More fundamentally, some children may have only a rudimentary familiarity with times of the day – 'home time' being a more fundamental idea than '3 pm'. Ask:

• *What time do you start school? What time do you go home?*

STRENGTHENING UNDERSTANDING

Some children will benefit from additional practice, perhaps using a model clock to indicate a range of times. This could easily be turned into a simple game: make some cards, with a time written on each one, deal the cards face down, then turn them up one card at a time, and score one point for each time that is made successfully on the model clock.

GOING DEEPER

Some children may already be able to tell the time to the nearest hour or better. Challenge confident children to draw their own clock faces to show particular times. This is a very useful exercise that encourages children to think carefully about the angular spacing of the numbers around the clock face.

KEY LANGUAGE

In lesson: o'clock, minute hand, hour hand, minute, hour

Other language used by the teacher: longer, shorter

STRUCTURES AND REPRESENTATIONS

Clock faces

RESOURCES

Optional: clock tool, printed clock faces, model clocks, real clocks, cards with times written on

 In the eTextbook of this lesson, you will find interactive links to a selection of teaching tools.

Quick recap
Give children a clock or display a child-friendly one on the whiteboard. Ask children to count around the clock with you from 1 to 12. Then ask them if they know what the big and small hands show.

Unit 16: Time, Lesson 4

Discover

WAYS OF WORKING Pair work

ASK

• Question 1 a): *Look at the picture. How are you going to work out what time it is now?*
• Question 1 a): *How many hands can you see on the clock? Do you know how to use them to work out what the time is now?*
• Question 1 b): *When the party starts, what will the hour hand point to? What will the minute hand point to?*

IN FOCUS This activity features a familiar scenario where children might be keen to tell the time. Point out the clock on the wall, and make sure that children understand that they will need to use both the information on the clock and the facts provided in the question to reach a solution.

PRACTICAL TIPS You could hand out model or analogue clocks to pairs of children – they can use them to make the times 3 o'clock and 5 o'clock.

ANSWERS

Question 1 a): The time is 3 o'clock.

Question 1 b): The hour hand will point to 5 and the minute hand will point to 12.

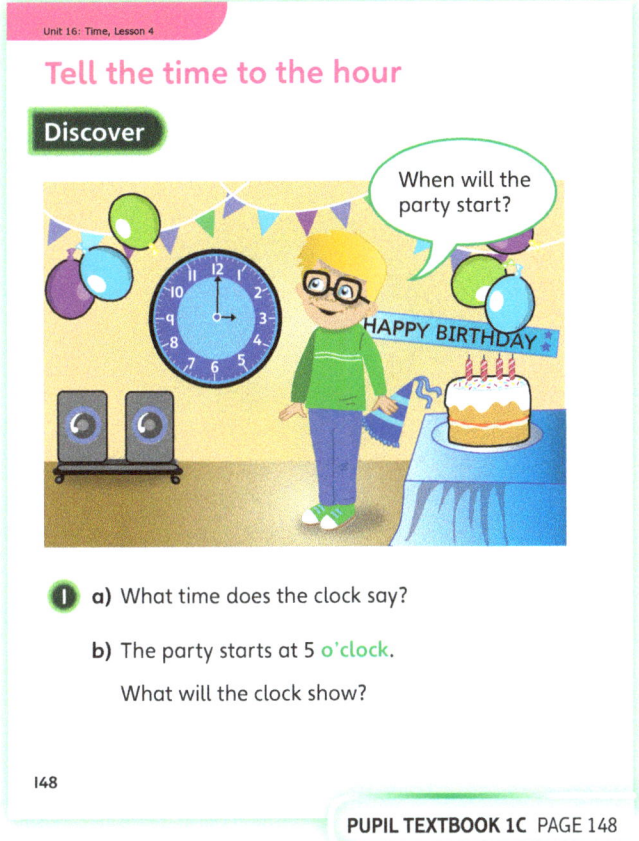

PUPIL TEXTBOOK 1C PAGE 148

Share

WAYS OF WORKING Whole class teacher led

ASK

• Question 1 a): *Look at what Ash says. Why are there two hands on the clock? What are they called?*
• Question 1 a): *Where will the minute hand be pointing when it is an o'clock time?*
• Question 1 a): *Astrid says that she knows how to draw a clock face for the other o'clock times. Would you know how to do it?*

IN FOCUS The key ideas to communicate to children in questions 1 a) and b) are:

• The clock has two hands – a minute hand and an hour hand. The hour hand is the shorter one.
• You do not need to worry too much about the minute hand at first. All you need to know is that when it is pointing straight up, it is one of the o'clock times.
• When the minute hand is pointing straight up, the hour hand tells us 'what o'clock' it is.

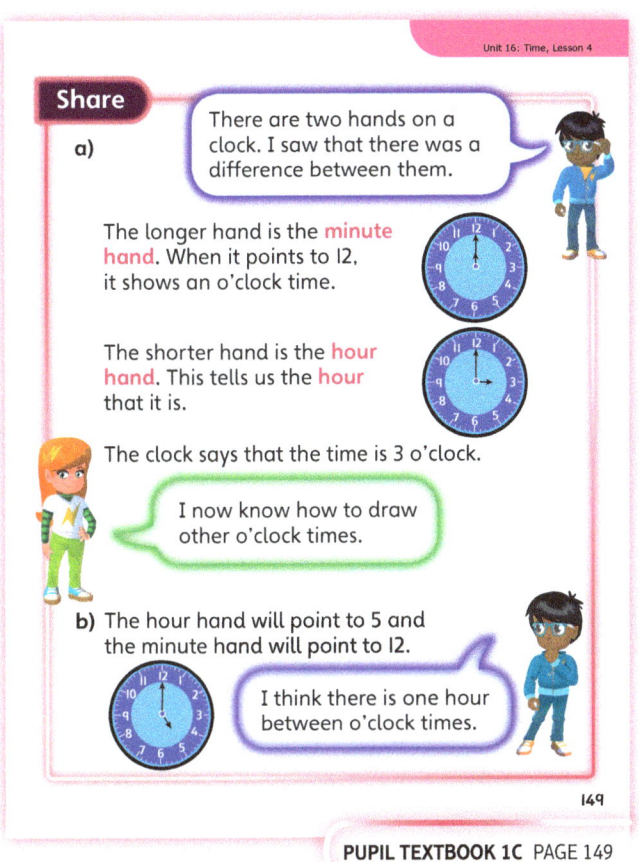

PUPIL TEXTBOOK 1C PAGE 149

187

Unit 16: Time, Lesson 4

Think together

WAYS OF WORKING Whole class teacher led (I do, We do, You do)

ASK
- Questions 1 and 2: *Both of these times are 'something o'clock'. How do you know that?*
- Questions 1 and 2: *Which of the hands on the clock is the hour hand? How can you use the hour hand to work out what the times are?*

IN FOCUS In questions 1 and 2, the key ideas are to distinguish between the minute and hour hand, then to recognise that the minute hand pointing straight up indicates 'something o'clock', and finally that the hour hand indicates 'what o'clock' it is.

STRENGTHEN If children find questions 1 and 2 diffcult, check each stage of the process for telling the time to the nearest hour. Can children consistently distinguish between the minute and hour hands? Can they associate the minute hand pointing straight up with 'something o'clock'? Can they use the hour hand to determine what the hour is?

DEEPEN The very simple clock faces used in these questions are designed to be as easy to read as possible. Real clocks may not be quite so simple. Try finding pictures of clocks with more complicated faces – for example, Big Ben. Can children explain how these work?

ASSESSMENT CHECKPOINT Check that children can correctly read the time on the clock faces shown in questions 1 and 2. If in doubt, set further examples of the same kind to help children practise each stage of the process for telling time carefully.

ANSWERS

Question 1: The time is 7 o'clock.

Question 2: The time is 2 o'clock.

Question 3: Harry has put the hour hand pointing to 11 instead of 10, so he has drawn 11 o'clock. Maya has mixed up the clock hands and put the minute hand pointing to 4 and the hour hand pointing to 12. They should be the other way round.

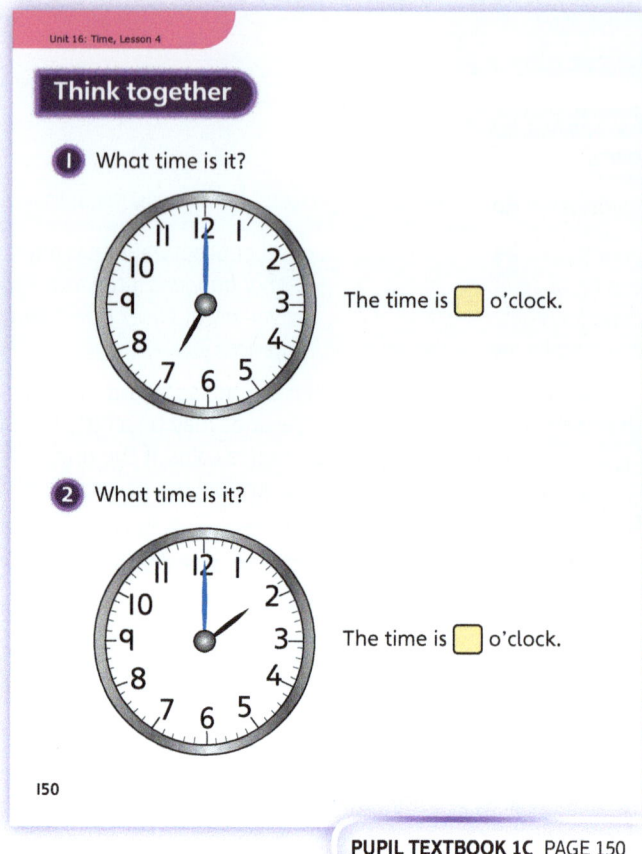

PUPIL TEXTBOOK 1C PAGE 150

PUPIL TEXTBOOK 1C PAGE 151

Unit 16: Time, Lesson 4

Practice

WAYS OF WORKING Independent thinking

IN FOCUS Question ❶ provides some straightforward examples that cover the main content of the lesson. Ensure that children know what is required. They can simply draw lines to show which clock face goes with each of the times.

STRENGTHEN Some children may benefit from additional practice with reading or writing the time to the nearest hour, using either printed clock faces or a demonstration clock.

DEEPEN Question ❷ features a wider variety of clock faces. Ask children to investigate what other watch and clock faces are available. Can they tell the time with other designs of clock face?

THINK DIFFERENTLY Question ❹ requires children to use reasoning to identify which child is correct and which child has made a mistake. To extend this further, you could ask children to explain the mistake that has been made.

ASSESSMENT CHECKPOINT Use question ❸ to check that children can draw their own clock faces to represent each of the times shown. Make sure that they draw the minute hand pointing straight up, and clearly longer than the hour hand, which should be pointing at the correct hour. If there are mistakes with any of these points, ask children to explain what they did in order to try to work out the exact step where they are going wrong.

ANSWERS Answers for the **Practice** part of the lesson can be found in the *Power Maths* online subscription.

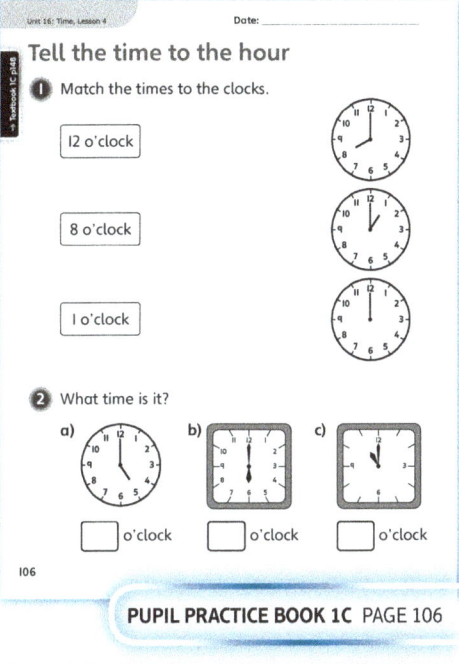

PUPIL PRACTICE BOOK 1C PAGE 106

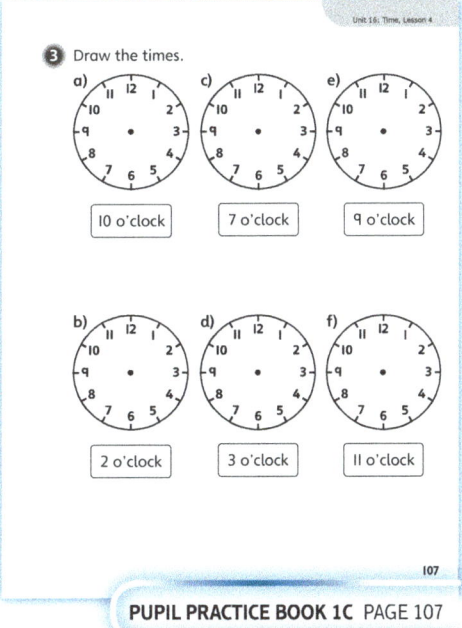

PUPIL PRACTICE BOOK 1C PAGE 107

Reflect

WAYS OF WORKING Independent thinking

IN FOCUS In this question, children need to sketch a clock face showing a time of 4 o'clock. They are required to draw recognisable minute and hour hands in the correct positions.

ASSESSMENT CHECKPOINT Look out for the main features of a correct answer: the minute hand pointing straight up, with a shorter hour hand pointing at the number 4, which should be in the right general position on the clock face.

ANSWERS Answers for the **Reflect** part of the lesson can be found in the *Power Maths* online subscription.

After the lesson ⏸

- How will you build further practice into the daily classroom routine, perhaps using the classroom clock?
- Can children already tell the time to greater levels of precision?

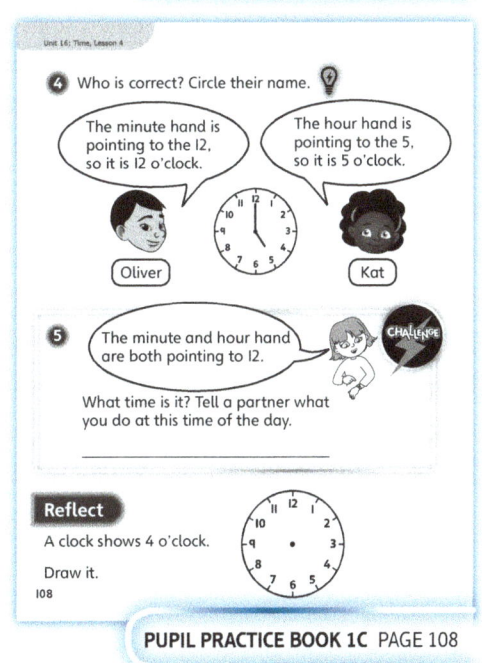

PUPIL PRACTICE BOOK 1C PAGE 108

189

Unit 17: Time, Lesson 5

Tell the time to the half hour

Learning focus
In this lesson, children will use an analogue clock face to tell the time to the nearest half hour ('half past something').

Before you teach
- Are children confident in telling the time to the hour?
- Are there any routine events in the day that happen at 'half past something' that could be used as examples?

NATIONAL CURRICULUM LINKS

Year 1 Measurement – time

Tell the time to the hour and half past the hour and draw the hands on a clock face to show these times.

ASSESSING MASTERY

Children can distinguish between the minute and hour hands of an analogue clock or watch. Children can use the position of the minute hand to identify times that are 'something o'clock' and 'half past something', and they can use the position of the hour hand to identify the hour.

COMMON MISCONCEPTIONS

Some children may have difficulty in deciding which number the hour hand has just passed. Ask:
- *Where is the hour hand? Which numbers is it between? Which number has it already passed? So, you know it is half past something – but half past what hour?*

STRENGTHENING UNDERSTANDING

Some children may be less familiar with the idea of 'half past the hour'. Try to use descriptive language to support understanding. For example, *The time is half past 8. That means that we have already gone past 8 o'clock, but it is not 9 o'clock yet; it is exactly half-way between 8 o'clock and 9 o'clock.*

GOING DEEPER

Some children may already know how to tell time to the nearest minute and may be able to read a wide range of analogue and digital clock faces. You could encourage them to think more deeply about the relative positions of the hands on a clock. Ask: *How can you tell whether the position of the hands on a clock is real or made up? Which positions are never allowed?*

KEY LANGUAGE

In lesson: **half-past**, hour, minute, hand, half-way, between

STRUCTURES AND REPRESENTATIONS

Analogue clock face, marked with the numbers 1 to 12

RESOURCES

Optional: clock tool marked with the numbers 1 to 12, printed clock faces for children to complete

 In the eTextbook of this lesson, you will find interactive links to a selection of teaching tools.

Quick recap

Check that children can read and draw or make clocks that show times to the hour. For example, show children some analogue clocks with o'clock times and ask them to write or say the time. Then ask them to draw or make 2 o'clock, 6 o'clock, 11 o'clock. Look out for them getting the hands in the correct position.

Unit 16: Time, Lesson 5

Discover

WAYS OF WORKING Pair work

ASK
- Question 1 a): *Can you see the minute hand on the classroom clock in the picture? Where is it pointing?*
- Question 1 b): *What time does the teacher say that the assembly starts?*
- Question 1 b): *How long does it take for the minute hand to go around once? What will happen to the hour hand in the same time?*

IN FOCUS This activity uses a familiar setting to introduce the idea of times that are 'half past something'. Children may already be familiar with the spoken form of this pattern; the activity asks them to relate this to the display on an analogue clock face.

PRACTICAL TIPS You could give pairs of children model or analogue clocks – they can use them to make the times half past 9 and half past 10.

ANSWERS

Question 1 a): The time is half past 9.

Question 1 b): At half past 10, the hour hand will be between 10 and 11 and the minute hand will point to 6.

Share

WAYS OF WORKING Whole class teacher led

ASK
- Question 1 a): *How do you tell which hand is the hour hand and which hand is the minute hand?*
- Question 1 a): *Which way are the hands of the clock moving? Is it this way [indicate clockwise], or this way [indicate anticlockwise]?*
- Question 1 a): *How can you use the minute hand to spot times that are 'half past something'?*

IN FOCUS The amount of information involved in this part of the lesson may be quite daunting. Ask children to focus on the key ideas in the following order:
- Children already know that the minute hand pointing straight up indicates 'something o'clock'.
- Now children are finding out what it means when the minute hand is pointing straight down: these are 'half-past something' times.
- Children find out the 'something' by looking at the hour hand – the time is half past the hour that the hour hand has just passed. They also need to know which way the hour hand is turning (clockwise).

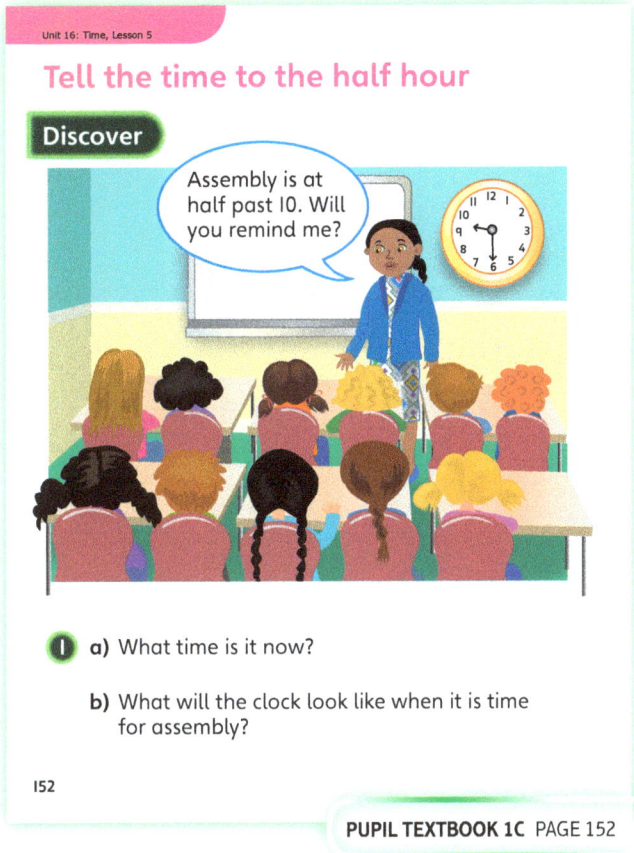

PUPIL TEXTBOOK 1C PAGE 152

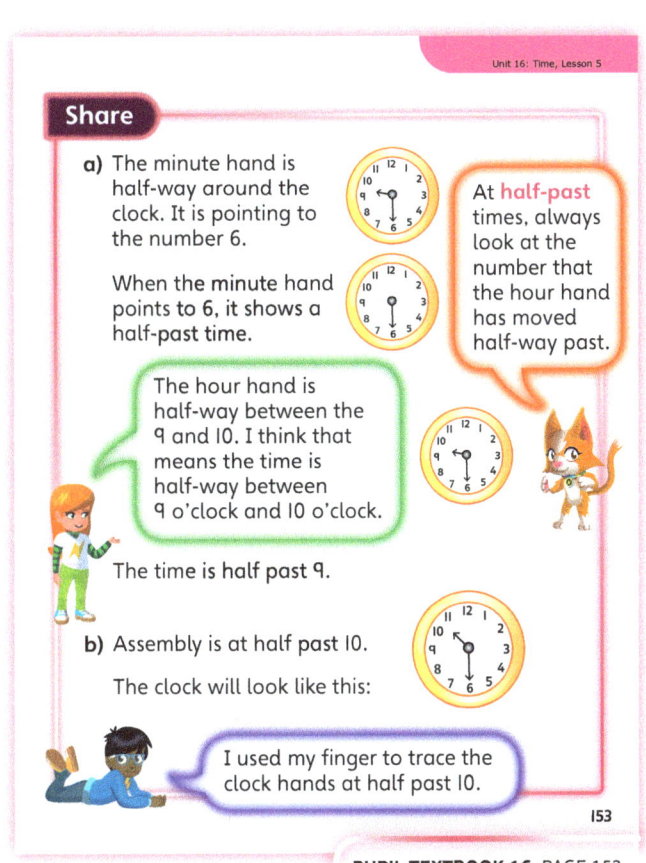

PUPIL TEXTBOOK 1C PAGE 153

Unit 16: Time, Lesson 5

Think together

WAYS OF WORKING Whole class teacher led (I do, We do, You do)

ASK
- Questions ❶ and ❷: *How do you know that these must be half-past times?*
- Questions ❶ and ❷: *It is half past an hour. How will you decide what the hour is?*

IN FOCUS Questions ❶ and ❷ provide opportunity for further discussion and consolidation where needed. Children are expected to be able to identify the half-past times shown on the analogue clocks.

STRENGTHEN Where children find the step to telling the time to the nearest half hour difficult, try to work out exactly where they are going wrong. Can they tell the time to the nearest hour? Are they correctly distinguishing between the hour and minute hands? Are they finding the correct hour?

The additional support or practice that children may need will depend on the stage where their difficulty arises. For example, if they are consistently choosing the wrong hour, try reinforcing the idea of 'clockwise', perhaps using a demonstration clock.

DEEPEN Discuss with children that very early clocks only had an hour hand and that the minute hand was a later invention. Ask children to think about how they would tell the time on a clock that only had an hour hand.

ASSESSMENT CHECKPOINT Question ❸ can be used to check understanding by analysing some common errors that arise when moving on to telling time to the half hour. Check that children do not confuse the hour and minute hands and that they use their understanding of 'clockwise' to correctly identify the hour.

ANSWERS

Question ❶: The time is half past 11.

Question ❷: The time is half past 1.

Question ❸: Myra has got the clock hands the wrong way round. The longer minute hand should be pointing to 6 and the shorter hour hand should be pointing between 8 and 9. Filip needed to draw the hour hand half-way between the 2 and the 3.

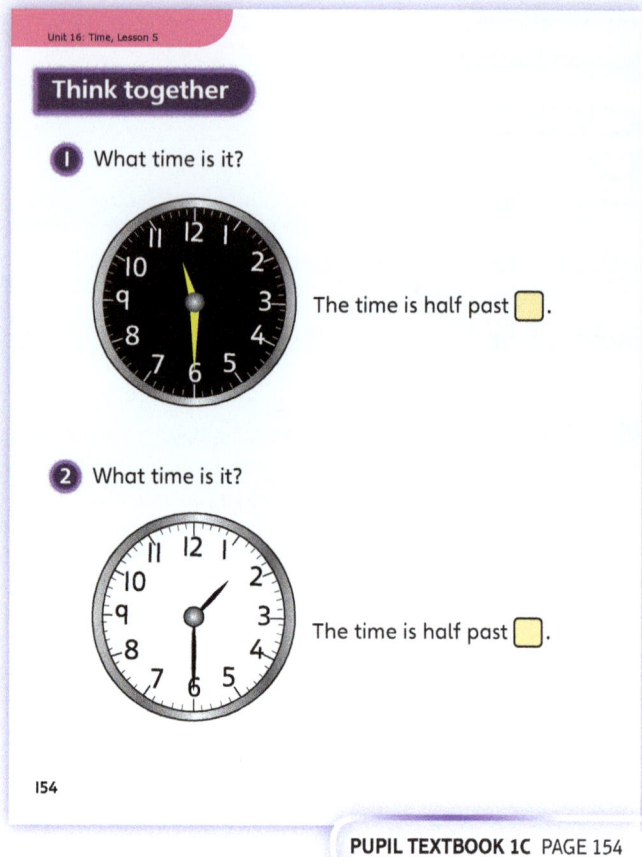

PUPIL TEXTBOOK 1C PAGE 154

PUPIL TEXTBOOK 1C PAGE 155

Unit 16: Time, Lesson 5

Practice

WAYS OF WORKING Independent thinking

IN FOCUS Question ❸ asks children to draw the stated times on a series of analogue clock faces. Remind children of all the key ideas so far; the minute hand is longer than the hour hand; the minute hand points straight down for a half-past time; the hour hand is correctly placed between two numbers for a half-past time.

STRENGTHEN Where children find it difficult to identify the correct hour, give them some printed clock faces and ask them to draw in the hour hand only, for a number of half-past times. This additional practice can be used as a means of reinforcing the idea of 'clockwise'. Ask: *Which way is the hour hand moving? So, which number has it gone past? Which number will it get to next?*

DEEPEN Question ❺ could be used to ask children questions such as: *Why can't the minute hand point to 6 and the hour hand point to 2?* Children could start to reason that the hour hand has to be pointing half-way between two numbers when the minute hand is pointing to 6.

THINK DIFFERENTLY Question ❹ requires children to use reasoning in order to decide whether Maya is correct or not. They should be able to recognise that Maya is wrong because the minute hand is pointing to 12, not 6. They may also begin to reason that the hour hand would need to be half-way between 6 and 7.

ASSESSMENT CHECKPOINT Use question ❸ to check children's understanding of the importance of distinguishing between the minute and hour hands. If this proves difficult, you could try looking at a variety of clock faces and identifying the hour and minute hands in each case.

ANSWERS Answers for the **Practice** part of the lesson can be found in the *Power Maths* online subscription.

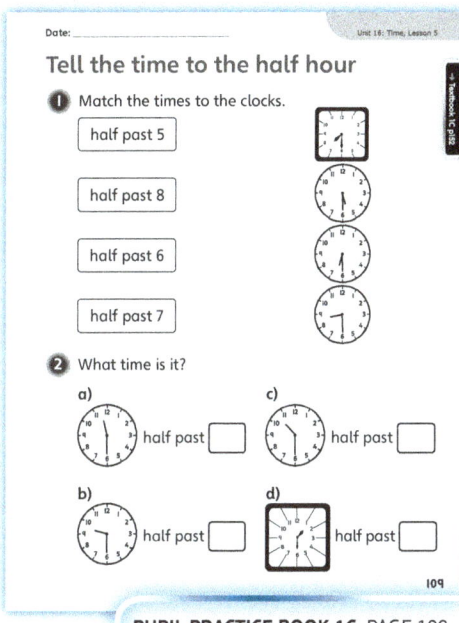

PUPIL PRACTICE BOOK 1C PAGE 109

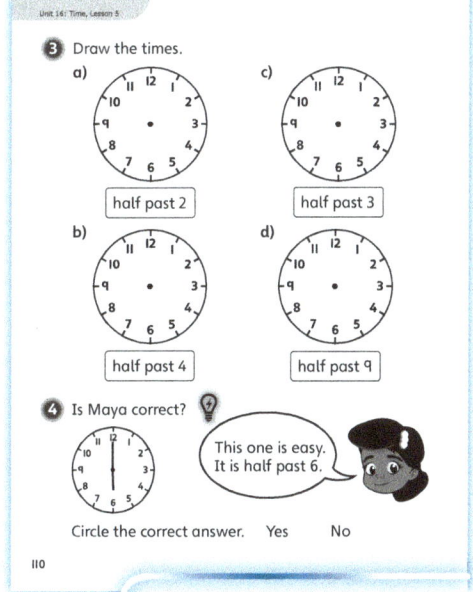

PUPIL PRACTICE BOOK 1C PAGE 110

Reflect

WAYS OF WORKING Pair work

IN FOCUS This question focuses on children knowing the main information required for a clock to be showing a half-past time.

ASSESSMENT CHECKPOINT Children should be able to draw the minute hand straight down and the hour hand in between the 7 and 8.

ANSWERS Answers for the **Reflect** part of the lesson can be found in the *Power Maths* online subscription.

- Does this lesson open up more possibilities to reinforce children's learning by talking about the timing of a wider variety of events in the school day?
- How will you use the classroom clock when talking about time?

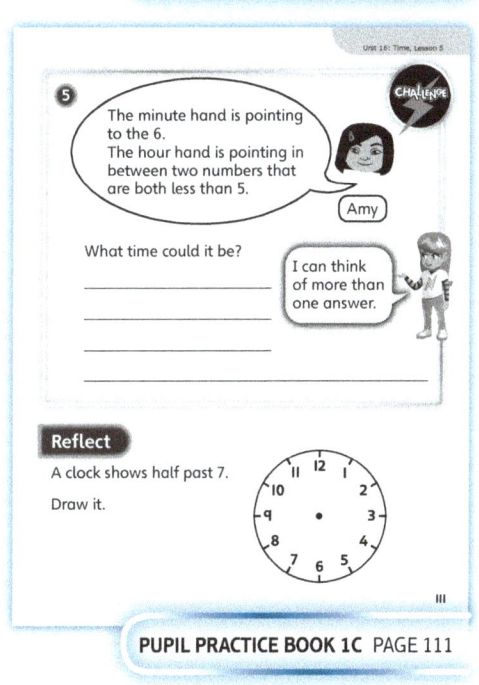

PUPIL PRACTICE BOOK 1C PAGE 111

Unit 16: Time

End of unit check

Don't forget the unit assessment grid in your *Power Maths* online subscription.

WAYS OF WORKING Group work adult led

IN FOCUS
- Question **1** assesses children's ability to recognise the months of the year and their chronological order.
- Question **2** assesses children's ability to recognise dates on a calendar and work out on which day of the week they fall, given a specific calendar.
- Question **3** assesses children's ability to recognise times on an analogue clock. It will require children to understand the terminology of 'X o'clock'.
- Question **4** assesses children's ability to find times earlier or later and tell the time to the half hour. It will require children to understand the terminology of 'X o'clock' and 'half past X'.
- Question **5** assesses children's ability to identify a given date on a calendar and work out what 7 days after this date would be.

Think!

WAYS OF WORKING Pair work or small groups

IN FOCUS Use this section to remind children of the importance of distinguishing carefully between the hour and minute hands on analogue clocks.

ANSWERS AND COMMENTARY Children who have mastered this unit will be able to work confidently within simple situations involving time, including understanding clocks and calendars. They will be able to tell the time to the half hour using an analogue clock, and use a calendar to say what day of the week a particular date falls on. They will be able to use a range of language to order familiar events, and they will be able to solve simple problems involving various units of time.

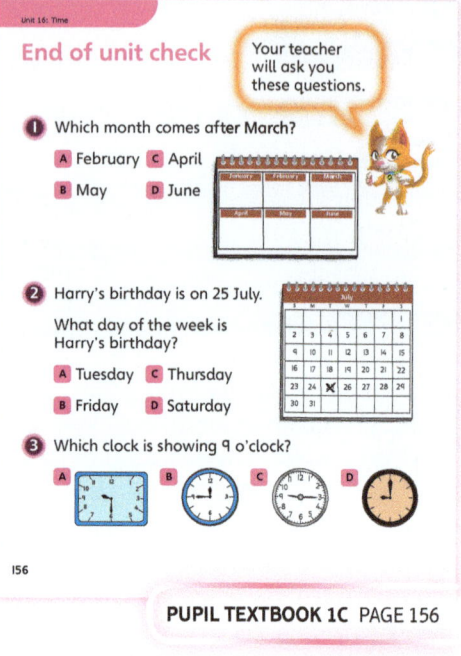

PUPIL TEXTBOOK 1C PAGE 156

PUPIL TEXTBOOK 1C PAGE 157

Q	A	WRONG ANSWERS AND MISCONCEPTIONS	STRENGTHENING UNDERSTANDING
1	C	Children may choose D if they simply read down the column, which could indicate a lack of familiarity with the sequence of month names.	Build clock-reading and calendar activities into the daily routine of the classroom wherever possible and ensure that you focus on the core techniques required. For example, when telling the time, ask: *Which hand on the clock should you look at first? So, what does that tell you? What do you need to do next?* When using a calendar, ask: *Where is today on the calendar? What day of the week is it? What will the date be in a week's time?*
2	A	Choosing C indicates that children do not appreciate the fact that two days' names start with T.	
3	D	Children may choose B if they do not correctly identify the hour and minute hands.	
4	C	Choosing D would indicate a confusion between the concepts of before and after.	
5	22 August	Any incorrect answer here might indicate that children have either not found the correct start date of the holiday or have made a mistake when finding 7 days after this date.	

Unit 16: Time

My journal

WAYS OF WORKING Independent thinking

ANSWERS AND COMMENTARY

What is the same?
- Both of the clocks show half past the hour.
- Both of the minute hands are pointing straight down at the 6.

What is different?
- The hour hands are in different positions.
- The first clock shows half past 4 and the second one shows half past 8.

Notice that children could point out features relating to the position of the hands on the clock faces, or to the times that are shown (or some combination of both aspects).

PUPIL PRACTICE BOOK 1C PAGE 112

Power check

WAYS OF WORKING Independent thinking

ASK
- Could you tell the time before you started this unit?
- What about now?
- How much do you think you have learnt?
- How confident do you feel about telling the time?
- How confident do you feel about reading dates on a calendar?

Power play

WAYS OF WORKING Pair work or small groups

IN FOCUS This puzzle will assess children's recognition of written times and their ability to convert them into times shown on a clock face.

Each pair of children will need a demonstration clock for this activity – make sure that it has a working gear mechanism that keeps the hands properly synchronised.

ANSWERS AND COMMENTARY Look out for any disagreements about what times are shown on the clocks. The most likely errors children may make are choosing the wrong hours with the half-past times and being confused between the hour and minute hands.

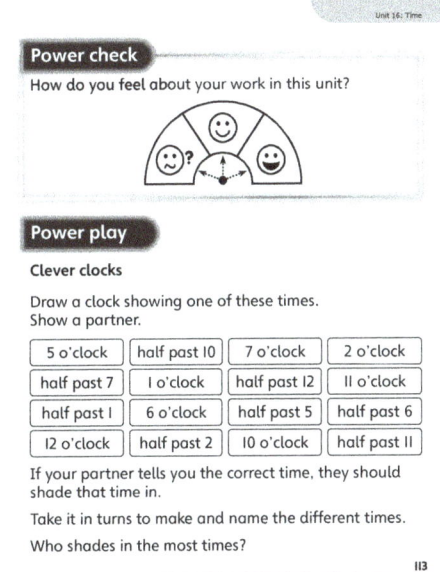

PUPIL PRACTICE BOOK 1C PAGE 113

After the unit

- Are children now telling the time and using a calendar confidently?
- Did the 'before and after' activities suggest that children had problems with sequencing events?

Strengthen and **Deepen** activities for this unit can be found in the *Power Maths* online subscription.

Published by Pearson Education Limited, 80 Strand, London, WC2R 0RL.

www.pearsonschools.co.uk

Text © Pearson Education Limited 2017, 2023
Edited by Pearson and Florence Production Ltd
First edition edited by Pearson, Little Grey Cells Publishing Services and Haremi Ltd
Designed and typeset by Pearson and PDQ Digital Media Solutions Ltd
First edition designed and typeset by Kamae Design
Original illustrations © Pearson Education Limited 2017, 2023
Illustrated by Fran and David Brylewski, Nigel Dobbyn, Adam Linley, Nadene Naude and Jorge Santillan at Beehive Illustration; Emily Skinner at Graham-Cameron Illustration; Paul Higgins at Hunter Higgins Illustrations; and Kamae Design
Images: The Royal Mint, 1971, 1982, 1990, 1992, 1998, 2017, 2023: 157, 159–161, 163–165, 167–171; Bank of England: 157, 159–161, 163–165, 167–171
Images Strengthen and Deepen: Bank of England: Y1 Unit 15: Money: Deepen Activities
Cover design by Pearson Education Ltd
Back cover illustration © Will Overton at Advocate Art and Nadene Naude at Beehive Illustration

Series editor: Tony Staneff; Lead author: Josh Lury
Authors (first edition): Tony Staneff, David Board, Julia Hayes, Derek Huby, Neil Jarrett and Timothy Weal
Consultants (first edition): Professor Liu Jian

The rights of Tony Staneff and Josh Lury to be identified as authors of this work have been asserted by them in accordance with the Copyright, Designs and Patents Act 1988.

This publication is protected by copyright, and permission should be obtained from the publisher prior to any prohibited reproduction, storage in a retrieval system, or transmission in any form or by any means, electronic, mechanical, photocopying, recording, or otherwise. For information regarding permissions, request forms and the appropriate contacts, please visit https://www.pearson.com/us/contact-us/permissions.html Pearson Education Limited Rights and Permissions Department.

First published 2017
This edition first published 2023

27 26 25 24 23
10 9 8 7 6 5 4 3 2 1

British Library Cataloguing in Publication Data
A catalogue record for this book is available from the British Library

ISBN 978 1 292 45049 0

Copyright notice
All rights reserved. No part of this publication may be reproduced in any form or by any means (including photocopying or storing it in any medium by electronic means and whether or not transiently or incidentally to some other use of this publication) without the written permission of the copyright owner, except in accordance with the provisions of the Copyright, Designs and Patents Act 1988 or under the terms of a licence issued by the Copyright Licensing Agency, Barnards Inn, 86 Fetter Lane, London EC4A 1EN (http://www.cla.co.uk). Applications for the copyright owner's written permission should be addressed to the publisher.

Printed in the UK by Ashford Press Ltd

For Power Maths online resources, go to:
www.activelearnprimary.co.uk

Note from the publisher
Pearson has robust editorial processes, including answer and fact checks, to ensure the accuracy of the content in this publication, and every effort is made to ensure this publication is free of errors. We are, however, only human, and occasionally errors do occur. Pearson is not liable for any misunderstandings that arise as a result of errors in this publication, but it is